Estimating for Residential Construction

Second Edition

Halsey Van Orman

◊ DELMAR PUBLISHERS INC.®

NOTICE TO THE READER

Publisher does not warrant or guarantee any of the products described herein or perform any independent analysis in connection with any of the product information contained herein. Publisher does not assume, and expressly disclaims, any obligation to obtain and include information other than that provided to it by the manufacturer.

The reader is expressly warned to consider and adopt all safety precautions that might be indicated by the activities described herein and to avoid all potential hazards. By following the instructions contained herein, the reader willingly assumes all risks in connection with such instructions.

The publisher makes no representations or warranties of any kind, including but not limited to, the warranties of fitness for particular purpose or merchantability, nor are any such representations implied with respect to the material set forth herein, and the publisher takes no responsibility with respect to such material. The publisher shall not be liable for any special, consequential or exemplary damages resulting, in whole or in part, from the readers' use of, or reliance upon, this material.

Delmar Staff

Senior Administrative Editor: David Anthony, Michael McDermott
Project Editor: Carol Micheli
Production Coordinator: Teresa Luterbach
Design Supervisor: Susan C. Mathews

Cover Design
Mary Beth Vought

Cover Photo
Tom Carney Photography

Printed in the United States of America
published simultaneously in Canada
by Nelson Canada,
a division of The Thomson Corporation

10 9 8 7 6 5 4 3 2 1

Library of Congress Cataloging-in-Publication Data

Van Orman, Halsey.
   Estimating for residential construction / Halsey Van Orman.–2nd ed.
      p.   cm.
   Includes bibliographical references.
   ISBN 0-8273-3643-8 (textbook)
      1. House construction–Estimates. 2. House construction–Costs.
   I. Title.
TH4815.8.V36   1991
692'.5–dc20   90-46970
                                                                          CIP

CONTENTS

Section 1 Introduction

Section 2 Architectural Drawings, Specifications, and Contracts

Section 3 Structural Work

Section 4 Interior and Exterior Finishing

PREFACE

ESTIMATING FOR RESIDENTIAL CONSTRUCTION, Second Edition presents the basic functions an éstimator performs in an easy-to-understand format. The author has written the text with the assumption that students often do not have the essential math skills to be an estimator, so the basic formulas are given and then applied so that their ease of use can be demonstrated.

This text provides the introductory information to prepare the student for the course on estimating — no one text can adequately cover all the aspects of the estimating process. Each facet of constructing a house is usually the function of the specialty subcontractor. The estimator for the plumbing sub-contractor, for instance, does not concern himself with the electrical estimating. The intent of this text is to provide the student with the basic, broad overview to begin working towards the specialty training.

ESTIMATING FOR RESIDENTIAL CONSTRUCTION, Second Edition, involves the student with the estimating process for residences. The first few units review prerequisite background material. After these beginning units the student is led through the process of doing a quantity takeoff. Each of these units presents a concise discussion of the materials and techniques involved in a certain aspect of construction, after which that portion of a house is estimated in the unit. A two-story colonial house has been chosen for this estimate, because it offers a wide variety of estimating experiences. The working drawings for this house are included in a packet at the back of the book.

Review questions at the end of each unit allow the instructor and the student to evaluate the student's progress. In many units these review questions include quantity-takeoff problems. After the last unit a comprehensive review is provided in which the student is given working drawings and specifications and is directed to do a quantity takeoff for the entire house.

A table of suggested man-hour requirements for the various phases of construction is shown in an appendix. This table should serve as a valuable reference for the student, even after the course is completed. The textbook also includes a glossary which defines 139 terms pertaining to building construction.

An Instructor's Guide is available which includes suggested lesson outlines for each unit in the textbook, answers to all review questions, and notes on how the answers to many questions were derived.

Halsey Van Orman has a solid background in the construction industry. His formal education includes a degree in Architectural Drafting from the State University of New York at Delhi, and courses in building construction and education at the State University of New York at Oswego and Russell Sage College. He has worked for more than 30 years as a mason, foreman, specifications writer, and estimator. He has been an instructor of Architectural Drafting for 12 years, a vocational supervisor for 7 years, and is now retired.

ACKNOWLEDGMENTS

The author wishes to thank the following companies for providing illustrations for this textbook:

American Institute of Architects, Washington, DC: 8-3

American Plywood Association, Tacoma, WA: 17-1, 19-6, 19-7, 23-2, 23-3

Anderson Corporation, Bayport, MN: 32-1, 32-2, 32-3, 32-6, 32-7, 32-9

Atlas Energy Products . . . . (Energy Shield)

Cahners Books International, Boston, MA: Appendix - Labor Estimating

Crandall Associates, Glens Falls, NY: Specifications for sample house

E. I. DuPont Inc.. . . . . (TYVEK)

Federal Housing Administration, Washington, DC: Specifications for review house

Georgia Pacafic Corporation: 23-4, 23-5

GAF Corporation, New York, NY: 13-5

Heatilator, Division of Vega Industries, Inc., Mount Pleasant, IA: 15-2

Kemper, Division of Tappan Company, Richmond, IN: 38-7

Kohler Company, Kohler, WI: 38-6, 39-1, 39-2

Richard T. Kreh Sr., Photographer: 1-1, 1-4, 9-2, 10-5, 11-1, 12-3, 16-3, 18-1, 22-2, 22-4, 23-1, 25-1, 25-2, 26-1, 29-1, 29-2, 29-4, 30-1, 30-3, 32-5, 32-8, 34-1, 35-2, 36-1, 38-4, 38-5, 40-3, 41-3, 41-6.

Louisiana-Pacific. . . . . (Oriented Waferboard)

Morgan Company, Oshkosh, WI: 31-1, 31-2, 31-3

Pease Industries Inc.: 31-1

Red Cedar Shingle & Handsplit Shake Bureau, Seattle, WA: 30-2

Reemay Inc.. . . . . (TYPAR Housewrap)

H.J. Scheirich Company, Louisville, KY: 38-2, 38-3

Sheldon Slate Products, Middle Granville, NY: 15-1

Joseph Tardi Associates, Photographer: 1-3, 16-1, 22-1, 22-3, 32-4, 35-1, 40-1, 40-2, 41-1

Weiser Company, South Gate, CA: 37-1

Western Wood Molding and Millwork Producers, Portland, OR: 27-1

York International Corp. . . . . (Heat Pumps)

The following individuals reviewed the material to insure its accuracy and appropriateness:

Dr. Stuart Bennett, Central Connecticut State

James Clark, Assistant Professor, Orange Coast College, Costa Mesa, CA

Dean Dressler, Western Iowa Tech.

Harry C. Huth, New Berlin, NY

Jerry Lawrence, Southern Utah State

Theodore W. Marotta, Assistant Professor, Hudson Valley Community College, Troy, NY

Frederick Platt, Tri-County Tech.

James Y. Robinson, Jr., AIA, Ferbee, Walters and Associates, Charlotte, NC

John Siegenthaler, Mohawk Valley Community College

C. Dennis Spring, California State University, Fresno

Alan R. Trellis, Director of Technical Services, National Association of Home Builders, Washington, DC

SECTION 1 INTRODUCTION

unit 1 professions in construction

OBJECTIVES

After completing this unit the student will be able to

- list several professions that use the knowledge and skills of an estimator.
- briefly describe the relationship between the architect, estimator, and general contractor.
- briefly describe the relationship between the general contractor and the subcontractors.

There are a number of jobs in the construction industry that require or benefit from the knowledge gained in a construction estimating course. It is important to consider long range plans in deciding on a career. Some careers such as construction estimator and superintendent, require a higher education and previous experience in construction. Knowledge of the techniques and procedures used in building construction estimating is sure to be very worthwhile to anyone who plans a career in the construction industry.

Before the construction of a building begins, a variety of professions play an important part in the planning stages. An estimator should be acquainted with the functions of these trade specialists and understand their influence on construction.

CONSTRUCTION ESTIMATOR

The construction estimator determines the costs and profit for a construction project. Estimators work indoors in an office using working drawings, specifications, building codes, and calculators, figure 1-1. Most estimators are employed by construction companies, but building materials suppliers also employ estimators as a service to their customers. An estimator for a supplier is generally concerned only with the quantity-takeoff portion of an estimate. The *quantity takeoff* is taken from the plans and specifications and is a listing of all the materials needed

to complete a building. This is only an estimate but it is the most reliable and accurate method for figuring costs.

Estimators use a variety of printed forms for recording their calculations. The form illustrated in figure 1-2, page 2, is used throughout this textbook. This form is typical of what an estimator might use in preparing a quantity takeoff. Notice that although spaces are included for several specific items describing the materials, it is sometimes more convenient to alter the method of recording data.

Fig. 1-1 Estimators work indoors to determine construction costs and material requirements.

1

ESTIMATING DATA SHEET

Building __SMITH HOUSE__

Location __Rt #2, CENTERVILLE__

Architect __JOHN DOE__

Estimator __BILL SMITH__

Sheet __1__ of __9__

Date _____

Description		Unit	Quantity	Unit Price	Total Material Cost	Labor	Total
EXCAVATION FOR FOUNDATION	28'x54'	CU. YDS	223				
FIR- CEILING JOISTS	2x4x16		22				
½" AC EXT. PLYWOOD PORCH CEILING			5				
FLAT WALL PAINT		GAL.	6				
FACE BRICKS- CHIMNEY			350				
½" COPPER PIPE, TYPE L		FT.	110				
#12 AWG ROMEX CABLE		FT.	750				
3,000 LB AIR ENTRAINED CONCRETE- DRIVEWAY	10'x50'x3"	CU. YDS.	6				

Fig. 1-2 A typical quantity-takeoff form

PURCHASING AGENT

The purchasing agent for a construction company is responsible for buying all of the materials, supplies, and equipment the company uses. This individual works in an office, meeting with salespersons and researching catalogs to determine which suppliers can best meet the needs of the construction company. The purchasing agent must also keep extensive records of orders and expenditures. While an overall knowledge of construction practices is important for a purchasing agent, it is also important to like to work with other people and to enjoy office work.

Fig. 1-3 Salesperson at a counter in a retail lumber outlet

BUILDING MATERIALS SALESPERSON

All of the materials used by the construction industry must be distributed to the builders and contractors. This is done by retail and wholesale salespersons, figure 1-3. These positions require people who are familiar with a wide range of construction materials. Retail salespersons normally work behind the counter in a retail outlet, such as a lumber company. Wholesale salespersons spend a great deal of their time meeting purchasing agents and contractors to secure large volumes of business. Knowledge of construction estimating is very valuable for people in these positions.

BUILDING INSPECTOR

Building inspectors are employed by several federal, state, and local government agencies. These people inspect buildings and construction sites to insure that they are safe and comply with existing building codes. Although building inspectors do no estimating, the overall knowledge of building construction that comes from estimating experience and study is valuable to inspectors.

ARCHITECTS

To communicate the intended design of the building to the various trade specialists, the architect works with building designers and drafters to prepare drawings that show the size and location of the various parts of the house. Not all of the information needed to describe materials used can be shown on these drawings, so specifications are written to describe materials and their use in more detail. The architect is responsible for the work produced by the drafters and specification writers.

CONTRACTORS

A contractor is anyone who enters into a legal agreement (contract) to perform certain work. In the building construction field, contractors may be grouped as either general contractors or subcontractors. The *general contractor* assumes overall responsibility for the construction of the building. This is the person who enters into the agreement with the future owner, figure 1-4.

General contractors vary in the amount of actual construction that they or their companies perform. The general contractor may employ

Fig. 1-4 General contractor on the job site explaining a problem to building inspectors and engineers

several carpenters, masons, painters, and specialists in other trades who perform a major portion of the work. Other general contractors may employ only a small number of carpenters to take care of details that would otherwise go unfinished. All of the work that is not performed by employees of the general contractor is performed by subcontractors and their employees.

Subcontractors specialize in a particular trade. such as masonry, concrete work, carpentry, plumbing, painting, cabinetry, flooring, and excavating. Before the general contractor agrees to construct a building for a certain price, bids must be obtained from one or more subcontractors in each of the required trades. These subcontractors estimate the cost of performing their portion of the construction, add allowances for overhead and profit, and submit their bid to the general contractor. A contract is then made between the subcontractor and the general contractors. Based on the bids received from the subcontractors, the general contractor then submits a bid for the total construction project.

SUMMARY

There are a number of occupations in the building construction industry that rely on knowledge of building materials and quantity-takeoff procedures. For the most part, buildings are designed by architects and engineers, who transmit their ideas to drafters. The drafter prepares working drawings from which the building is constructed. The construction estimator uses the information from these drawings and specially prepared specifications to calculate the quantity of materials, labor requirements, overhead, and profit for the construction project. Many large construction companies employ purchasing agents to contact building materials salespersons and check catalogs to obtain materials and supplies for the company. Building inspectors are employed by government agencies and some large construction companies to ensure that work is properly completed and meets the requirements set forth in the building codes.

A contractor is one who agrees to complete certain work for a fixed price. Contractors in building construction fall into two categories; general contractors and subcontractors. General contractors agree to have a building constructed for a future owner. Subcontractors specialize in one phase of construction and agree to perform that part of the work for the general contractor. The general contractor has the final responsibility for seeing that each of the subcontractors' portion of the work is completed.

REVIEW QUESTIONS

Select the letter preceding the best answer.

1. Who is responsible for buying all of the supplies and equipment for a company?

   a. Building materials salesperson   c. General contractor
   b. Purchasing agent   d. Architect

2. Who works with the drafter in preparing the drawings that show the size and location of the parts of a building?

   a. Architect   c. General contractor
   b. Building inspector   d. Purchasing agent

3. Who assumes the overall responsibility for the construction of a building?

   a. Subcontractor   c. Construction estimator
   b. General contractor   d. Architect

4. Who makes a formal agreement with the general contractor to perform certain construction work?

   a. Subcontractor   c. Architect
   b. Building inspector   d. Building materials salesperson

5. On what basis does the general contractor submit the bid for the total building project?

   a. The cost of the land
   b. The owner's ability to pay
   c. The bids from the subcontractors
   d. None of the above

6. The construction estimator's major concern is

   a. Where a building should be constructed
   b. The costs and profit of a project
   c. What belongs in the specifications for the construction of a house
   d. None of the above

unit 2 basic arithmetic review

OBJECTIVES

After studying this unit the student will be able to

- find the perimeter of an object.
- find the areas of common geometric figures.
- apply the principles of mensuration to estimating.

To assist the student who has no knowledge of geometry and *mensuration* (measurement of areas and volumes) definitions and explanations of new terms and propositions are given in this unit.

A *point* indicates position only; it has no length, thickness, or width.

A *line* has one dimension only—length. It may be straight, curved, or irregular, figure 2-1. If the direction of the line does not change, it is a straight line. If the direction changes continually, it is a curved line. An irregular line is a combination of straight and curved lines.

A *perpendicular line* is a line that is drawn at right angles (90 degrees) to another line, figure 2-2.

Plane Figures

A *surface* is the exterior part of anything that has length and width.

A *plane surface* is a flat surface with no depressions or high points. It is usually referred to as a plane, and has only two dimensions—length and width.

The *perimeter* of a surface is the distance around its sides, or the sum of the length of all the sides.

Triangles

All triangles have three sides and three angles. An *isosceles triangle* is a triangle having two of its sides equal, figure 2-3. An *equilateral triangle* is a triangle having all three sides equal, figure 2-4. A *scalene triangle* is a triangle having no two sides equal, figure 2-5.

The *base* of any triangle is the side on which the triangle is considered to rest. The *altitude* or height of a triangle is the perpendicular distance between the base and the *vertex* (the point at

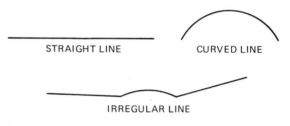

STRAIGHT LINE   CURVED LINE

IRREGULAR LINE

Fig. 2-1

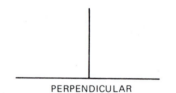

PERPENDICULAR

Fig. 2-2

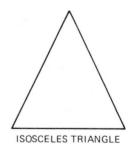

ISOSCELES TRIANGLE

Fig. 2-3

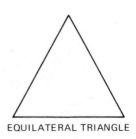

EQUILATERAL TRIANGLE

Fig. 2-4

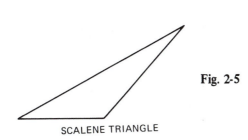

Fig. 2-5

SCALENE TRIANGLE

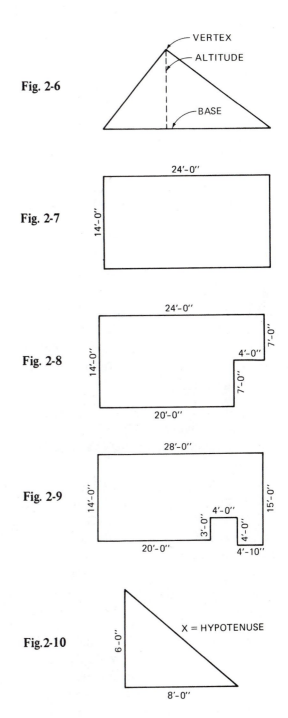

Fig. 2-6

Fig. 2-7

Fig. 2-8

Fig. 2-9

Fig. 2-10

which the two sides of an angle come together) of the angle opposite the base, figure 2-6.

A *right triangle* is a triangle having one right angle, (90 degrees). The side opposite the right angle is called the *hypotenuse.*

One must know linear measurement rules to find the *perimeter* (distance around) of buildings, and the length of walls, baseboards, and ceiling moldings. One must also know area measurements to find the areas of floors, walls, ceilings, gables, and roofs.

Perimeter

The distance around a square or a rectangle is found by adding the length and width and multiplying by 2, or by adding the length of all four sides. The same rule applies to a building. The distance around the building is found by adding the length and width of all the sides.

In figure 2-7, 24'-0" plus 14'-0" equals 38'-0". 38 times 2 equals 76 linear feet of perimeter.

In figure 2-8, add all of the dimensions shown. 24'-0" plus 14'-0" plus 20'-0" plus 4'-0" plus 7'-0" plus 7'-0" equals 76 lin. ft. perimeter.

What is the perimeter of figure 2-9?

The correct answer is shown at the end of this unit.

Hypotenuse

In a right triangle, the square of the hypotenuse equals the sum of the squares of the other two sides. This formula, known as the Pythagorean theorem, is useful in determining the length of a common rafter and in laying out a square corner of a foundation.

To find the length of X in figure 2-10

1. Square 6. (6 x 6) equals 36.

2. Square 8. (8 x 8) equals 64.

3. Add these figures. 100.

4. Find the square root of this total. $\sqrt{100}$ equals 10.

Answer: X equals 10

7

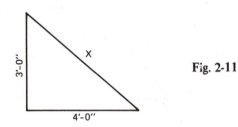

Fig. 2-11

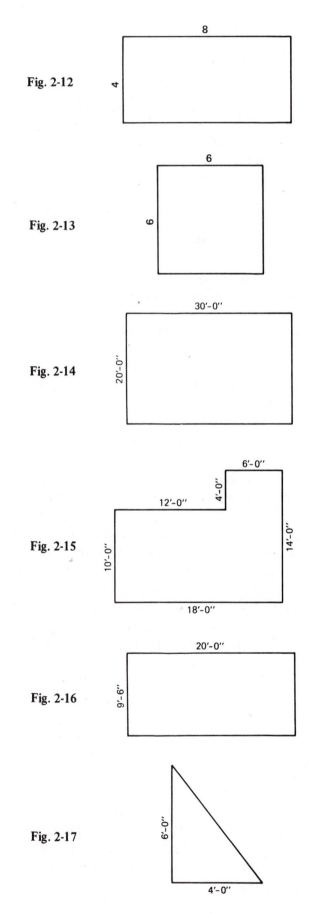

Fig. 2-12

Fig. 2-13

Fig. 2-14

Fig. 2-15

Fig. 2-16

Fig. 2-17

What is the length of X in figure 2-11?
The correct answer is shown at the end
of this unit.

Areas

The area of a square or a rectangle is found by
multiplying the length by the width, or A = ℓ x w.

In figure 2-12 the area equals 8 multiplied by
4, or 32 square feet.

In figure 2-13, find the area by multiplying
one side (6) by the other side (6) to get 36 square
feet for the area (A = s x s or s^2).

What is the area of figure 2-14? What is
the area of figure 2-15?

The correct answers are shown at the
end of this unit.

Notice that in figures 2-12 and 2-13 the numbers 8 by 4 and 6 by 6 are given. It is not indicated
if they are inches, feet, or yards. One must be careful in estimating to be sure that feet are multiplied
by feet, inches by inches, and yards by yards. If the
measurement has feet and inches, make the inches
fractional parts of a foot and then multiply.

In figure 2-16, a building is 20 feet long and 9
feet 6 inches wide. To find the area, multiply
20'-0'' by 9'-6''. 6 inches is one-half of 12 inches
or one-half (0.5) of a foot. Multiply 20 by 9-1/2
(9.5) to get an area of 190 square feet.

If the measurements in figure 2-16 were
changed so that the 9'-6'' (9.5) were to become 9'-8'' (9.66), what would the area of
figure 2-16 be?

The correct answer is shown at the end of
this unit.

The area of a triangle is found by multiplying
the length of the base by 1/2 of the altitude
(known as height), A = bh/2.

In figure 2-17, multiply 1/2 by 6'-0'' which
equals 3, and multiply this figure by the base
(4'-0'') to get an area of 12 square feet.

Another method is to multiply the altitude by the base and then divide by 2.

What is the area of figure 2-18? What is the area of figure 2-19?
The correct answers are shown at the end of this unit.

The area of a circle may be found by using any of these three formulas:

Area equals the square of the radius times π (3.1416), or $A = \pi r^2$

Area equals the square of the diameter times 0.7854, or $A = \pi d$.

Area equals the square of the circumference times 0.07958, or $A = C \times 0.07958$.

Using the above formulas, find the areas of the circles shown in figures 2-20, 2-21, 2-22.
The correct answers are shown at the end of this unit.

To find the area of an irregular polygon, divide the figure into triangles, parallelograms, or other shapes, and then find the area of each. The sum of the partial areas is the area of the irregular polygon. To find the area of figure 2-23, divide the polygon into triangles, find the area of each and then add them together.

1. 10 times 4 equals 40.
2. 40 divided by 2 equals 20.
3. 12 times 5 equals 60.
4. 60 divided by 2 equals 30.
5. 20 plus 30 equals 50 square inches.
   Answer: Area equals 50 square inches.

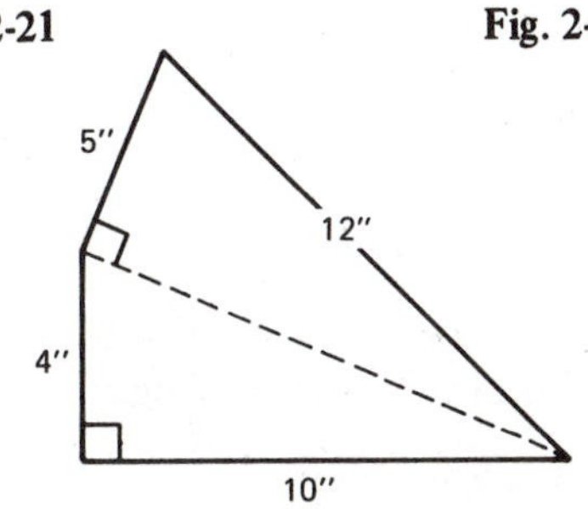

What is the area of figure 2-24?

The correct answer is shown at the end of this unit.

To find the area of a parallelogram, multiply the length of either side by the altitude, figure 2-25. (The altitude equals the perpendicular distance to the opposite side.) In figure 2-26 multiply the length (14) by the altitude (8) to get an area of 112 square inches.

What is the area of figure 2-26?

The correct answer is shown at the end of this unit.

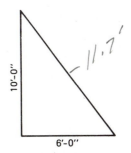

Fig. 2-18

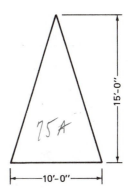

Fig. 2-19

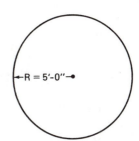

Fig. 2-20

Fig. 2-21   **Fig. 2-22**

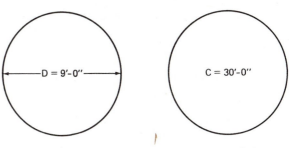

Fig. 2-23

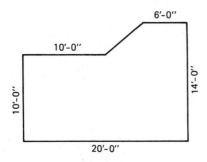

Fig. 2-24

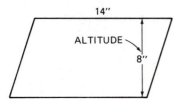

Fig. 2-25

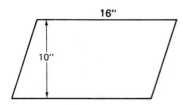

Fig. 2-26

SUMMARY

Quantity takeoff requires a practical knowledge of the principles of measurement. The estimator should know linear measurement and area measurement rules in order to determine the perimeter of buildings and the square-foot areas of floors, walls, ceilings, and roofs.

ANSWERS TO SAMPLE PROBLEMS

Figure number	Answers
2-9	92 feet –10 inches
2-11	5 feet
2-14	600 square feet
2-15	204 square feet
2-16	193 1/3 square feet
2-18	30 square feet
2-19	75 square feet
2-20	78.54 square feet
2-21	63.6 square feet
2-22	716 square feet
2-24	232 square feet
2-26	160 square inches

REVIEW QUESTIONS

Select the letter preceding the best answer.

1. What is the distance around the outside of a building called?

   a. Length   c. Altitude
   b. Perimeter   d. Circumference

2. Finding the length of a common rafter is the same as finding the_____ of a right triangle.

   a. Base   c. Area
   b. Altitude   d. Hypotenuse

3. Which two factors must be multiplied to find the area of a building?

   a. Height   d. Length
   b. Width   e. Circumference
   c. Perimeter   f. Diameter

4. Finding the area of a gable end on a house is the same as finding the area of a triangle. What is the area of figure 2-27 on page 11?

   a. 29 sq. feet   c. 120 sq. feet
   b. 600 sq. feet   d. 60 sq. feet

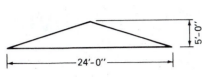

Fig. 2-27

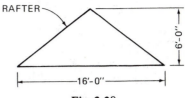

Fig. 2-28

5. A farmer is building a silo with a radius of 10'-0". What will be the square area of the floor?

    a. 314 sq. ft.   b. 400 sq. ft.

    b. 3140 sq. ft.   d. 53 sq. ft.

6. A man wants to finish one wall of his garage. The garage is 24'-0" long and it is 7'-6" from the floor to the ceiling. What is the area of this wall?

    a. 180 sq. ft.   c. 200 sq. ft.

    b. 1800 sq. ft.   d. 18 sq. ft.

7. What would be the length of the rafter in figure 2-28?

    a. 17 ft.   c. 100 ft.

    b. 28 ft.   d. 10 ft.

SECTION 2 ARCHITECTURAL DRAWINGS, SPECIFICATIONS AND CONTRACTS
unit 3 working drawings

OBJECTIVES

After studying this unit the student will be able to

- explain how to read working drawings in plans, elevations, sections, and special details.
- tell how three-dimensional objects are drawn.
- describe orthographic projection as it applies to architectural drawings.
- explain why different scales are used on working drawings.

Plans are drawn by the architect or drafter to give the builder exact information about how to build a particular structure. Such information cannot be easily conveyed in words. Working drawings, together with specifications, contain all of the necessary information. This information is presented in the form of drawings, notes, dimensions, and any other indications necessary to guide the builder in the construction of the building.

A set of working drawings shows the size and thickness of every part of the building. On working drawings, every part is drawn to scale. Working drawings and specifications often indicate the material to be used and the finish to be applied. To make drawings easy to understand, architects have developed certain symbols, signs, and abbreviations to represent various materials, fixtures, and forms of construction.

SCALE

Since the drawings cannot be made the full size of the building, they are scaled down. A fraction of an inch on the drawing stands for a foot on the actual building. The plans of an average-size house are usually drawn to the 1/4-inch-to-the-foot scale. This is written 1/4″ = 1′-0″. This is the scale normally used for floor plans and elevations. Drawings made to this scale are 1/48th the size of the building.

On wall sections and details, however, a larger scale, such as 3/4″ = 1′-0″, is used to show the drawings in greater detail. The architect or draftsman chooses a scale depending upon the space available for the drawings and the nature of the part to be described as well as complying with F.H.A. if required. Simple details do not have to be drawn as large as more complex ones.

KINDS OF DRAWINGS

Plans, elevations, sections, and special details are all needed to understand a set of drawings. In other words, plans of every floor and the basement are needed, as well as the elevations of the four sides of the building and a section showing the construction of the wall from the basement to the attic. In addition, details which give more precise information on particular sections or parts are needed.

The architect or drafter has two general ways of presenting three-dimensional objects on two-dimensional paper: (1) pictoral drawings, which include perspective, isometric, and oblique; and (2) orthographic-projection drawings.

An *orthographic-projection* drawing of a block of wood would include a drawing as seen by looking straight at the front of the block, a drawing as seen by looking straight down from the top, and a drawing as seen by looking straight at one end. This method, sometimes referred to as third-angle projection, is standard practice for all forms of architectural drawing.

ELEVATIONS

A building is represented in much the same way as the block of wood. To show a building in orthographic projection, a drawing is made of the building as it is seen by looking straight at the front, the rear, and the sides. These drawings are known as *elevations*.

An elevation is the easiest part of a drawing to visualize. *Elevations* are drawn as one looks at the building. Different elevations are indicated by different names. For example, in figure 3-1, the front of the house would be drawn as the front elevation, the right-hand side would be drawn as the right-side elevation, the side on the left would be drawn as the left-side elevation, and the back of the building would be drawn as the rear elevation. Sometimes the elevations are referred to as north elevation, east elevation, etc.

PLAN VIEWS

Next, assume that a horizontal cut is made through the building about four feet from the floor and the top section is removed. The view looking straight down from above at the lower part of the building is what is known as the *floor plan*. The cut through the building is made at the proper height so that it passes through doors, windows, and other wall openings. This permits them to be drawn as visible edge or object lines.

In figure 3-2, page 14, the view of the cut through the building shown on the right becomes the floor plan shown on the left. This allows each floor and basement plan to be drawn separately. Only the layout for that particular floor is shown. This provides plans that can be easily read and understood. The floor plan shows the location of the outside walls, the centers of all door and window openings and their sizes.

SECTIONAL VIEWS

Now, looking at the same building as it was before the horizontal line was cut through it to make the floor plans, assume that a vertical cut is made through the building and one part is removed. Looking straight at one end of the building, one sees a sectional view, figure 3-3, page 14.

SUMMARY

Working drawings give the builder exact information about how to build a particular structure. Working drawings, together with specifications, contain all of the necessary information to guide the builder in the construction of the building. This information is given in the form of drawings, notes, dimensions, and other indications.

Working drawings show the size and thickness of every part of the building, with each part drawn to scale. In order to read and understand a set of working drawings, plans, elevations, sections, and special detail drawings are needed.

Orthographic projection, sometimes referred to as third-angle projection, is standard practice for all forms of architectural drawing. The use of orthographic projection helps to show the elevations, floor plans, and sectional views of a building.

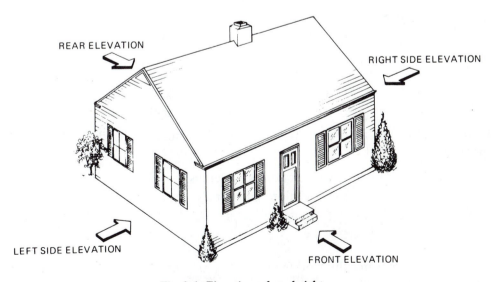

REAR ELEVATION

RIGHT SIDE ELEVATION

LEFT SIDE ELEVATION

FRONT ELEVATION

Fig. 3-1 Elevations show height.

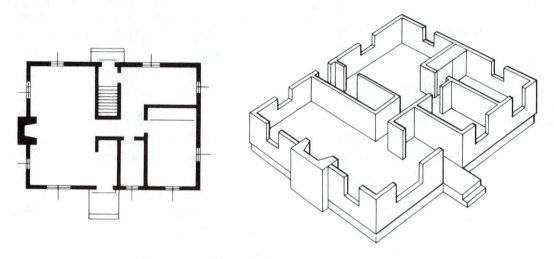

Fig. 3-2 The floor plan shows the layout of rooms.

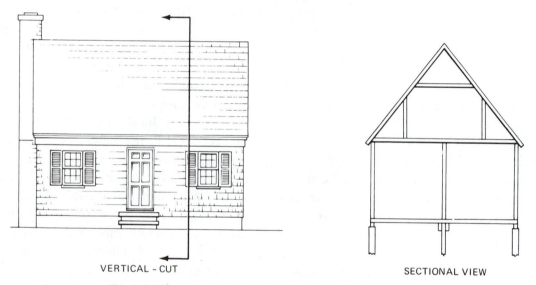

VERTICAL - CUT

SECTIONAL VIEW

Fig. 3-3 The sectional view shows the inside of a building.

REVIEW QUESTIONS

Select the letter preceding the best answer.

1. What is the scale of most floor plans?

   a. 3/4″ = 1′-0″   c. 1/4″ = 1′-0″
   b. 1/2″ = 1′-0″   d. 3/8″ = 1′-0″

2. How does the scale for wall sections and details compare with the scale for the rest of the plans?

   a. Smaller scale is used on wall sections and details.
   b. Larger scale is used on wall sections and details.
   c. No scale is used on wall sections and details.
   d. They both use the same scale.

3. What must accompany the working drawings to contain all of the information necessary to build the building?

4. List the types of pictorial drawings that may be used by the architect.

5. Which drawing looks the most like what one actually sees, looking at the front of a building?

   a. Section
   b. Elevation

   c. Floor plan
   d. Detail

6. Which type of drawing shows what would actually be seen if a horizontal cut were made through a building and the top portion was removed?

   a. Section
   b. Elevation

   c. Floor plan
   d. Perspective

7. Which type of drawing shows what would be seen if a vertical cut was made through a building and one portion was removed?

   a. Elevation
   b. Floor plan

   c. Section
   d. Oblique

8. What type of drawing would be used to show how the entrance floor is joined to the living room floor?

   a. Floor plan
   b. Special detail

   c. Perspective
   d. Pictorial

unit 4 the architect's scale

OBJECTIVES

After studying this unit the student will be able to

- list the eleven scales found on the architect's triangular scale.
- calculate dimensions given at various scales.
- tell how the scales are compatible with each other.

Triangular scales are made of boxwood or plastic. The plastic scale works best because it has sharp edges and distinct machine markings which are necessary to accurately read measurements. The triangular scale combines eleven different scales and is easy to handle while estimating.

The architect's scale is *open divided*, figure 4-1. This means the scales have the main units undivided, and a fully subdivided extra unit placed at the zero end of the scales.

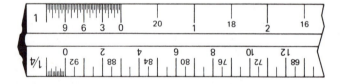

Fig. 4-1 Architect's triangular scale

Listed below are the eleven scales found on the architect's triangular scale:

Full scale

1/8″ = 1'-0″	1/4″ = 1'-0″
3/8″ = 1'-0″	3/4″ = 1'-0″
1/2″ = 1'-0″	1″ = 1'-0″
1 1/2″ = 1'-0″	3″ = 1'-0″
3/32″ = 1'-0″	3/16″ = 1'-0″

Two scales are combined on each face, except the full-size scale which is fully divided into sixteenths. The combined scales are compatible because one is twice as large as the other and their zero points and extra-divided units are on opposite ends of the scale.

Architectural drawings use feet and inches as the major units of measurement. The architect's scale is calibrated into these units in conventionally reduced scales so that large buildings and details can be conveniently drawn on paper. This makes the drawings smaller and more easily handled.

The fraction, or number, near the zero at each end of the scale indicates the unit length in inches that is used on the drawing to represent one foot of the actual building. The extra unit near the zero end of the scale is subdivided into twelfths of a foot, or inches, as well as fractions of inches on the larger scales.

Most plans for houses and small buildings are drawn to the 1/4-inch scale. This means that each quarter of an inch on the plans equals one foot of the actual size of the building. The scale of the drawing is noted on the plans and is usually given in the title box on each page of the plans. Sometimes when special details are given, the scale is placed directly under the detail.

To read the architect's triangular scale, turn it to the 1/4-inch scale. The scale is divided on the left from the zero towards the 1/4 mark so that each line represents one inch. Counting the marks from the zero toward the 1/4 mark, there are twelve lines marked on the scale. Each one of these lines is one inch on the 1/4″ = 1'-0″ scale.

The fraction 1/8 is on the opposite end of the same scale, figure 4-2, page 17. This is the 1/8-inch scale and is read from the right to the left. Notice that the dividend unit is only half as large as the one on the 1/4-inch end of the scale. Counting the lines from the zero toward the 1/8 mark, there are only six lines. This means that each line represents two inches at the 1/8-inch scale.

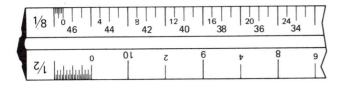

Fig. 4-2

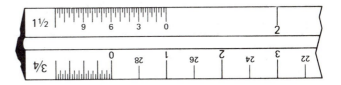

Fig. 4-3

Now look at the 1 1/2-inch scale, figure 4-3. The divided unit is broken into twelfths of an inch and also a fractional part of an inch. Reading from the zero toward the number 1 1/2, notice the figures 3, 6, and 9. These represent the measurements of 3 inches, 6 inches, and 9 inches at the 1 1/2″ = 1'-0″ scale. From the zero to the first long mark that represents one inch (which is the same length as the mark shown at 3) are 4 lines. This means that each line on the scale is equal to 1/4 of an inch. Reading the zero to the 3, read each line as follows: 1/4, 1/2, 3/4, 1, 1 1/4, 1 1/2, 1 3/4, 2, 2 1/4, 2 1/2, 2 3/4, and 3 inches.

SUMMARY

The architect's triangular scale combines eleven different scales. The architect's scale is open divided; this means the scales have the main units undivided, and a fully subdivided extra unit placed at the zero end of the scales. All scales are related to 12 inches.

Two scales are combined on each face except the full-size scale which is fully divided into sixteenths. The combined scales are compatible because one is twice as large as the other, and their zero points and extra-divided units are on opposite ends of the scale.

REVIEW QUESTIONS

Select the letter preceding the best answer.

1. How many scales are included on the architect's scale?

    a. 9   c. 11

    b. 12   d. 4

2. On an open-divided scale what units are undivided?

    a. The end units   c. The open units

    b. The main units   d. The smaller units

3. Where is the fully-divided extra unit located on an architect's scale?

    a. At the end opposite zero   c. At the zero end

    b. In the middle   d. On the top

4. What are the major units of measurement on architectural drawings?

    a. Feet and inches   c. Decimals and fractions

    b. Feet and yards   d. None of these

5. The fraction or number at the zero end of the scale indicates that the units on the scale represent:

    a. Yards   c. Feet

    b. Inches   d. None of these

B. Give the exact dimension in feet and inches at the following scales.

    1. 1/4″ = 1'-0″

        a. |————————————————————————|

  b. _____

  c. _____

  d. _____

  e. _____

  f. _____

2. 1/8″ = 1′-0″

  a. _____

  b. _____

  c. _____

  d. _____

  e. _____

3. 3/4″ = 1′-0″

  a. _____

  b. _____

  c. _____

4. 3/8″ = 1′-0″

  a. _____

  b. _____

  c. _____

5. 1/2″ = 1′-0″

  a. _____

  b. _____

  c. _____

6. 3″ = 1′-0″

  a. _____

  b. _____

  c. _____

unit 5 the alphabet of lines

OBJECTIVES

After studying this unit the student will be able to

- describe the types of lines used on working drawings.
- tell how to recognize the lines on working drawings.

To read a set of plans it is important that the estimator know the different types of lines commonly used in making a set of working drawings. The mechanical drafter and the architectural drafter use the same alphabet of lines, mainly object lines, hidden lines, centerlines, extension lines, and dimension lines.

On architectural drawings the following types of lines are most commonly used:

Border Line. This is a solid line drawn medium-dark to give an even border around the drawing. The title box in the lower right-hand corner of the drawing is made the same weight and becomes a part of the border line.

Object Line. The shape of the drawing is always drawn with a dark, solid line to show the outline of the building.

Extension Line. This is a light, solid line used to show the exact size or dimension. These lines extend away from the object lines at the exact points between which the dimensions are given.

Dimension Line. Dimension lines are light lines, with arrowheads or diagonal lines at their ends. They can be solid, with the dimension given in a break. The arrowheads touch the extension lines or centerlines and give the exact distance referred to by the dimension line in feet and inches, figure 5-1A and 5-1B.

Equipment Line. This is a light, solid line that is used to show the outline of equipment on a floor plan. Equipment lines are used to show sinks, tubs, toilets, and cabinets, to name a few.

Symbol Section Line. Two light, solid lines are drawn with a space between them. The type of

material to be used is then drawn between these lines, figure 5-2. Various materials are now standardized and are easily found in most reference books, both in plans and elevations.

Break Line. This is a solid line with a quick up-and-down break in a straight line, figure 5-3. This indicates that either parts have been left out or that the full length of some part is not drawn.

Invisible Line. To be complete, a drawing must include all lines that represent walls and surfaces. Sometimes these lines cannot be seen because they are covered by other sections of the object. To show that a line or surface is hidden, the drafter uses a series of short dashes, figure 5-4, page 20. Invisible lines are used mostly on basement and foundation plans to show footings.

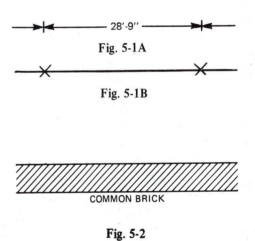

28'-9''

Fig. 5-1A

Fig. 5-1B

COMMON BRICK

Fig. 5-2

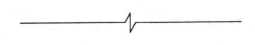

Fig. 5-3

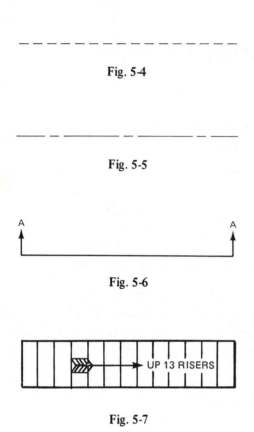

Fig. 5-4

Fig. 5-5

Fig. 5-6

Fig. 5-7

Centerline. A centerline is drawn as a light, broken line of long and short dashes spaced alternately, figure 5-5. Centerlines are used to indicate the center of door and window openings and the center of partitions.

Cutting-plane Line. A cutting-plane line is a solid line turned at right angles on each end with an arrowhead showing the direction of the section cut, figure 5-6. The letter at the end of the arrowhead identifies the section that is referred to in the details. The cutting-plane line is used to refer to a detail that shows exactly what is to be done on that part of the construction.

Stair Indicator. A stair indicator is found on any plan that has stairs going from one level to another. It is a straight line with an arrowhead on one end showing the direction of the stairs, figure 5-7. The other end is closed, to represent the tail of an arrow. Over the top of the stair indicator the words up or down are given, depending on the direction. Sometimes the number of risers is also given.

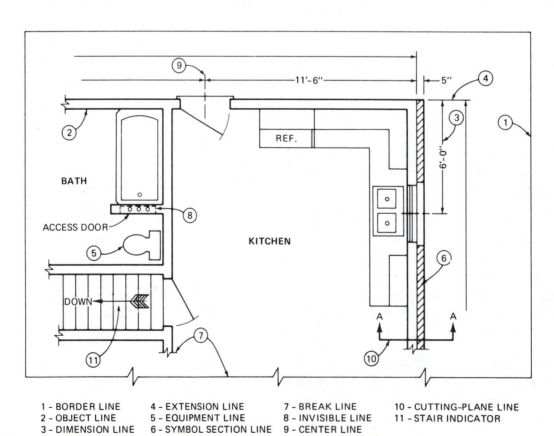

1 - BORDER LINE	4 - EXTENSION LINE	7 - BREAK LINE	10 - CUTTING-PLANE LINE
2 - OBJECT LINE	5 - EQUIPMENT LINE	8 - INVISIBLE LINE	11 - STAIR INDICATOR
3 - DIMENSION LINE	6 - SYMBOL SECTION LINE	9 - CENTER LINE	

Fig. 5-8

Figure 5-8 shows where the various lines mentioned in this unit might be found on a set of plans.

SUMMARY

An estimator must know how to read working drawings. It is important to know the different types of lines used in making a set of working drawings in order to be able to read those drawings. These lines have become standardized throughout the industry, so that learning the types of lines and where they are used makes it easier to read and understand working drawings.

REVIEW QUESTIONS

Select the letter preceding the best answer.

1. What is the heavy solid line on a drawing showing the shape of the building?

    a. Hidden line

    b. Extension line

    c. Object line

    d. Centerline

    e. Border line

2. What line is used to indicate the middle of a window or door opening?

    a. Object line

    b. Centerline

    c. Hidden line

    d. Extension line

    e. Border line

3. What are the light lines used to show the precise point a dimension refers to?

    a. Object lines

    b. Dimension lines

    c. Extension lines

    d. Centerlines

    e. Hidden lines

4. What are the light lines, with arrowheads at their ends, which give exact distances?

    a. Hidden lines

    b. Extension lines

    c. Object lines

    d. Dimension lines

    e. Centerlines

5. What is a solid line, turned at right angles on the ends, with arrowheads showing direction?

    a. Dimension line

    b. Centerline

    c. Object line

    d. Extension line

    e. Cutting-plane line

6. What is a light line made by a series of short dashes?

    a. Centerline

    b. Hidden line

    c. Object line

    d. Cutting-plane line

    e. Break line

7. What line is used to show that a part of an object has not been drawn?

    a. Break line

    b. Object line

    c. Centerline

    d. Cutting-plane line

    e. Border line

unit 6 architectural symbols and conventions

OBJECTIVES

After studying this unit the student will be able to

- explain how architectural symbols and conventions are shown on working drawings.
- tell how various materials are shown on drawings.
- explain how dimensions are shown on working drawings.

ARCHITECTURAL SYMBOLS

In order for the architect to make drawings useful, there must be a way to show materials and the type of construction procedures used so they are easily understood and drawn. This is done by the use of symbols.

Figure 6-1 lists some examples of the symbols most commonly found on a set of working drawings. The estimator should also be acquainted with the abbreviations most commonly used on drawings. These abbreviations are shown in figure 6-2, page 23.

When two symbols are very similar, knowledge of the materials involved helps make the distinction between them. For example, the symbols for gypsum plaster and concrete are almost identical. From experience, it is apparent that gypsum plaster is not used on the exterior of a building, where it would be exposed to the sun, rain, snow,

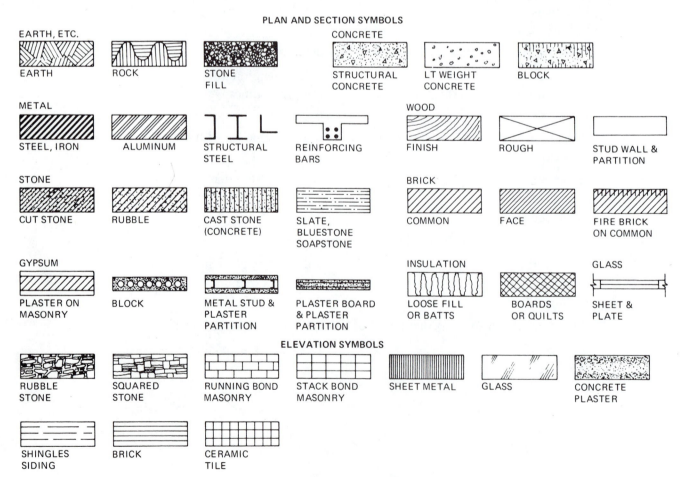

Fig. 6-1 Symbols found on architectural drawing

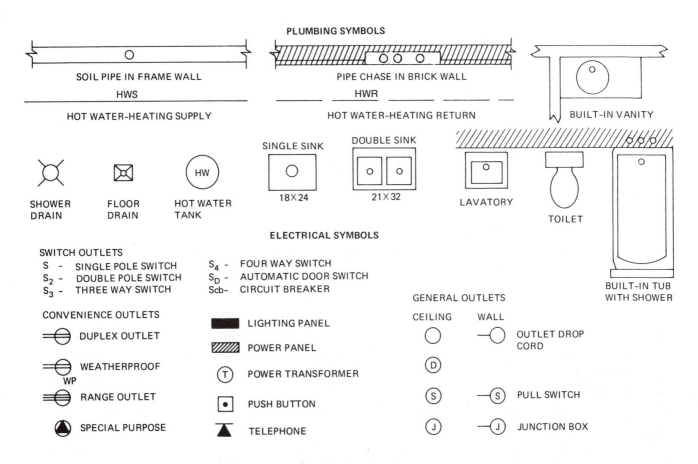

Fig. 6-1 Continued

AWG	American Wire Gauge	GL	Glass
B	Bathroom	HB	Hose Bibb
BR	Bedroom	C	Hundred
BD	Board	INS	Insulation
BM	Board Measure	INT	Interior
BTU	British Thermal Unit	KD	Kiln Dried
BLDG	Building	K	Kitchen
CLG	Ceiling	LAV	Lavatory
C to C	Center to Center	LR	Living Room
CL or ℄	Centerline	MLDG	Molding
CLO	Closet	OC	On Center
COL	Column	REF	Refrigerator
CONC	Concrete	R	Riser
CFM	Cubic feet per minute	RM	Room
CU YD	Cubic Yard	SPEC	Specification
DR	Dining Room	STD	Standard
ENT	Entrance	M	Thousand
EXT	Exterior	T & G	Tongue and
			Groove
FIN	Finish	UNFIN	Unfinished
FL	Floor	WC	Water Closet
FTG	Footing	WH	Water Heater
FDN	Foundation	WP	Waterproof
GA	Gauge	WD	Wood

Fig. 6-2 Abbreviations used on working drawings

and freezing temperatures. Concrete might be used on the exterior of a building where it is in contact with the elements.

CONVENTIONS

In drawing a set of plans or working drawings, the architect must have some way of showing the sizes and location of equipment, doors, windows, rooms, kitchen cabinets, appliances, and so forth. These are shown by using centerlines, extension lines, dimension lines, object lines, and symbols in a conventional or standard way. The architect uses conventions to show such things as the direction doors swing, their size, and their location. Conventions are used by the architect to draw windows in plan, showing their exact location, size, and style.

All windows and doors on exterior walls are shown. The dimensions are given to the center of all door and window openings. Doors are shown on plans as in figure 6-3, page 24. Notice that the swing of the door is shown. On some plans the architect designates the door

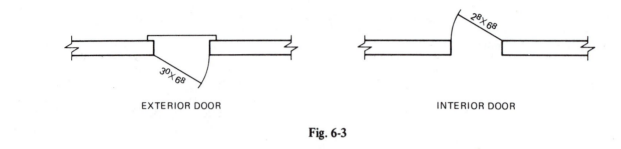

EXTERIOR DOOR INTERIOR DOOR

Fig. 6-3

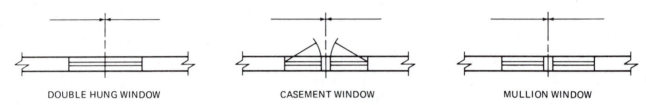

DOUBLE HUNG WINDOW CASEMENT WINDOW MULLION WINDOW

Fig. 6-4

size on the line showing the door swing. On some plans a circled number is shown at each door opening. This indicates that there is a schedule which gives the specifications for the door that corresponds to the number in the circle. If the doorway does not have a door, it may be designated with the abbreviation CO, which means cased opening.

Also, notice that the exterior door has a line extending across the front parallel with the line of the wall. This represents the threshold and extends on each side of the door, the width of the outside door casing.

Windows are shown on the plan by a centerline drawn to the center of the opening, figure 6-4. Window sizes are usually given on the plans just under the symbol designating the location of the window. Here, also, a circled number or letter may appear. This indicates that there is a schedule located on the plans which lists the window size and style.

In veneer construction and in frame construction, dimensions are given from the outside face of the exterior sheathing to the center of the partitions or openings. All dimensions must add up to equal the overall dimensions, figure 6-5, page 25.

Figure 6-5 shows doors and windows in a brick-veneer exterior wall. Notice that the dimension lines are drawn from the face of the exterior sheathing to the centerline of the door and window openings.

Figure 6-6, page 25, shows doors and windows in frame construction on exterior walls. Notice that

the dimensions for the location of the picture window are given to the center of the opening. In architectural drawing, the width is always given first and the height second. This applies to all windows and doors. For example, the picture window in figure 6-6 is a five foot wide center sash with a two foot wide double-hung window on each side. All the windows in this unit are four feet six inches high. The same conventions are used for interior wall and door openings with only minor changes. In figure 6-7, page 25, notice that the distance from the face of the nearest partition to the edge of the openings is given. The opening is dimensioned from one side to the other with a dimension line.

On wood frame partitions, figure 6-8, page 25, the dimensions are given from the face of the exterior sheathing to the center of the door opening or the center of the partition. The size of the door is noted in the same manner as for an exterior door.

SUMMARY

It is impossible to convey all of the information necessary to construct a building without the use of symbols and conventions. Architectural symbols are used on drawings to designate certain materials and equipment for the various tradespersons. Conventions are used by architects to show dimensions, centerlines, and exact locations of parts on a drawing. It is important for an estimator to be able to recognize these symbols and conventions.

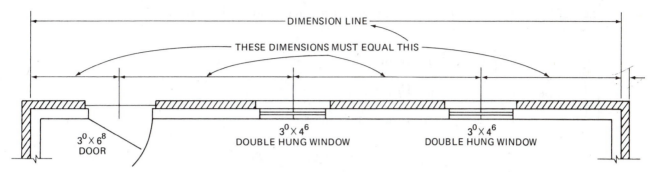

Fig. 6-5

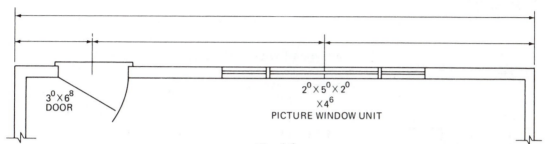

Fig. 6-6

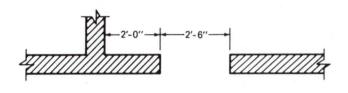

Fig. 6-7

Fig. 6-8

REVIEW QUESTIONS

A. Identify each of the following architectural symbols.

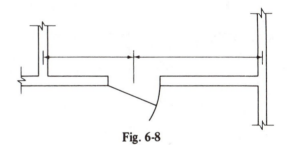

1.

2.

3.

4.

5.

6.

7.

8.

9. S_3

10.

B. What do each of the following abbreviations stand for?

1. O.C.

2. G.I.

3. Gl.

4. Flr.

5. ₵L

C. In each of the following cases, use the drawing provided to answer the related questions.

1.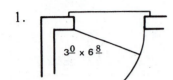

    a. Is this an interior or exterior door?

    b. How high is the door?

    c. How wide is the door?

2.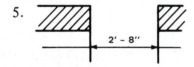

    a. What type of window is this?

    b. What type of wall is this?

    c. How wide is the window?

3.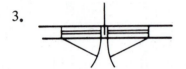

    a. What type of window is this?

    b. What type of wall is this?

4.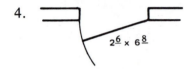

    a. Is this an interior or exterior door?

    b. How wide is the door?

    c. How high is the door?

5.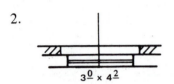

    a. What type of wall is this?

    b. Does 2'-8" indicate the height or width of the opening?

unit 7 specifications

OBJECTIVES

After studying this unit the student will be able to

- define specifications.
- describe the relationship between working drawings and specifications.
- discuss the legal aspects of building construction in relation to proposals, performance bonds, and contracts.

PURPOSE OF SPECIFICATIONS

When a set of working drawings for a building is made, it is impossible to include all of the information necessary to build the structure. For example, if the plans show wood floors, they might be oak, maple, or vertical grain fir. Tile flooring on the plans might be ceramic tile, asphalt tile, or vinyl tile. Where the plans show flashing, it could be galvanized steel, aluminum, or copper. Roof shingles can be asphalt or wood. The types of materials used and the quality of these materials must be described in some way.

Information that cannot be clearly shown on the drawings is conveyed to the builder by written specifications. The working drawings for a building give the shape, size, and location. Specifications describe the quality and type of materials, colors, finishes, and workmanship required. Inspections, rejections, and approvals are stated so that the contractor knows what is to be expected.

The legal aspects of building construction may be confusing to the unacquainted reader. The estimator should understand these documents, because they affect the cost of building construction.

A *contract* is an agreement in which one party agrees to perform certain work and the other party agrees to pay for the work and services.

A *performance bond* is a guarantee issued by a bonding company stating that the work will be done according to the plans and specifications. If the contractor fails to finish the contract, the bond insures that money will be provided to hire another contractor to complete the job.

The specifications become a distinct part of the contract. Once the contract has been signed, the specifications cannot be changed. Any corrections that must be made in the specifications after this point, must be accompanied by a *change order*. In the case of a dispute between the drawings and the specifications, that which is indicated by the specifications is binding.

Specifications are prepared by the architect or contractor and cover the entire project. The amount of detail and the exact form of the specifications may vary. Specifications serve several purposes:

- They make up a legal document that gives instructions for bids, owner-contractor agreements, insurance, and bond forms that are necessary.
- They help prevent disputes between the builder and the owner, or between the contractor and the architect.
- They eliminate conflicting opinions about the grade and quality of the material to be used.
- They help the contractor estimate the material and labor.
- Together with the working drawings, they are necessary to complete the contract.

- Specifications are part of a legal document. Combined with the drawings, they may be used in a court of law as binding evidence. To make a complete set of specifications for each new job would be unnecessarily time consuming. Instead, specification writers rely on various references for standard specifications from which they compile a set for each new job.

DIVISIONS OF SPECIFICATIONS

In general, specifications for a residence are broken down into divisions that cover work by the

different trades. It is standard practice to write the specifications in the order in which the house will be constructed. This makes it easier for the estimator to write a material list and for the contractor and subcontractors to locate specifications for their particular trade, material, or work.

The following are typical divisions of specifications for a residence:

General Conditions
General Requirements
Excavating & Backfill
Grading
Concrete
Masonry
Carpentry & Millwork
Sheet Metal & Roofing
Glass
Painting
Hardware
Heating & Air Conditioning
Plumbing
Electrical

The General Conditions section is usually first. In the specifications for house construction, the A.I.A. (American Institute of Architects) Short Form of Agreement Between the Owner and the Contractor is sometimes included as part of the General Conditions.

SUMMARY

To avoid misunderstandings between the owner and the contractor concerning what is to be built, how it is to be built, and what materials are to be used, the house is described in plans and working drawings. Items that cannot be clearly shown on the drawings are conveyed to the builder by specifications.

The working drawings for a building give the shape, size, and location. Specifications give an explanation of the quality and type of materials, colors, and finishes to be used, as well as the workmanship that is to be expected.

REVIEW QUESTIONS

A. Select the letter preceding the best answer.

1. What document describes the quality of the materials to be used in the construction of a building?

   a. Working drawings
   b. Specifications
   c. Performance bond
   d. Insurance

2. What document describes the size and shape of a building?

   a. Working drawings
   b. Specifications
   c. Performance bond
   d. Insurance

3. What document guarantees that the work will be completed on the building?

   a. Contract
   b. Specifications
   c. Performance bond
   d. Insurance

4. What is the legal term for an agreement between parties?

   a. Contract
   b. Specifications
   c. Performance bond
   d. Insurance

5. The divisions of a set of specifications are normally written according to:

   a. The order in which the work is performed.
   b. The cost of the materials involved.
   c. The order in which the drawings are prepared.
   d. The various trades involved in the construction of the building.

unit 8 general conditions and general requirements of specifications

OBJECTIVES

After studying this unit the student will be able to

- list the items most commonly covered in the general conditions of a set of specifications.
- list the items most commonly covered in the general requirements of a set of specifications.
- explain the purpose of the A.I.A. Short Form of Agreement Between Owner and Contractor.

The first parts of a set of specifications provide general information about the construction project, the procedure for bidding on the project, and the agreement between the parties involved. In order to better understand the various parts of the specifications, each will be discussed in the order in which it appears.

TITLE PAGE AND INDEX

The first page of the specifications is usually a title page. The title page, figure 8-1, tells what the title of the job is, who it is for (the owner), the architect's name, and the date. The second page is the index, figure 8-2, page 30. The index helps the contractor, estimator, and all who must read the specifications locate the specifications for the various trades.

CONTRACTS

A contract is a legally binding agreement made between two or more parties, in which one party agrees to perform a service for another party. Although it is not always necessary for a contract to be in writing, it is more practical to use a written contract in the case of building construction. In this manner, all parties concerned, such as banks, contractors, the owner, and the architect, are assured of receiving the same information. In as much as the specifications must describe all aspects of the construction project, they normally

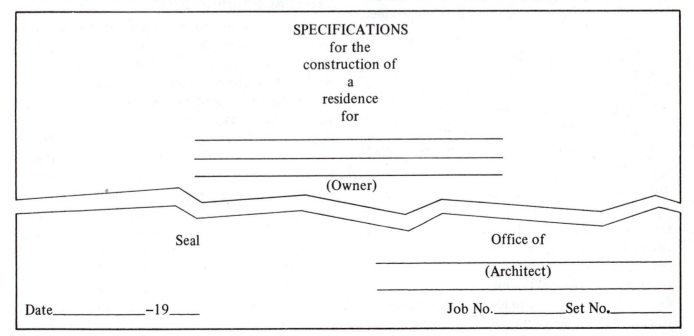

Fig. 8-1 Title page of specifications

```
┌─────────────────────────────────────────┐
│                 INDEX                     │
│  Instructions to Bidders                  │
│  General Conditions                       │
│  Division I - General Requirements        │
│  Division II - Site Work                  │
│  Division III - Concrete                  │
│  Division IV - Masonry                    │
│  Division V - Metals                      │
│  Division VI - Carpentry & Millwork       │
│  Division VII - Moisture Protection       │
│  Division VIII - Doors, Windows, & Glass  │
│  Division IX - Finishes                   │
│  Division X - Specialties                 │
│  Division XI - Mechanical                 │
│  Division XII - Electrical                │
└─────────────────────────────────────────┘
```

Fig. 8-2 Specifications Index

indicate the form of the contract to be used. A typical contract form for residential construction is the Standard Form of Agreement Between Owner and Contractor, from the American Institute of Architects, figure 8-3. This form, together with the working drawings and the detailed specifications, constitutes the contract for a given construction project.

GENERAL CONDITIONS

In order for the contract to provide complete protection to the parties involved, the specifications include General Conditions, such as the following:

- The contract form
- Supervision of the contract
- The architect's responsibility
- The contractor's responsibility for furnishing the lot lines and any restrictions
- Protection of the work in progress
- The following of manufacturers' instructions
- Quality of workmanship

This section usually mentions the fact that where manufacturers' trade names are listed, it is done to serve as a guide for quality and is not to restrict competitive bidding. Any work or materials not covered elsewhere in the specifications are normally covered in the General Conditions.

If the Instructions to Bidders has not been included as a separate section, they are included with the General Conditions. Bidders may be required

to obtain a bid bond. This is a guarantee that the contractor will not withdraw the bid before a given date. Bidders are usually required to have certain types of insurance. Compensation insurance is required to protect all of the contractor's employees and for any employees of subcontractors. Liability insurance protects the contractor and the owner from suits resulting from accidents arising out of any aspect of the construction. In general, the Instructions to Bidders provides all of the information a contractor needs in order to bid on the job.

GENERAL REQUIREMENTS

The General Requirements division of the specifications gives the location of the work, the owner's name and address, and the scope of the work to be performed. It discusses temporary provisions for such items as toilet facilities, storage, water, electric power supply, and heat, if necessary. It describes the requirements for the contractor's insurance. This is required to protect the workers and the owner from any law suit that may result from an accident on the job. It also requires that the contractor furnish the owner with certificates for specific amounts of property damage. The General Requirements also require the owner to maintain builders' risk insurance.

Frequently the General Conditions are omitted. In this case many of the items normally covered in that section are included in the General Requirements.

AIA Document A107

Abbreviated Form of Agreement Between Owner and Contractor

For CONSTRUCTION PROJECTS OF LIMITED SCOPE where the Basis of Payment is a STIPULATED SUM

1987 EDITION

THIS DOCUMENT HAS IMPORTANT LEGAL CONSEQUENCES; CONSULTATION WITH AN ATTORNEY IS ENCOURAGED WITH RESPECT TO ITS COMPLETION OR MODIFICATION.

This document includes abbreviated General Conditions and should not be used with other general conditions. It has been approved and endorsed by The Associated General Contractors of America.

AGREEMENT

made as of the   day of   in the year of
Nineteen Hundred and

BETWEEN the Owner:
(Name and address)

and the Contractor:
(Name and address)

The Project is:
(Name and location)

The Architect is:
(Name and address)

The Owner and Contractor agree as set forth below.

ARTICLE 1
THE WORK OF THIS CONTRACT

1.1   The Contractor shall execute the entire Work described in the Contract Documents, except to the extent specifically indicated in the Contract Documents to be the responsibility of others, or as follows:

ARTICLE 2
DATE OF COMMENCEMENT AND SUBSTANTIAL COMPLETION

2.1   The date of commencement is the date from which the Contract Time of Paragraph 2.2 is measured, and shall be the date of this Agreement, as first written above, unless a different date is stated below or provision is made for the date to be fixed in a notice to proceed issued by the Owner.

(Insert the date of commencement, if it differs from the date of this Agreement or, if applicable, state that the date will be fixed in a notice to proceed.)

2.2   The Contractor shall achieve Substantial Completion of the entire Work not later than

(Insert the calendar date or number of calendar days after the date of commencement. Also insert any requirements for earlier Substantial Completion of certain portions of the Work, if not stated elsewhere in the Contract Documents.)

, subject to adjustments of this Contract Time as provided in the Contract Documents.

(Insert provisions, if any, for liquidated damages relating to failure to complete on time.)

ARTICLE 3
CONTRACT SUM

3.1   The Owner shall pay the Contractor in current funds for the Contractor's performance of the Contract the Contract Sum of
Dollars
($), subject to additions and deductions as provided in the Contract Documents.

3.2 The Contract Sum is based upon the following alternates, if any, which are described in the Contract Documents and are hereby accepted by the Owner:

(State the numbers or other identification of accepted alternates. If decisions on other alternates are to be made by the Owner subsequent to the execution of this Agreement, attach a schedule of such other alternates showing the amount for each and the date until which that amount is valid.)

3.3 Unit prices, if any, are as follows:

ARTICLE 4
PROGRESS PAYMENTS

4.1 Based upon Applications for Payment submitted to the Architect by the Contractor and Certificates for Payment issued by the Architect, the Owner shall make progress payments on account of the Contract Sum to the Contractor as provided below and else-where in the Contract Documents. The period covered by each Application for Payment shall be one calendar month ending on the last day of the month, or as follows:

4.2 Payments due and unpaid under the Contract shall bear interest from the date payment is due at the rate stated below, or in the absence thereof, at the legal rate prevailing from time to time at the place where the Project is located.

(Insert rate of interest agreed upon, if any.)

(Usury laws and requirements under the Federal Truth in Lending Act, similar state and local consumer credit laws and other regulations at the Owner's and Contractor's principal places of business, the location of the Project and elsewhere may affect the validity of this provision. Legal advice should be obtained with respect to deletions or modifications, and also regarding requirements such as written disclosures or waivers.)

ARTICLE 5
FINAL PAYMENT

5.1 Final payment, constituting the entire unpaid balance of the Contract Sum, shall be made by the Owner to the Contractor when the Work has been completed, the Contract fully performed, and a final Certificate for Payment has been issued by the Architect.

ARTICLE 6
ENUMERATION OF CONTRACT DOCUMENTS

6.1 The Contract Documents are listed in Article 7 and, except for Modifications issued after execution of this Agreement, are enumerated as follows:

6.1.1 The Agreement is this executed Abbreviated Form of Agreement Between Owner and Contractor, AIA Document A107, 1987 Edition.

6.1.2 The Supplementary and other Conditions of the Contract are those contained in the Project Manual dated , and are as follows:

Document	Title	Pages

6.1.3. The Specifications are those contained in the Project Manual dated as in Subparagraph 6.1.2, and are as follows:

(Either list the Specifications here or refer to an exhibit attached to this Agreement.)

Section	Title	Pages

6.1.4 The Drawings are as follows, and are dated unless a different date is shown below:
(Either list the Drawings here or refer to an exhibit attached to this Agreement.)

Number	Title	Date

6.1.5 The Addenda, if any, are as follows:

Number	Date	Pages

Portions of Addenda relating to bidding requirements are not part of the Contract Documents unless the bidding requirements are also enumerated in this Article 6.

6.1.6 Other documents, if any, forming part of the Contract Documents are as follows:
(List any additional documents which are intended to form part of the Contract Documents.)

Fig. 8-3 (Continued)

ARTICLE 7
CONTRACT DOCUMENTS

7.1 The Contract Documents consist of this Agreement with Conditions of the Contract (General, Supplementary and other Conditions), Drawings, Specifications, addenda issued prior to the execution of this Agreement, other documents listed in this Agreement and Modifications issued after execution of this Agreement. The intent of the Contract Documents is to include all items necessary for the proper execution and completion of the Work by the Contractor. The Contract Documents are complementary, and what is required by one shall be as binding as if required by all; performance by the Contractor shall be required only to the extent consistent with the Contract Documents and reasonably inferable from them as being necessary to produce the intended results.

7.2 The Contract Documents shall not be construed to create a contractual relationship of any kind (1) between the Architect and Contractor, (2) between the Owner and a Subcontractor or Sub-subcontractor or (3) between any persons or entities other than the Owner and Contractor.

7.3 Execution of the Contract by the Contractor is a representation that the Contractor has visited the site and become familiar with the local conditions under which the Work is to be performed.

7.4 The term "Work" means the construction and services required by the Contract Documents, whether completed or partially completed, and includes all other labor, materials, equipment and services provided or to be provided by the Contractor to fulfill the Contractor's obligations. The Work may constitute the whole or a part of the Project.

ARTICLE 8
OWNER

8.1 The Owner shall furnish surveys and a legal description of the site.

8.2 Except for permits and fees which are the responsibility of the Contractor under the Contract Documents, the Owner shall secure and pay for necessary approvals, easements, assessments and charges required for the construction, use or occupancy of permanent structures or permanent changes in existing facilities.

8.3 If the Contractor fails to correct Work which is not in accordance with the requirements of the Contract Documents or persistently fails to carry out the Work in accordance with the Contract Documents, the Owner, by a written order, may order the Contractor to stop the Work, or any portion thereof, until the cause for such order has been eliminated; however, the right of the Owner to stop the Work shall not give rise to a duty on the part of the Owner to exercise this right for the benefit of the Contractor or any other person or entity.

ARTICLE 9
CONTRACTOR

9.1 The Contractor shall supervise and direct the Work, using the Contractor's best skill and attention. The Contractor shall be solely responsible for and have control over construction means, methods, techniques, sequences and procedures and for coordinating all portions of the Work under the Contract, unless Contract Documents give other specific instructions concerning these matters.

9.2 Unless otherwise provided in the Contract Documents, the Contractor shall provide and pay for labor, materials, equipment, tools, construction equipment and machinery, water, heat, utilities, transportation, and other facilities and services necessary for the proper execution and completion of the Work, whether temporary or permanent and whether or not incorporated or to be incorporated in the Work.

9.3 The Contractor shall enforce strict discipline and good order among the Contractor's employees and other persons carrying out the Contract. The Contractor shall not permit employment of unfit persons or persons not skilled in tasks assigned to them.

9.4 The Contractor warrants to the Owner and Architect that materials and equipment furnished under the Contract will be of good quality and new unless otherwise required or permitted by the Contract Documents, that the Work will be free from defects not inherent in the quality required or permitted, and that the Work will conform with the requirements of the Contract Documents. Work not conforming to these requirements, including substitutions not properly approved and authorized, may be considered defective. The Contractor's warranty excludes remedy for damage or defect caused by abuse, modifications not executed by the Contractor, improper or insufficient maintenance, improper operation, or normal wear and tear under normal usage. If required by the Architect, the Contractor shall furnish satisfactory evidence as to the kind and quality of materials and equipment.

9.5 Unless otherwise provided in the Contract Documents, the Contractor shall pay sales, consumer, use, and other similar taxes which are legally enacted when bids are received or negotiations concluded, whether or not yet effective or merely scheduled to go into effect, and shall secure and pay for the building permit and other permits and governmental fees, licenses and inspections necessary for proper execution and completion of the Work.

9.6 The Contractor shall comply with and give notices required by laws, ordinances, rules, regulations, and lawful orders of public authorities bearing on performance of the Work. The Contractor shall promptly notify the Architect and Owner if the Drawings and Specifications are observed by the Contractor to be at variance therewith.

9.7 The Contractor shall be responsible to the Owner for the acts and omissions of the Contractor's employees, Subcontractors and their agents and employees, and other persons performing portions of the Work under a contract with the Contractor.

9.8 The Contractor shall review, approve and submit to the Architect Shop Drawings, Product Data, Samples and similar submittals required by the Contract Documents with reasonable promptness. The Work shall be in accordance with approved submittals. When professional certification of performance criteria of materials, systems or equipment is required by the Contract Documents, the Architect shall be entitled to rely upon the accuracy and completeness of such certifications.

9.9 The Contractor shall keep the premises and surrounding area free from accumulation of waste materials or rubbish caused by operations under the Contract. At completion of the Work the Contractor shall remove from and about the Project waste materials, rubbish, the Contractor's tools, construction equipment, machinery and surplus materials.

9.10 The Contractor shall provide the Owner and Architect access to the Work in preparation and progress wherever located.

9.11 The Contractor shall pay all royalties and license fees; shall defend suits or claims for infringement of patent rights and shall hold the Owner harmless from loss on account thereof, but shall not be responsible for such defense or loss when a particular design, process or product of a particular manufacturer or manufacturers is required by the Contract Documents unless the Contractor has reason to believe that there is an infringement of patent.

9.12 To the fullest extent permitted by law, the Contractor shall indemnify and hold harmless the Owner, Architect, Architect's consultants, and agents and employees of any of them from and against claims, damages, losses and expenses, including but not limited to attorneys' fees, arising out of or resulting from performance of the Work, provided that such claim, damage, loss or expense is attributable to bodily injury, sickness, disease or death, or to injury to or destruction of tangible property (other than the Work itself) including loss of use resulting therefrom, but only to the extent caused in whole or in part by negligent acts or omissions of the Contractor, a Subcontractor, anyone directly or indirectly employed by them or anyone for whose acts they may be liable, regardless of whether or not such claim, damage, loss or expense is caused in part by a party indemnified hereunder. Such obligation shall not be construed to negate, abridge, or reduce other rights or obligations of idemnity which would otherwise exist as to a party or person described in this Paragraph 9.12.

9.12.1 In claims against any person or entity indemnified under this Paragraph 9.12 by an employee of the Contractor, a Subcontractor, anyone directly or indirectly employed by them or anyone for whose acts they may be liable, the indemnification obligation under this Paragraph 9.12 shall not be limited by a limitation on amount or type of damages, compensation or benefits payable by or for the Contractor or a Subcontractor under workers' or workmen's compensation acts, disability benefit acts or other employee benefit acts.

9.12.2 The obligations of the Contractor under this Paragraph 9.12 shall not extend to the liability of the Architect, the Architect's consultants, and agents and employees of any of them arising out of (1) the preparation or approval of maps, drawings, opinions, reports, surveys, Change Orders, Construction Change Directives, designs or specifications, or (2) the giving of or the failure to give directions or instructions by the Architect, the Architect's consultants, and agents and employees of any of them provided such giving or failure to give is the primary cause of the injury or damage.

ARTICLE 10
ADMINISTRATION OF THE CONTRACT

10.1 The Architect will provide administration of the Contract and will be the Owner's representative (1) during construction, (2) until final payment is due and (3) with the Owner's concurrence, from time to time during the correction period described in Paragraph 18.1

10.2 The Architect will visit the site at intervals appropriate to the stage of construction to become generally familiar with the progress and quality of the completed Work and to determine in general if the Work is being performed in a manner indicating that the Work, when completed, will be in accordance with the Contract Documents. However, the Architect will not be required to make exhaustive or continuous on-site inspections to check quality or quantity of the Work. On the basis of on-site observations as an architect, the Architect will keep the Owner informed of progress of the Work and will endeavor to guard the Owner against defects and deficiencies in the Work.

10.3 The Architect will not have control over or charge of and will not be responsible for construction means, methods, techniques, sequences or procedures, or for safety precautions and programs in connection with the Work, since these are solely the Contractor's responsibility as provided in Paragraphs 9.1 and 16.1. The Architect will not be responsible for the Contractor's failure to carry out the Work in accordance with the Contract Documents.

10.4 Based on the Architect's observations and evaluations of the Contractor's Applications for Payment, the Architect will review and certify the amounts due the Contractor and will issue Certificates for Payment in such amounts.

10.5 The Architect will interpret and decide matters concerning performance under and requirements of the Contract Documents on written request of either the Owner or Contractor. The Architect will make initial decisions on all claims, disputes or other matters in question between the Owner and Contractor, but will not be liable for results of any interpretations or decisions rendered in good faith. The Architect's decisions in matters relating to aesthetic effect will be final if consistent with the intent expressed in the Contract Documents. All other decisions of the Architect, except those which have been waived by making or acceptance of final payment, shall be subject to arbitration upon the written demand of either party.

10.6 The Architect will have authority to reject Work which does not conform to the Contract Documents.

10.7 The Architect will review and approve or take other appropriate action upon the Contractor's submittals such as Shop Drawings, Product Data and Samples, but only for the limited purpose of checking for conformance with information given and the design concept expressed in the Contract Documents.

10.8 All claims or disputes between the Contractor and the Owner arising out or relating to the Contract, or the breach thereof, shall be decided by arbitration in accordance with the Construction Industry Arbitration Rules of the American Arbitration Association currently in effect unless the parties mutually agree otherwise and subject to an initial presentation of the claim or dispute to the Architect as required under Paragraph 10.5. Notice of the demand for arbitration shall be filed in writing with the other party to this Agreement and with the American Arbitration Association and shall be made within a reasonable time after the dispute has arisen. The award rendered by

AIA DOCUMENT A107 • ABBREVIATED OWNER-CONTRACTOR AGREEMENT • NINTH EDITION • AIA® • ©1987
THE AMERICAN INSTITUTE OF ARCHITECTS, 1735 NEW YORK AVENUE, N.W., WASHINGTON, D.C. 20006

Fig. 8-3 (Continued)

the arbitrator or arbitrators shall be final, and judgment may be entered upon it in accordance with applicable law in any court having jurisdiction thereof. Except by written consent of the person or entity sought to be joined, no arbitration arising out of or relating to the Contract Documents shall include, by consolidation, joinder or in any other manner, any person or entity not a party to the Agreement under which such arbitration arises, unless it is shown at the time the demand for arbitration is filed that (1) such person or entity is substantially involved in a common question of fact or law, (2) the presence of such person or entity is required if complete relief is to be accorded in the arbitration, (3) the interest or responsibility of such person or entity in the matter is not insubstantial, and (4) such person or entity is not the Architect or any of the Architect's employees or consultants. The agreement herein among the parties to the Agreement and any other written agreement to arbitrate referred to herein shall be specifically enforceable under applicable law in any court having jurisdiction thereof.

ARTICLE 11
SUBCONTRACTS

11.1 A Subcontractor is a person or entity who has a direct contract with the Contractor to perform a portion of the Work at the site.

11.2 Unless otherwise stated in the Contract Documents or the bidding requirements, the Contractor, as soon as practicable after award of the Contract, shall furnish in writing to the Owner through the Architect the names of the Subcontractors for each of the principal portions of the Work. The Contractor shall not contract with any Subcontractor to whom the Owner or Architect has made reasonable and timely objection. The Contractor shall not be required to contract with anyone to whom the Contractor has made reasonable objection. Contracts between the Contractor and Subcontractors shall (1) require each Subcontractor, to the extent of the Work to be performed by the Subcontractor, to be bound to the Contractor by the terms of the Contract Documents, and to assume toward the Contractor all the obligations and responsibilities which the Contractor, by the Contract Documents, assumes toward the Owner and Architect, and (2) allow to the Subcontractor the benefit of all rights, remedies and redress afforded to the Contractor by these Contract Documents.

ARTICLE 12
CONSTRUCTION BY OWNER OR BY SEPARATE CONTRACTORS

12.1 The Owner reserves the right to perform construction or operations related to the Project with the Owner's own forces, and to award separate contracts in connection with other portions of the Project or other construction or operations on the site under conditions of the contract identical or substantially similar to these, including those portions related to insurance and waiver of subrogation. If the Contractor claims that delay or additional cost is involved because of such action by the Owner, the Contractor shall make such claim as provided elsewhere in the Contract Documents.

12.2 The Contractor shall afford the Owner and separate contractors reasonable opportunity for the introduction and storage of their materials and equipment and performance of their activities, and shall connect and coordinate the Contractor's construction and operations with theirs as required by the Contract Documents.

12.3 Costs caused by delays, improperly timed activities or defective construction shall be borne by the party responsible therefor.

ARTICLE 13
CHANGES IN THE WORK

13.1 The Owner, without invalidating the Contract, may order changes in the Work consisting of additions, deletions or modifications, the Contract Sum and Contract Time being adjusted accordingly. Such changes in the Work shall be authorized by written Change Order signed by the Owner, Contractor and Architect, or by written Construction Change Directive signed by the Owner and Architect.

13.2 The Contract Sum and Contract Time shall be changed only by Change Order.

13.3 The cost or credit to the Owner from a change in the Work shall be determined by mutual agreement.

ARTICLE 14
TIME

14.1 Time limits stated in the Contract Documents are of the essence of the Contract. By executing the Agreement the Contractor confirms that the Contract Time is a reasonable period for performing the Work.

14.2 The date of Substantial Completion is the date certified by the Architect in accordance with Paragraph 15.3.

14.3 If the Contractor is delayed at any time in progress of the Work by changes ordered in the Work, by labor disputes, fire, unusual delay in deliveries, abnormal adverse weather conditions not reasonably anticipatable, unavoidable casualties or any causes beyond the Contractor's control, or by other causes which the Architect determines may justify delay, then the Contract Time shall be extended by Change Order for such reasonable time as the Architect may determine.

ARTICLE 15
PAYMENTS AND COMPLETION

15.1 Payments shall be made as provided in Articles 4 and 5 of this Agreement.

15.2 Payments may be withheld on account of (1) defective Work not remedied, (2) claims filed by third parties, (3) failure of the Contractor to make payments properly to Subcontractors or for labor, materials or equipment, (4) reasonable evidence that the Work cannot be completed for the unpaid balance of the Contract Sum, (5) damage to the Owner or another contractor, (6) reasonable evidence that the Work will not be completed within the Contract Time and that the unpaid balance would not be adequate to cover actual or liquidated damages for the anticipated delay, or (7) persistent failure to carry out the Work in accordance with the Contract Documents.

15.3 When the Architect agrees that the Work is substantially complete, the Architect will issue a Certificate of Substantial Completion.

15.4 Final payment shall not become due until the Contractor has delivered to the Owner a complete release of all liens arising out of this Contract or receipts in full covering all labor, materials and equipment for which a lien could be filed, or a bond satisfactory to the Owner to indemnify the Owner against such

lien. If such lien remains unsatisfied after payments are made, the Contractor shall refund to the Owner all money that the Owner may be compelled to pay in discharging such lien, including all costs and reasonable attorneys' fees.

15.5 The making of final payment shall constitute a waiver of claims by the Owner except those arising from:

.1 liens, claims, security interests or encumbrances arising out of the Contract and unsettled;

.2 failure of the Work to comply with the requirements of the Contract Documents; or

.3 terms of special warranties required by the Contract Documents.

Acceptance of final payment by the Contractor, a Subcontractor or material supplier shall constitute a waiver of claims by that payee except those previously made in writing and identified by that payee as unsettled at the time of final Application for Payment.

ARTICLE 16
PROTECTION OF PERSONS AND PROPERTY

16.1 The Contractor shall be responsible for initiating, maintaining, and supervising all safety precautions and programs in connection with the performance of the Contract. The Contractor shall take reasonable precautions for safety of, and shall provide reasonable protection to prevent damage, injury or loss to:

.1 employees on the Work and other persons who may be affected thereby;

.2 the Work and materials and equipment to be incorporated therein; and

.3 other property at the site or adjacent thereto.

The Contractor shall give notices and comply with applicable laws, ordinances, rules, regulations and lawful orders of public authorities bearing on safety of persons and property and their protection from damage, injury or loss. The Contractor shall promptly remedy damage and loss to property at the site caused in whole or in part by the Contractor, a Subcontractor, a Sub-subcontractor, or anyone directly or indirectly employed by any of them, or by anyone for whose acts they may be liable and for which the Contractor is responsible under Subparagraphs 16.1.2 and 16.1.3, except for damage or loss attributable to acts or omissions of the Owner or Architect or by anyone for whose acts either of them may be liable, and not attributable to the fault or negligence of the Contractor. The foregoing obligations of the Contractor are in addition to the Contractor's obligations under Paragraph 9.12.

16.2 The Contractor shall not be required to perform without consent any Work relating to asbestos or polychlorinated biphenyl (PCB).

ARTICLE 17
INSURANCE

17.1 The Contractor shall purchase from and maintain in a company or companies lawfully authorized to do business in the jurisdiction in which the Project is located insurance for protection from claims under workers' or workmen's compensation acts and other employee benefit acts which are applicable, claims for damages because of bodily injury, including death, and from claims for damages, other than to the Work

itself, to property which may arise out of or result from the Contractor's operations under the Contract, whether such operations be by the Contractor or by a Subcontractor or anyone directly or indirectly employed by any of them. This insurance shall be written for not less than limits of liability specified in the Contract Documents or required by law, whichever coverage is greater, and shall include contractual liability insurance applicable to the Contractor's obligations under Paragraph 9.12. Certificates of such insurance shall be filed with the Owner prior to the commencement of the Work.

17.2 The Owner shall be responsible for purchasing and maintaining the Owner's usual liability insurance. Optionally, the Owner may purchase and maintain other insurance for self-protection against claims which may arise from operations under the Contract. The Contractor shall not be responsible for purchasing and maintaining this optional Owner's liability insurance unless specifically required by the Contract Documents.

17.3 Unless otherwise provided, the Owner shall purchase and maintain, in a company or companies lawfully authorized to do business in the jurisdiction in which the Project is located, property insurance upon the entire Work at the site to the full insurable value thereof. This insurance shall be on an all-risk policy form and shall include interests of the Owner, the Contractor, Subcontractors and Sub-subcontractors in the Work and shall insure against the perils of fire and extended coverage and physical loss or damage including, without duplication of coverage, theft, vandalism and malicious mischief.

17.4 A loss insured under Owner's property insurance shall be adjusted with the Owner and made payable to the Owner as fiduciary for the insureds, as their interests may appear, subject to the requirements of any applicable mortgagee clause.

17.5 The Owner shall file a copy of each policy with the Contractor before an exposure to loss may occur. Each policy shall contain a provision that the policy will not be cancelled or allowed to expire until at least 30 days' prior written notice has been given to the Contractor.

17.6 The Owner and Contractor waive all rights against each other and the Architect, Architect's consultants, separate contractors described in Article 12, if any, and any of their subcontractors, sub-subcontractors, agents and employees, for damages caused by fire or other perils to the extent covered by property insurance obtained pursuant to this Article 17 or any other property insurance applicable to the Work, except such rights as they may have to the proceeds of such insurance held by the Owner as fiduciary. The Contractor shall require similar waivers in favor of the Owner and the Contractor by Subcontractors and Sub-subcontractors. The Owner shall require similar waivers in favor of the Owner and Contractor by the Architect, Architect's consultants, separate contractors described in Article 12, if any, and the subcontractors, sub-subcontractors, agents and employees of any of them.

ARTICLE 18
CORRECTION OF WORK

18.1 The Contractor shall promptly correct Work rejected by the Architect or failing to conform to the requirements of the Contract Documents, whether observed before or after Substantial Completion and whether or not fabricated, installed or completed, and shall correct any Work found to be not in accordance with the requirements of the Contract Documents within a period of one year from the date of Substantial Com-

AIA DOCUMENT A107 • ABBREVIATED OWNER-CONTRACTOR AGREEMENT • NINTH EDITION • AIA® • ©1987
THE AMERICAN INSTITUTE OF ARCHITECTS, 1735 NEW YORK AVENUE, N.W., WASHINGTON, D.C. 20006

Fig. 8-3 (Continued)

pletion of the Contract or by terms of an applicable special warranty required by the Contract Documents. The provisions of this Article 18 apply to Work done by Subcontractors as well as to Work done by direct employees of the Contractor.

18.2 Nothing contained in this Article 18 shall be construed to establish a period of limitation with respect to other obligations which the Contractor might have under the Contract Documents. Establishment of the time period of one year as described in Paragraph 18.1 relates only to the specific obligation of the Contractor to correct the Work, and has no relationship to the time within which the obligation to comply with the Contract Documents may be sought to be enforced, nor to the time within which proceedings may be commenced to establish the Contractor's liability with respect to the Contractor's obligations other than specifically to correct the Work.

ARTICLE 19
MISCELLANEOUS PROVISIONS

19.1 The Contract shall be governed by the law of the place where the Project is located.

19.2 As between the Owner and the Contractor, any applicable statute of limitations shall commence to run and any alleged cause of action shall be deemed to have accrued:

.1 not later than the date of Substantial Completion for acts or failures to act occurring prior to the relevant date of Substantial Completion;

.2 not later than the date of issuance of the final Certificate for Payment for acts or failures to act occurring subsequent to the relevant date of Substantial Completion and prior to issuance of the final Certificate for Payment; and

.3 not later than the date of the relevant act or failure to act by the Contractor for acts or failures to act occurring after the date of the final Certificate for Payment.

ARTICLE 20
TERMINATION OF THE CONTRACT

20.1 If the Architect fails to recommend payment for a period of 30 days through no fault of the Contractor, or if the Owner fails to make payment thereon for a period of 30 days, the Contractor may, upon seven additional days' written notice to the Owner and the Architect, terminate the Contract and recover from the Owner payment for Work executed and for proven loss with respect to materials, equipment, tools, and construction equipment and machinery, including reasonable overhead, profit and damages applicable to the Project.

20.2 If the Contractor defaults or persistently fails or neglects to carry out the Work in accordance with the Contract Documents or fails to perform a provision of the Contract, the Owner, after seven days' written notice to the Contractor and without prejudice to any other remedy the Owner may have, may make good such deficiencies and may deduct the cost thereof, including compensation for the Architect's services and expenses made necessary thereby, from the payment then or thereafter due the Contractor. Alternatively, at the Owner's option, and upon certification by the Architect that sufficient cause exists to justify such action, the Owner may terminate the Contract and take possession of the site and of all materials, equipment, tools, and construction equipment and machinery thereon owned by the Contractor and may finish the Work by whatever method the Owner may deem expedient. If the unpaid balance of the Contract Sum exceeds costs of finishing the Work, including compensation for the Architect's services and expenses made necessary thereby, such excess shall be paid to the Contractor, but if such costs exceed such unpaid balance, the Contractor shall pay the difference to the Owner.

ARTICLE 21
OTHER CONDITIONS OR PROVISIONS

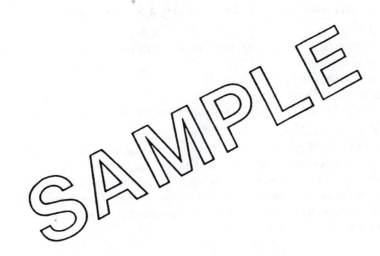

This Agreement entered into as of the day and year first written above.

OWNER

CONTRACTOR

_____
(Signature)

_____
(Signature)

_____
(Printed name and title)

_____
(Printed name and title)

The following are the General Requirements for the sample house, discussed in this textbook.

Specifications

DIVISION 1: GENERAL REQUIREMENTS

A. **ARCHITECT'S SUPERVISION**

The architect will have continual supervisory responsibility for this job.

B. **TEMPORARY CONVENIENCES**

The general contractor shall provide suitable temporary conveniences for the use of all workers on this job. Facilities shall be within a weather tight, painted enclosure complying with legal requirements. The general contractor shall maintain all temporary toilet facilities in a sanitary condition.

C. **PUMPING**

The general contractor shall keep the excavation and the basement free from water at all times and shall provide, maintain, and operate at his own expense such pumping equipment as shall be necessary.

D. **PROTECTION**

The general contractor shall protect all existing driveways, parking areas, sidewalks, curbs, and existing paved areas on, or adjacent to, the owner's property.

E. **GRADES, LINES, LEVELS, AND SURVEYS**

The owner shall establish the lot lines.

The general contractor shall:

1. Establish and maintain bench marks.
2. Verify all grades, lines, levels, and dimensions as shown on the drawings, and report any errors or inconsistencies before commencing work.
3. Lay out the building accurately under the supervision of the architect.

F. **FINAL CLEANING**

In addition to the general room cleaning, the general contractor shall do the following special cleaning upon completion of the work:

1. Wash and polish all glass and cabinets.
2. Clean and polish all hardware.
3. Remove all marks, stains, fingerprints, and other soil or dirt from walls, woodwork, and floors.

G. **GUARANTEES**

The general contractor shall guarantee all work performed under the contract against faulty materials or workmanship. The guarantee shall be in writing with duplicate copies delivered to the architect. In case of work performed by subcontractors where guarantees are required, the general contractor shall secure written guarantees from those subcontractors. Copies of these guarantees shall be delivered to the architect upon completion of the work. Guarantees shall be signed by both the subcontractor and the general contractor.

H. **FOREMAN**

The general contractor shall have a responsible foreman at the building site from the start to the completion of construction. The foreman shall be on duty during all working hours.

I. **FIRE INSURANCE**

The owner shall effect and maintain builder's risk completed-value insurance on this job.

SUMMARY

In a set of specifications the General Conditions and the General Requirements division describe the responsibilities of the contractor and the owner. These divisions discuss the clearing of the site, the location of lot lines, the location and scope of the work to be performed, insurance required by the contractor and the owner, the contract, and any restrictions. The contract is the agreement signed by the contrator and the owner. The contract indicates what is to be done, the time limitation, the contract amount, the payment schedule, and the final payment. The contract, with the specifications and working drawings, becomes a legal document. It can be used in a court of law in the case of a legal dispute.

REVIEW QUESTIONS

Select the letter preceding the best answer.

1. Which of the following is covered by the General Conditions section of a set of specifications?

   a. The form of the contract

   b. Supervisory responsibilities

   c. Order of precedence of drawings and specifications

   d. All of these

2. Where would the requirement of temporary provisions be found?

   a. A.I.A. Short Form of Agreement
   b. Drawings
   c. General Conditions section of the specifications
   d. General Requirements division of the specifications

3. The order of precedence is listed in the General Conditions section of the specifications to help solve conflicts between the drawings and specifications. Number the following items in the order of their precedence.

   _____ Detail specifications
   _____ Detail (large scale) drawings
   _____ Small scale drawings
   _____ General Conditions

4. Where is the responsibility for supervision of the contract discussed?

   a. Title page
   b. General requirements

   c. General conditions
   d. Mortgage

5. Who is responsible for establishing lot lines?

   a. Contractor
   b. Owner

   c. Architect
   d. None of these

6. Who is responsible for protecting the work?

   a. Contractor
   b. Owner

   c. Architect
   d. All of these

7. Who is responsible for maintaining builder's risk insurance?

   a. Contractor
   b. Owner

   c. Architect
   d. Bank

8. What is the purpose of a bid bond?

   a. To insure that the bid is not too high
   b. To insure that the bidder is a qualified contractor
   c. To insure that the bid will not be withdrawn for a certain period of time
   d. To insure that all work is properly performed

9. Who is the owner's representative during the construction of the building?

   a. Bank
   b. Building inspector

   c. Architect
   d. Contractor

10. Where is the progress of payments indicated?

    a. On the mortgage
    b. On the working drawings

    c. In the General Conditions
    d. In the General Requirements

SECTION 3 STRUCTURAL WORK
unit 9 site work

OBJECTIVES

After completing this unit the student will be able to

- explain the specifications regarding site work.
- discuss plot plans.
- determine, in cubic yards, the amount of earth to be excavated.

The first construction work on a new building is usually the excavation of the foundation and cellar. This may not require the ordering of materials, but the estimator must determine the quantity of material to be moved, as this determines the cost of the excavation.

Estimating the excavation is done in much the same manner as estimating other phases of construction. The estimator receives a complete set of specifications in one unit and a complete set of working drawings in another. Each of these sets is referred to for each part of the construction process.

SPECIFICATIONS

The specifications for site work are found in Division 2 of the specifications for the sample house.

Specifications

DIVISION 2: SITE WORK

A. WORK INCLUDED

This work shall include, but shall not be limited by the following:
1. Clearing site
2. Excavating, backfilling, grading, and related items
3. Removal of excess earth
4. Protection of existing trees to remain on the site

All excavation and backfill required for heating, plumbing, and electrical work will be done by the respective contractors and are not included under site work.

B. CLEARING THE SITE

1. Clear the area within the limits of the building of all trees, shrubs, or other obstructions as necessary.
2. Within the limits of grading work as shown on the drawings, remove such trees, shrubs, or other obstructions as are indicated on the drawings to be removed, without injury to trunks, interfering branches, and roots of trees to remain. Do cutting and trimming only as directed. Box and protect all trees and shrubs in the construction area to remain; maintain boxing until finished grading is completed.
3. Remove all debris from the site; do not use it for fill.

C. EXCAVATION

1. Carefully remove all sod and soil throughout the area of the building and where finish grade levels are changed. Pile on site where directed. This soil is to be used later for finished grading. Do not strip below the topsoil.

2. Do all excavation required for footings, piers, walls, trenches, areas, pits, and foundations. Remove all materials encountered in obtaining indicated lines and grades required.

   Beds for all foundations and footings must have solid, level, and undisturbed bed bottoms. No backfill will be allowed and all footings shall rest on unexcavated earth.

3. The general contractor shall notify the architect when the excavation is complete so that he or she may inspect the soil before the concrete is placed.

4. Excavate to elevations and dimensions indicated, leaving sufficient space to permit erection of walls, drain tile, waterproofing, masonry, and the inspection of foundations. Protect the bottom of the excavation from frost. Do not place the foundation, footings, or slabs on frozen ground.

D. BACKFILL

1. All outside walls shall be backfilled to within 6 inches of the finished grade with clean fill. Backfill shall be thoroughly puddled and tamped solid.

2. Backfill under basement floor slabs and elsewhere as required to bring the earth to proper levels and grades for subsequent work. Use only gravel or stone fill properly tamped. All fill shall be well tamped and puddled to prevent settling.

3. Unless otherwise directed by the architect, no backfill shall be placed until after the first floor framing is in place. No backfill shall be placed until all walls have developed such strength to resist thrust due to filling operations.

E. INTERIOR GRADING

1. Furnish and place graded gravel or bank run gravel fill as approved under all basement floor slabs.

2. Fill under all slabs shall be sand not less than 6 inches deep. Where existing grades are lower than 6 inches below the bottom of the slabs, sand shall be used for the entire depth of fill.

3. All fill shall be shaped to line and grade and brought to proper elevations without voids and depressions to receive the concrete slabs.

F. EXTERIOR GRADING

See Plot Plan for area limits under contract.

1. Do all excavating, filling, and rough grading to bring entire area outside of the building to levels shown on the Plot Plan and at the building as shown on the drawings.

2. Where existing trees are to remain, if the new grade is lower than the natural grade under the trees, a sloping mound shall be left under the base of the tree extending out as far as the branches; if the grade is higher, a dry well shall be constructed around the base of the tree to provide the roots with air and moisture.

   Sand and, or gravel, removed from the excavation for the building, may be used for grading.

3. After rough grading has been completed and approved, spread the topsoil evenly to the previously stripped area. Prepare the topsoil to receive grass seed by removing

stone, debris, and unsuitable materials. Hand rake to remove water pockets and irregularities.

Seeding will be done by the owner.

PLOT PLANS

The plot plan for a residence accompanies the specifications and is usually found on the first page of the working drawings. It may be drawn with an engineer's scale in feet and decimal parts of a foot or with an architect's scale at 1/16" = 1'-0".

In most areas, zoning laws require a survey and a plot plan to be submitted before a building permit is issued. Most banks and lending institutions also require a survey and a plot plan before any loans can be approved on the property. Zoning laws define or restrict the type of building that may be built in specific areas of the community. They may, for example, order commercial buildings to be separated from residential buildings. Zoning laws vary from one community to another.

The following is a list of features shown on most plot plans:

- Scale and title of the drawing
- Dimensions of the plot
- Compass bearings
- Locations and dimensions of all buildings
- Locations and dimensions of walks, driveways, fences, etc.
- Elevation of main floor
- Finish grade levels of main corners of the building
- Locations of a bench mark and its elevation
- Location of utilities, such as electric, water, gas, and sewage

In reference to plot plans, the terms location, elevation, and size have special meanings. The location of the lot is determined by a surveyor, who may be either government employed or privately employed. Surveyors get information from municipal land records, deeds, and markers established by other surveyors. The size of a plot is determined by its boundary lines.   The survey indicates the length and compass bearing of each boundary line. The elevation is determined from a known coastal sea level. Almost every city or town has one or more permanent points established from this sea level. These points are known as city or county

datum. These are marked by a metal rod fixed in a concrete base which is set in the ground. These points, from which elevations are determined, are referred to as *bench marks.*

CONTOUR LINES

On some plot plans, where the ground is irregular, the surveyor may show *contour lines.* These are continuous lines running through all points of the same elevation of a plot of land, figure 9-1. The vertical distance between the contour lines is called the *vertical contour interval.* These intervals can be any measurement, but normally, they are one foot. For example, if two contour lines are marked 100' and 99', the vertical contour interval is one foot. This means that the ground is one foot higher along the 100' line than it is at any point along the 99' line. This one foot difference in elevation is the same regardless of the spacing between the contour lines. Therefore, the closer the contour lines are together, the steeper the slope is at that point.

The plan view of figure 9-1 shows how the contour lines might appear on a plot plan. The

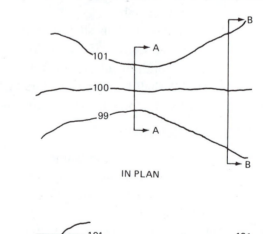

IN PLAN

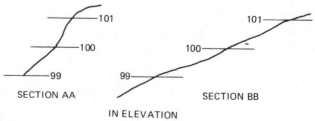

SECTION AA   SECTION BB

IN ELEVATION

Fig. 9-1

elevation view provides a better understanding of how the slope appears. Notice that the slope is steeper in section AA than it is in section BB. This is because the contour lines are much closer on the plan view at the point where section AA is taken.

On most plot plans the surveyor establishes the reference elevation mark in feet and inches for the convenience of the contractor. Frequently, however, the surveyor measures in feet and decimal parts of a foot with an engineer's scale. An engineer's scale has each foot divided into ten parts and each of these tenths divided again into ten parts. An example of a measurement taken from an engineer's scale might be 112.129 feet. To convert such a dimension into feet and inches use the following procedure:

    a. Multiply the decimal part by 12 to find the number of inches and decimal parts of an inch. (.129 x 12 = 1.548 inches)

    b. Multiply the remaining decimal by 16 to find 16ths of an inch. (.548 x 16 = 8.768/16 or 9/16 inch)

    c. Add the fractional part of an inch to the whole feet and inches. (112 feet, 1 inch + 9/16 inch = 112 feet, 1-9/16 inches)

    d. To convert to 64ths of an inch instead of 16ths of an inch multiply the remaining decimal by 64 instead of 16. (.548 x 64 = 35.072/64 or 35/64 inch)

The excavation is measured in cubic yards of earth to be removed. To find this amount, make a section view of the plot to figure from.

Excavations for the main footings should extend at least two feet beyond the foundation walls allowing room to work around the footing. To determine the cubic yards to be excavated, first multiply the area to be excavated in square feet by the depth of the excavation in feet. Divide this product by 27 to find the number of cubic yards. There are 27 cubic feet in a cubic yard.

For trench work, multiply the linear feet by the width of the trench in feet. Multiply this number by the depth in feet and divide by 27. The result is the number of cubic yards to be excavated. Excavations of sloping ground for buildings are computed by using the average depth.

Normally, an excavating contractor is hired for the excavation and backfill, figure 9-2. Both operations are paid for in one payment. The operator strips the topsoil and excavates the cellar to the specified depth. After the foundation wall is either waterproofed or dampproofed the earth is pushed back and graded so that the land slopes away from the foundation. The topsoil is then spread over the area to complete the backfill operation.

Before the final contract price is agreed upon, the contractor should visit the construction site. This is to determine whether any additional fill is needed, or if earth and trees must be removed from the site.

The plot plan for the sample house is the one for which the specifications shown in this book were written.

SUMMARY

The specification division for site work discusses permits, protection of adjoining property, how the land is to be cleared, and the excavation. It also gives information about backfilling, grading, and clearing the site.

The plot plan shows the elevation of the land, the location of the building on the lot, and the boundary lines. It shows the shape or contour of the land and any trees that are to be saved. The plot plan also gives the finished elevation and layout of all sidewalks and curbs. In other words, the plot plan provides all of the information necessary to clear the lot, the size of the lot, and the location of the building.

Fig. 9-2 Excavating for a foundation

REVIEW QUESTIONS

Select the letter preceding the best answer.

1. Who is responsible for repairing damage to adjoining property?

    a. Owner of the adjoining property
    b. Owner of the property on which the construction is being done
    c. Contractor
    d. Individual worker doing the damage

2. Where is the plot plan normally found?

    a. In the specifications
    b. On the first page of the working drawings
    c. On the last page of the working drawings
    d. Plot plans are usually separate from other documents.

3. If the plot plan is drawn to an engineer's scale, how are the dimensions given?

    a. Feet and decimal parts of a foot   c. Feet and inches
    b. Yards and feet   d. Rods and fractions of a rod

4. List five features normally shown on a plot plan.

5. What is a bench mark?

    a. A survey point giving elevation
    b. A reference point from which the location of a building can be measured
    c. A metal rod driven in the ground to indicate that a zoning law is in effect
    d. None of the above

6. What is indicated if the contour lines on a plot plan are close together?

    a. A steep slope   c. Deep topsoil
    b. A gradual slope   d. A zoning law is in effect

7. Convert 59.579 feet to feet and inches. Round off to the nearest 1/16 inch.

    a. 6 feet 9 3/16 inches   c. 59 feet 15/16 inches
    b. 59 feet 6 9/16 inches   d. 59 feet 6 15/16 inches

8. Approximately how many cubic yards of earth must be removed for an excavation measuring 26 feet x 40 feet x 6 feet?

    a. 308 cubic yards   c. 231 cubic yards
    b. 1040 cubic yards   d. 130 cubic yards

9. How many cubic yards of earth must be excavated for a trench measuring 24 inches wide x 162 feet long x 8 inches deep?

    a. 324 cubic yards   c. 8 cubic yards
    b. 192 cubic yards   d. 213 cubic yards

10. Referring to figure 9-3 on page 49, what is the relationship between the elevation of the first floor of the building and the sidewalk?

    a. They are the same.

    b. The sidewalk is lower.

    c. The sidewalk is higher.

    d. It is impossible to tell from this plot plan.

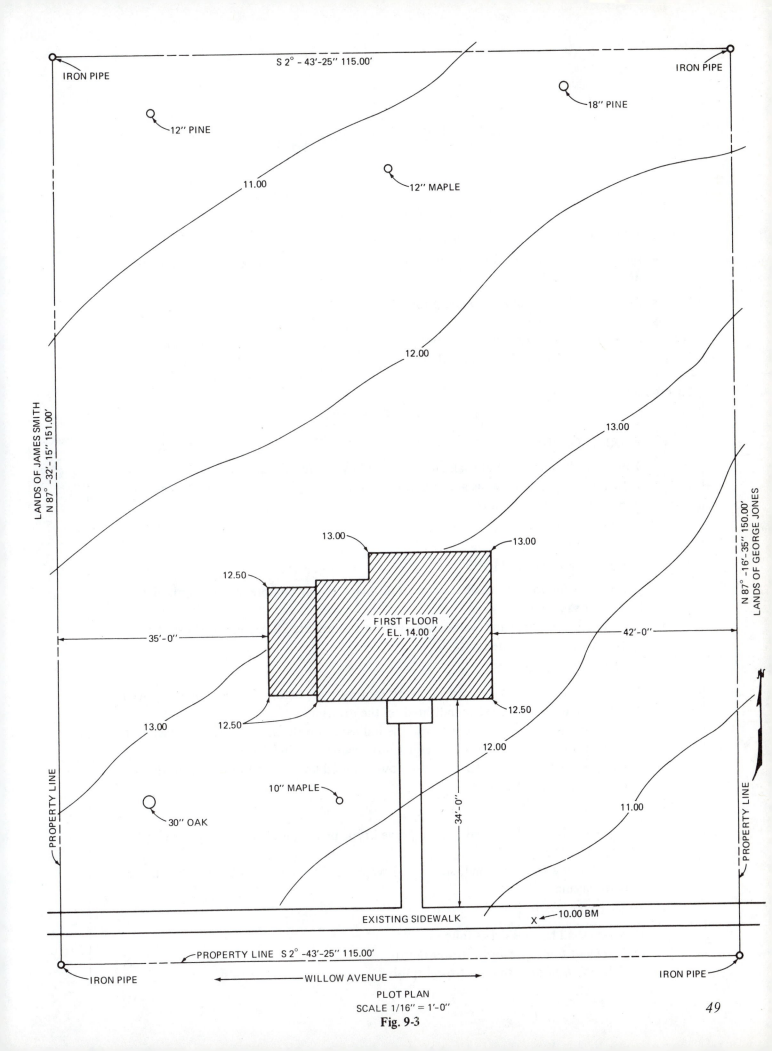

PLOT PLAN
SCALE 1/16" = 1'-0"

Fig. 9-3

49

unit 10 footings

OBJECTIVES

After studying this unit the student will be able to

- explain the specifications relating to concrete work.

- tell how to recognize concrete symbols and where to find them on a set of work. drawings.

- determine the quantity of concrete for footings.

The next two units on concrete cover both the footings for the house and the concrete floors and slabs. The following specifications describe the concrete and how it is to be placed and finished.

Specifications

DIVISION 3: CONCRETE

A.   WORK INCLUDED

This work shall include but shall not be limited by the following:
1. Concrete footing for walls and support posts
2. Basement concrete floors
3. Porch floors
4. Sidewalks

B.   CONCRETE PROPORTIONS

1. Concrete for footings shall develop an ultimate 28-day compressive strength of 3000 lbs. per square inch.
2. Concrete for floors and walks shall develop an ultimate 28-day compressive strength of 2500 lbs. per square inch.

C.   CONCRETE FOOTINGS

1. Notify the architect for inspection of footing beds before any concrete is placed.
2. All footings shall be of sizes indicated on the drawings.
3. All footings shall start at the elevations indicated on the drawings and at least 3'-0" below finish grade. Colder climates require deeper footing depths.
   Where excavation has been carried below required level, start footings from the bottom of the excavation.

D.   MIXING CONCRETE

1. All concrete shall be mixed in an approved type power batch mixer equipped with a water measuring device.
2. For freezing weather; sand, stone, and water for concrete shall be properly heated before using.

E.   CONCRETE FINISH

Concrete shall be finished as follows:
1. *Basement floor* — smooth trowel finish
2. *Side porch floor* — smooth trowel finish

---

3. *Entrance porch floors* — float finish with steel trowel
4. *Sidewalks* — wood float finish
5. Slabs on fill:

   After the concrete footings have been placed, remove the forms and backfill to proper level to receive the floor construction.

   Compact backfill to provide a solid bearing.

   Under all floor areas place a 6-inch layer of graded gravel fill; puddle, tamp, and level properly to a reasonable true, even surface.

FOOTINGS

Every building rests upon a footing or foundation of some sort. The footing distributes the weight of the building evenly on the earth. They vary in width and height according to the weight they have to support and the bearing capacity of the soil. For practical purposes, in house construction and small buildings construction, the footing is usually twice as wide as the thickness of the wall that is to be built on top of it. The height of the footing is usually the same as the thickness of the wall, figure 10-1.

For example, a footing to support an 8 inch thick wall of either concrete or concrete block is 16 inches wide and 8 inches high. If the wall is 12 inches thick, then the footings are 24 inches wide and 12 inches high.

These dimensions may be changed if unforeseen circumstances are found, when digging the foundation and footings. Footings must be placed on solid ground or rock. Never place footings on fresh earthen fill.

The footing is shown on the basement and foundation plan as dotted lines, figure 10-2. The dotted line is approximately four inches on each side of the solid object line that marks the foundation wall.

Footings for posts and fireplaces are also shown with a hidden line at the exact location where they are to be placed. They are usually accompanied by notes explaining the size and depth, figure 10-3. The size of the footing may be on the wall section view of the plans. This drawing includes a side view of the footings. Figure 10-4 is an example of the footing, concrete floor, and wall in a wall section view on the plans.

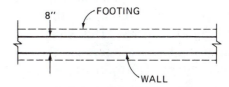

Fig. 10-2 Foundation wall footing

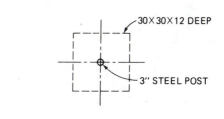

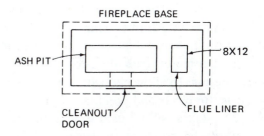

Fig. 10-3 Footings for posts and fireplace

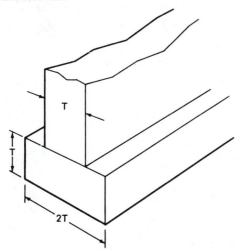

Fig. 10-1 The footing should be twice as wide as the thickness of the foundation wall and the same height as the thickness of the wall.

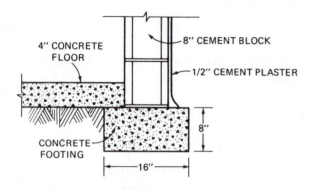

4" CONCRETE FLOOR

8" CEMENT BLOCK

1/2" CEMENT PLASTER

8"

CONCRETE FOOTING

16"

WALL SECTION DETAIL

Fig. 10-4 Wall section detail

It is standard practice in building construction that any time two dimensions are given, the first is the width. For example if the opening for a cellar window is 32" x 16", the width is 32 inches and the height is 16 inches.

With the site cleared and the excavation made in the correct location as indicated on the Plot Plan, work is ready to begin on the footings. The Basement and Foundation Plan is used to find the exact size and location of the footings and foundation walls.

Looking at the Basement and Foundation Plan, for the sample house, the thickness of the walls is given as 8 inches. This is indicated by the 8" notation that appears between the arrowheads at the object lines. The footings are 8 inches wider than on the wall, so the footings on this building are 16 inches wide. The footings are the same thickness as the wall so the footings here are 8 inches thick.

Estimating Footings

Forms

Forms for the sides of the footings are normally constructed of 2 x 8 lumber. Although the specifications indicate that the footings may be poured in trenches of undisturbed earth, it may be necessary to have side forms, figure 10-5. To find the amount of lumber that is needed for the forms, double the perimeter (the distance around the building). Be sure to allow enough for the footings for the posts, fireplace, and porches.

Wooden stakes will be used to hold the forms in place until the concrete is placed. When the concrete is in place, the stakes are pulled out. After

Fig. 10-5 Although it is sometimes possible to place concrete directly in trenches it is generally necessary to use side forms.

the concrete is set, the 2 x 8 lumber is removed and set aside.

Concrete

To figure the amount of concrete necessary for the footings, find the number of linear feet around the outside walls of the house. Include the length of any other walls, such as porches. Multiply the width of the footings in feet by the thickness in feet, then multiply this product by the length of the footings in feet. This is the number of cubic feet in the footings. Divide the number of cubic feet by 27 to find the number of cubic yards. The amount of concrete to be used for supporting posts and fireplace footings must be added to this amount. To figure these, multiply the length in feet times the width in feet times the thickness in feet and divide by 27.

The sample house is 30' x 24', which means that there is a total of 108 feet around the house. The side porch is 8' x 18' so there are 34 linear feet in the side porch footings. The front entrance is 6' x 4' and the rear entrance is 5' x 4'. This makes a total of 169 linear feet of footings for the house. If the footings are 8 inches thick by 16 inches wide, multiply 169 linear feet by 2/3 of a foot thick by 1 1/3 feet wide, for a total volume of approximately 150 cubic feet. Divide this by 27 to get 5 15/27 yards, or 6 yards when rounded off. The fireplace footing is 7' x 3' and there are two 30" x 30" x 12" deep footings for the supporting posts. This makes 33 1/2 cubic feet of footings for

the fireplace and the posts. Divide 33 1/2 by 27 to get approximately 1 1/3 yards of concrete for the footings. Most estimators allow 5 to 10 percent for waste to make up for varying conditions.

List the materials figured so far, on the material list as follows.

SUMMARY

The specifications describe how concrete is poured for footings and concrete slabs. They also state the proportions of the mix, the standards which the concrete must meet, and the finish to be used in specific areas.

The working drawings give the exact location and size of the foundation walls and the footings. To find the amount of concrete needed for footings, multiply the linear distance around the house in feet by the width in feet. Multiply this number by the depth in feet and then divide by 27. This gives the total number of cubic yards necessary for the footings.

FORMS FOR FOOTINGS	2 x 8	lin. ft.	373		
WOODEN STAKES	12 STAKES PER	bndls	4		
TRANSIT - MIX CONCRETE -					
FOOTINGS					
		yds.	8		

REVIEW QUESTIONS

Select the letter preceding the best answer.

1. Which of the following specifications indicates how wet the concrete should be mixed?

   a. 2500 lb. Building-Code-Design Concrete
   b. Number 57 aggregate
   c. 5-inch slump
   d. 1-2-4 mix

2. How wide should the footing be under an 8 inch thick wall?

   a. 8 inches   c. 16 inches
   b. 12 inches   d. 20 inches

3. How high should the footing be under an 8 inch thick wall?

   a. 8 inches   c. 16 inches
   b. 12 inches   d. 20 inches

4. How are the footings shown on the working drawings?

   a. On the floor plan with a solid line   c. On the basement plan with a solid line
   b. On the floor plan with a broken line   d. On the basement plan with a broken line

5. How many linear feet of lumber are required to build the forms for the footing for the foundation shown in figure 10-6?

   a. 144 feet   c. 56 feet
   b. 72 feet   d. 112 feet

6. How much concrete should be ordered for the footings in figure 10-6?

   a. 2 1/2 cubic yards   c. 5 cubic yards
   b. 64 cubic yards   d. 3 1/2 cubic yards

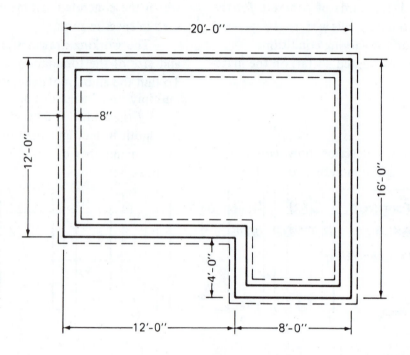

Fig. 10-6

7. How much concrete should be ordered for a fireplace footing 1 foot thick x 3 feet wide x 8 feet long?

   a. 24 cubic yards

   b. 2 cubic yards

   c. 1 cubic yard

   d. 1 1/3 cubic yards

8. Which of the following finishes should be applied on the basement floor?

   a. Trowel

   b. Steel float

   c. Wood float

   d. Screed

9. Which of the following finishes should be applied on the sidewalk?

   a. Trowel

   b. Steel float

   c. Wood float

   d. Screed

10. Which of the following finishes should be applied on the entrance porch?

   a. Trowel

   b. Steel float

   c. Wood float

   d. Screed

unit 11 concrete floors and slabs

OBJECTIVES

After studying this unit the student will be able to

- explain the specifications for placing and finishing concrete.
- tell where to look for concrete on working drawings.
- determine the quantity of concrete needed for a specific job.

The specifications division on concrete work indicates that the cement must be a domestic portland cement type 1. The concrete is to be ready mixed and delivered to the site in transit-mix trucks. The basement floor is to be a smooth, metal-trowel finish, as is the side porch floor. The front and rear entrance porch floors are to be metal-float finished with a steel trowel. The sidewalks are to be wood-float finish.

CEMENT

Portland cement is a type of cement, not a brand name. Every manufacturer of cement makes portland cement. Portland cements are made to meet the American Society for Testing and Materials' standard specifications ASTM Designation. The most common portland cements are divided into five types. The most commonly used in residential construction are C150 portland cement, types I and III and C175 air-entraining portland cement.

Type I, normal portland cement, is a general-purpose cement suitable for all uses when the special properties of the other types are not required. It is used in pavement and sidewalk construction, reinforced concrete buildings and bridges, railway structures, tanks and reservoirs, culverts, water pipe, masonry units, and for all uses of cement or concrete not subject to sulfate attack from soil or water. Since cement always gives off a certain amount of heat as it reacts to outside chemicals, it must also be used where the heat that is generated while the cement cures will not cause an objectionable rise in temperature.

Type III, high-early-strength portland cement, is used when high strengths are desired at very early periods—from one to three days. It is used when it is desired to remove forms as soon as possible or to put the concrete into service quickly, or in cold weather construction to reduce the period of protection against low temperatures.

Air-entraining portland cements are covered by ASTM C175 and are designated as Types IA, IIA, and IIIA. These correspond to Types I, II and III, respectively, in ASTM C150. In these cements, small quantities of air-entraining materials are incorporated during manufacture. These cements have been developed to produce concrete that will resist severe frost action as well as the effects of chemicals applied for snow and ice removal. Concrete made with these cements contains tiny, well-distributed, completely separated air bubbles. The bubbles are so minute that there are billions of them in a cubic foot of air-entrained concrete.

When water is added to cement, it makes a paste which binds fine and coarse materials, called *aggregates,* into a solid, rock-like mass known as concrete. Cement hardens through the chemical action between cement and water. This water should be clean enough to drink.

Aggregates

Aggregates are classified as *fine* or *coarse,* depending upon their size. Sand which passes through a 1/4-inch wire screen is called a fine aggregate. Coarse aggregates, which range in size from 1/4 inch to 3 inches, are usually crushed stone or cinders.

Both fine and coarse aggregates are used in concrete so that the spaces between the coarse aggregates (stone) are filled by the smaller ones (sand).

In addition to the aggregates already mentioned, there are *lightweight aggregates* which are

manufactured from vermiculite, pumice, and other minerals. These have been developed to meet the needs for less weight required in certain types of construction, such as roofs or walls.

Placing Concrete

A thickness of 4 inches is recommended for most floors, sidewalks, steps, porch floors, and most other slabs made of concrete.

Concrete is placed in forms as quickly as possible after it has been thoroughly mixed. It must be *consolidated* (worked to remove voids) when placed in the forms.

Concrete floors are leveled off evenly with a straight board called a *screed*. The concrete is then *floated* (smoothed with a tool called a float) to bring a film of cement and sand to the top, figure 11-1. After the concrete has been floated, it may be troweled to a smooth, flat finish.

The amount of concrete needed is determined by multiplying the length by the width by the depth or height, all in feet, and then dividing the total by 27. This gives the number of cubic yards of concrete needed to do the job. Be very careful to check drawing for any changes in thickness of slab such as slabs with integral footings (monolithic).

ESTIMATING CONCRETE

Find the total square-foot area of all places to be covered with concrete. This includes the basement floor, side porch floor, and the front and rear entrance porches. For the sample house in the back of the book there is a total of 908 square feet. The floors are 4 inches thick, so multiply 908 by 1/3 (4 inches is 1/3 of a foot) and get 303. Divide by 27 and get 11 6/27 cubic yards, or rounded off — 11 1/2 cubic yards.

The walk is not shown on the plans, but is called for by the specifications. Most walks are 3 feet wide. Looking at the plot plan for the basement and foundation in the sample house, notice that the house has a 30-foot setback. To figure the

Fig. 11-1 After the concrete is placed, the surface is floated.

amount of concrete needed for the sidewalk, multiply the length (30 feet) by the width (3 feet) to get 90 square feet. Sidewalks are normally 4 inches thick. Four is 1/3 of 12, so multiply 90 by 1/3 and get 30 cubic feet. Now divide 30 by 27 and get 1 3/27 cubic yards, or, rounded off, 1 1/2 cubic yards of concrete needed for the walk.

List this on the material list.

SUMMARY

Cement is a finely pulverized material made from limestone. When water is added, it makes a paste which binds fine and coarse aggregates into a solid rock-like mass called concrete. Sand which passes through a 1/4-inch screen is called a fine aggregate and crushed stone is known as a coarse aggregate. The sand, stone, and cement, when mixed with water, becomes concrete.

Concrete is placed in forms as quickly as possible after it has been thoroughly mixed. The concrete is leveled off evenly, then floated. This brings a film of cement and sand to the surface to allow troweling, which produces a smooth, flat surface.

CONCRETE - ALL FLOORS				CU. YD.	11 1/2																						
CONCRETE - SIDEWALK				CU.YD.	1 1/2																						

REVIEW QUESTIONS

Select the answer preceding the best answer.

1. What is the main ingredient of portland cement?
   a. Sand
   b. Aggregate
   c. Limestone
   d. All of these

2. What is the largest size of fine aggregates?
   a. fine sand
   b. 1/4 inch
   c. 1/2 inch
   d. 3 inches

3. What are the ingredients of concrete?
   a. Aggregates
   b. Portland cement
   c. Water
   d. All of these

4. Which of the following is normally used for concrete footings?
   a. White cement
   b. Gray cement
   c. Masonry cement
   d. None of these

5. What is the most common thickness for floors and sidewalks?
   a. 6 inches
   b. 3 inches
   c. 4 inches
   d. 8 inches

6. How much concrete should be ordered for the floor in figure 11-2?
   a. 3 1/3 cubic yards
   b. 3 cubic yards
   c. 1 cubic yard
   d. 10 cubic yards

7. How much concrete should be ordered for a standard sidewalk 40 feet long?
   a. 6 cubic yards
   b. 4 1/2 cubic yards
   c. 2 2/3 cubic yards
   d. 2 cubic yards

8. Which of the following makes the coarsest finish?
   a. Metal trowel
   b. Wood float
   c. Metal float
   d. All are about the same

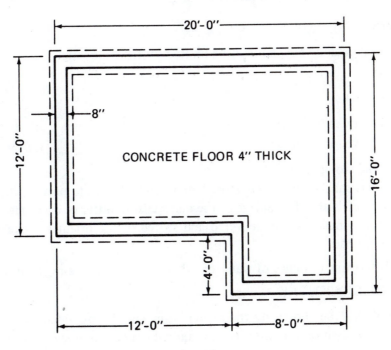

Fig. 11-2

unit 12 concrete block

OBJECTIVES

After studying this unit the student will be able to

- explain the masonry division of the specifications.
- describe the various types of concrete blocks.
- determine the number of blocks needed for a particular job.

CONCRETE BLOCK

Concrete blocks are made by mixing portland cement and water with a dense aggregate such as sand, gravel, crushed stone, or cinders. These blocks are formed under pressure, or a combination of vibration and pressure.

Concrete blocks are made 4, 6, 8, 10, and 12 inches thick, 8 inches high, and 16 or 18 inches long. These are *nominal* dimensions. Actually, each dimension is 3/8 of an inch smaller, to allow for the mortar joint. They are commonly called hollow masonry units or c.m.u.

Special sizes and forms of concrete blocks are made for particular uses such as half blocks, corner blocks, chimney blocks, pier blocks, jamb blocks, and decorative blocks.

Concrete blocks are widely used for foundations of residences and for foundations where brick veneer work is fastened to frame construction. Also, blocks are used extensively in small commercial construction in combination with brick. Modular layout is very important.

The specifications for block and brick work are combined here, as they normally are in residential specifications. All masonry is specified in one section of the specifications. However, this unit of the textbook discusses only block work.

Specifications

DIVISION 4: MASONRY

A. **WORK INCLUDED**

This work shall include but shall not be limited by the following:

1. Brickwork

2. Concrete blockwork

3. Mortar for brick and blockwork

B. **MASONRY MATERIALS**

1. Delivery and storage:
   All materials shall be delivered, stored, and handled so as to prevent the inclusion of foreign materials and the damage of the materials by water or breakage.
   Packaged materials shall be delivered and stored in the original packages until they are ready for use.

2. Materials showing evidence of water or other damage shall be rejected.

C. **BRICK AND BLOCKWORK**

1. The *brick* shall be chosen by the owner from samples provided by the contractor. Brick is to be used on the fireplace face and where the chimney is exposed above the roof.

Common Brick shall be hard-burned and uniform in size. No soft brick will be allowed. The brick shall be reasonably free of cracks, pebbles, particles of lime, and other substances which will affect their serviceability or strength. The brick shall be carefully protected during transportation and shall be unloaded by hand and carefully piled; dumping is not permitted.

2. *Mortar* used for laying brick and concrete block should consist of one (1) part portland cement, one (1) part hydrated mason's lime, and six (6) parts sand. The mortar ingredients shall comply with the following requirements:

a. *Aggregates:* ASTM C-144-52T

b. *Water:* Clean, fresh, free from acid, alkali, sewage, or organic material.

c. *Portland cement:* ASTM C-150-55 Type 1

d. *Masonry cement:* ASTM C-91-55T Type 2

e. *Hydrated lime:* ASTM C-207-49T Type S

3. *Concrete block* shall be load bearing, hollow, concrete masonry units and shall conform to the standard specifications of ASTM C-90.

4. *Wall Reinforcing;* reinforcing material for masonry walls shall be prefabricated welded steel.

5. *Installation;* erect all block and brickwork in accordance with the following requirements.

All work shall be laid true to dimensions, plumb, square, and in bond, and properly anchored. All courses shall be level and joints shall be of uniform width; no joints shall exceed the size specified.

Joints shall be finished as follows:

All brick shall be laid on full mortar bed with a shoved joint. All joints shall be completely filled with mortar. All horizontal and vertical joints shall be raked 3/8 of an inch deep.

All mortar joints for concrete block masonry shall have full mortar coverage on vertical and horizontal face shells.

Vertical joints shall be shoved tight. Full mortar bedding shall have ruled joints.

Horizontal reinforcement shall be placed in every other bed joint of block work. Reinforcement shall be placed in the first and second bed joints above and below all openings.

Concealed work shall have joints cut flush.

Protection: cover the wall each night and when the work is discontinued due to rain or snow.

Do not work during freezing weather without permission from the architect.

E. CLEANING AND POINTING

1. Concrete Block: Point up all the voids and open the joints with mortar. Remove all of the excess mortar and dirty spots from the entire surface.

2. Brickwork: Upon completion, all brickwork shall be thoroughly cleaned with clean water and stiff fiber brushes and then rinsed with clear water. The use of acids or wire brushes is not permitted.

Figure 12-1 shows several of the types and sizes of blocks. Corner blocks are used to make a square, straight corner. Pier blocks are closed on each end and are used to make a finished support. Half and whole sash blocks are used on each side of the basement windows. Solid blocks are used to cap the foundation wall. These cover the holes in the block wall and make a solid top for the foundation. Chimney blocks come in two styles; whole and half blocks. Whole chimney blocks are used where an 8-inch x 8-inch flue liner is installed in a single chimney. If a larger flue, such as 8 inches x 13 inches, is to be used, then use half chimney blocks and reverse the direction of the block on every other course.

Estimating Concrete Block

To find the number of concrete blocks required, first find the surface area of the wall. Each 8-inch x 16-inch block covers 0.888 square feet, so allow 112 1/2 blocks for every 100 square feet of wall. Add 5 to 10 percent for waste.

The foundation wall for the sample house is 7 feet, or 84 inches high. Blocks are 8 inches high, so 10 courses of full size or regular blocks and one course of solid 4-inch cap blocks are needed.

It is not considered good practice to pour footings and lay block walls on loose fill. Also, when the excavation is finished, enough space must be left around the outside of the foundation walls to allow workers to move easily. Therefore, when the excavation is finished, the foundation is usually made large enough to include room for the footings for the porches. This means that, although there is no basement under them, the foundations for the porches have the same number of block courses as the rest of the foundation.

Adding all of the walls, there is a total of 169 linear feet in the perimeter. Multiply this figure by 7 feet and get a total of 1183 square feet of wall. Allowing 10 percent for waste, the wall requires 1463 blocks. Check pilasters and wider blocks being used for deep foundation walls.

The foundation walls are to be 10 1/2 courses high, or a total height of 84 inches. One course around the foundation requires 127 blocks, so include 127 four-inch cap blocks.

Two half and two whole sash blocks are needed for each basement window. The basement windows are 32 inches (2 blocks) long and 20 inches (2 1/2 blocks) high. There are 5 basement windows, so 10 half sash blocks and 10 whole sash blocks are required.

Add these items to the material list shown at the top of page 61.

CONCRETE WALLS

If the foundation is to be of poured concrete instead of concrete block, the number of cubic yards of concrete in the wall must be computed. To do this, multiply the length of the foundation times its thickness in feet. Multiply this times its

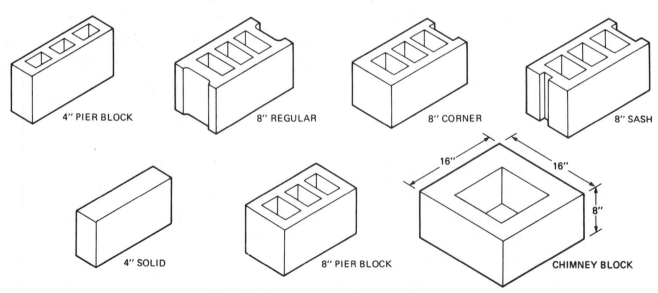

Fig. 12-1 Common types of concrete blocks

REGULAR CONCRETE BLOCKS	8×8×16		1463																							
REGULAR CAP BLOCKS	4×8×16		127																							
HALF SASH BLOCKS			10																							
WHOLE SASH BLOCKS			10																							

height in feet. This is the number of cubic feet in the wall. Divide this by 27 to find the number of cubic yards of concrete needed to do the job.

FOUNDATION FOR BRICK-VENEER WALLS

If the house is to have brick veneer, 12-inch concrete blocks must be used for the foundation. The thicker blocks require a larger footing, so more concrete must be included in the footings estimate.

When using brick veneer, it is common practice to lay 12-inch concrete blocks for nine courses. Then lay one course of 8-inch blocks, and one course of 4-inch blocks, with the 8-inch and 4-inch blocks flush with the inside face of the foundation wall. The holes in the top of the 12-inch wall are filled with concrete. The bricks are laid on top of the top course of the 12-inch block, figure 12-2. This places the brick below the finished grade and improves the appearance of the brickwork.

HORIZONTAL REINFORCEMENT

The specifications call for horizontal reinforcing, figure 12-3, in every second bed joint. The foundation is 84 inches high, which means there are 10 full courses and 1 course of solid cap block. The total distance around the foundation walls is 169 linear feet. Divide 10 (courses) by 2 (every second course) and get a total of 5. Multiply 5 times 169 and get 845 linear feet of reinforcing.

The specifications also indicate that reinforcing shall be placed in the first and second bed joint above and below the openings. There is no masonry above the basement windows, so reinforcement is only needed below the window openings. There are 5 basement windows, which are each 32 inches long. To be effective, the reinforcing should extend at least 1 1/2 blocks on each side of the opening. If there are 3 blocks for every 4 feet and the window openings are added, then the reinforcing under the openings is 80 inches long. Round this off by adding 4 inches and figure a total of 7 feet for each course under each window. Since there are 5 basement windows, multiply 14 by 5 and get 70 linear feet, which is added to 845 linear feet (for the walls) for a total of 915 linear feet of 8-inch horizontal reinforcement.

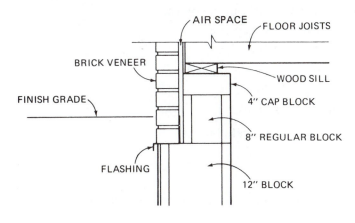

Fig. 12-2 When the building is to have brick veneer, the foundation is laid with larger blocks.

Fig. 12-3 Horizontal Reinforcement in a block wall

	HORIZONTAL REINFORCEMENT	8"		lin.ft.	915																					

Add this to the material list.

SUMMARY

Concrete blocks are made by mixing cement and water with a dense aggregate such as sand, gravel, or cinders. They are formed under pressure, or a combination of vibration and pressure. Concrete blocks are made 4, 6, 8, 10, and 12 inches thick and 8 inches high and 16 inches long. Special sizes and shapes are manufactured for the particular uses.

Concrete blocks are popular for foundations of residences and for foundations where brick-veneer work is fastened to frame construction.

Concrete blocks including the mortar joint are 16 inches long. This means there are 3 blocks for every 4 linear feet of wall. In estimating blocks, multiply the linear distance around the wall by three fourths. The blocks are 8 inches high, so for a 7-foot wall, figure 10 full courses and one course of 4-inch solid cap blocks. Multiply the number of blocks for one course by the number of courses to find the total number of blocks needed. Count the basement windows and allow 2 half sash and 2 whole sash blocks for each window.

REVIEW QUESTIONS

Select the letter preceding the best answer.

1. According to the specifications, how is the mortar to be mixed?
   a. 1 part line, 1 part sand, and 6 parts cement
   b. 1 part lime, 6 parts sand, and 1 part cement
   c. 6 parts lime, 1 part sand, and 1 part cement
   d. None of these

2. How is horizontal reinforcement to be placed around window openings in the foundation wall?
   a. Every other course above the opening and every course below the opening
   b. Every course above the opening and every other course below the opening
   c. In the first and second course above the opening and every other course below the opening
   d. None of the above

3. How is horizontal reinforcement to be placed in walls without openings?
   a. Every course
   b. Every other course
   c. Every third course
   d. None of these

4. What kind of blocks are used at the sides of cellar window openings?
   a. Sash blocks
   b. Corner blocks
   c. Pier blocks
   d. Regular blocks

5. Which of the following is a common nominal size for concrete blocks?
   a. 8" x 8" x 24"
   b. 5" x 8" x 16"
   c. 4" x 8" x 20"
   d. 8" x 8" x 16"

6. How many 8-inch blocks are required to build the foundation in figure 12-4 on page 63? Allow 10 percent for waste.
   a. 726
   b. 968
   c. 72
   d. 798

7. How many solid 4-inch blocks are required for the top of the foundation in figure 12-4?
   a. 88
   c. 66
   b. 44
   d. 112

8. How much concrete is required for the foundation in figure 12-4? Allow 5 percent for waste.
   a. 61 cubic yards
   c. 16 cubic yards
   b. 93 cubic yards
   d. 13 1/2 cubic yards

9. How many whole and half sash blocks are required for 4 openings 32 inches by 16 inches?
   a. 4 whole and 4 half blocks
   c. 4 whole and 8 half blocks
   b. 8 whole and 8 half blocks
   d. 16 whole and 16 half blocks

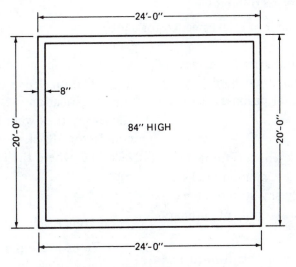

Fig. 12-4

unit 13 brick

OBJECTIVES

After studying this unit the student will be able to

- explain the specifications for brickwork.
- describe several types of bonds.
- calculate the number of bricks needed for a job.

Bricks are small, solid units made in rectangular form from clay or shale, and hardened by heat. Bricks used for structural purposes fall into three general classifications:

Common brick is a clay brick which is formed and then baked to a hard material. Common bricks are porous and are used primarily as a structural material where the appearance is not of primary importance.

Face brick is a shale brick which is formed and then baked to a hard material. Face bricks are dense shale and will not absorb water. The surface can have a smooth or matte finish, or it can be finished with a wire brush to give a textured finish. They are used for exterior facing and come in various colors.

Special types are bricks made for a particular purpose, such as glazed bricks, acid-resistant bricks, paving bricks, and firebricks.

The advantages of using brick are its durability, beauty, low maintenance cost, and higher resale value for houses on which they are used.

The nominal size of common and standard face brick is 2 1/4 inches thick, 3 3/4 inches wide, and 8 inches long. *Firebrick* is 2 1/2 inches thick, 4 1/2 inches wide, and 9 inches long. *Roman brick* is 1 5/8 inches thick, 3 3/4 inches wide, and 12 inches long. These are the most common types of bricks in residential construction. Bricks are also named according to their position in the wall, figure 13-1. By using bricks in various combinations of these positions a variety of bond patterns can be achieved, figure 13-2.

Division 4, Masonry, Section C, number 1 of the specifications states that the *Brick:* Shall be chosen by the owner from samples provided by the contractor. Brick is to be used on the fireplace face and where the chimney is exposed above the roof.

Estimating Brick

To estimate the quantity of common or standard face brick in veneer work, 675 bricks are needed for every 100 square feet of wall. An allowance of 5 to 10 percent should be included for waste. Brick veneer refers to one pier or wythe of brick as a face pier, which is backed by other masonry units or attached to a frame wall. If the bricks are laid on edge *(rowlock)* or standing on end *(soldier)* allow 4 1/2 bricks for each linear foot.

To estimate the number of bricks in a solid pier, with 3/8-inch joints, see figure 13-3.

To estimate bricks for brick steps with treads 10 inches wide and 7 1/2-inch rise, allow 8 1/2 bricks on edge for each linear foot of step, figure 13-4.

Because of the weight of most masonry products (120 to 140 pounds per cubic foot) it is necessary to provide additional support where they are to be used in a house. Many houses are not built to support this weight; however, many people desire the atmosphere that bricks create in the house, figure 13-5. To meet the demand, manufacturers produce special materials, such as

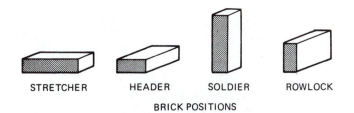

STRETCHER   HEADER   SOLDIER   ROWLOCK

BRICK POSITIONS

Fig. 13-1 Brick positions

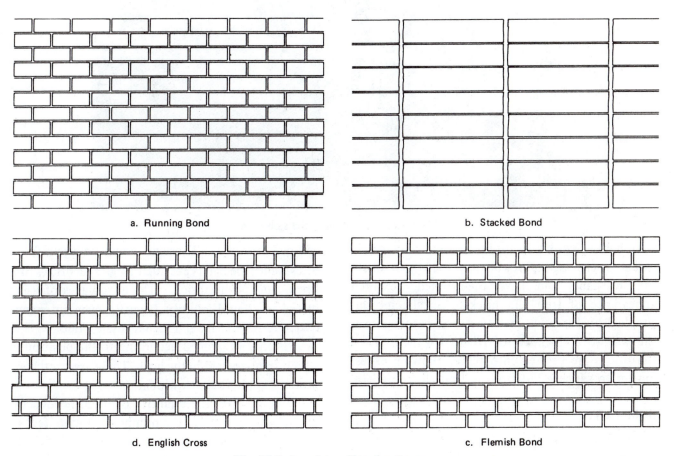

a. Running Bond   b. Stacked Bond

d. English Cross   c. Flemish Bond

Fig. 13-2 A variety of bond patterns

PIER SIZE	NUMBER OF BRICKS PER FOOT OF HEIGHT
8 × 8	13½
8 × 12	18
12 × 12	20½
16 × 16	36

SOLID PIER

Fig. 13-3 Estimating the bricks required for a solid pier

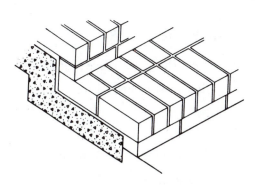

Fig. 13-4 Brick steps

Fig. 13-5 Special lightweight bricks on interior wall

thin, lightweight bricks for interior use. These individual bricks are approximately 3/16 inch thick, fire resistant, and have the rugged, textured appearance of true kiln bricks. Special mortar, supplied by the brick manufacturers is spread over the wall to a thickness of 1/8 to 1/4 inch. Then the bricks are pressed into place, allowing excess mortar to squeeze out from behind all four edges.

SUMMARY

A brick is a solid, rectangular unit made from clay or shale which is hardened by heat. Brick, used, for structural purposes, falls into three general classifications. Those are common brick, face brick, and special types such as glazed brick, firebrick, and paving brick. The advantages of using bricks are their durability, beauty, low maintenance cost, fire-resistance, and higher resale value.

Bricks are named according to the position in which they are laid in the wall. These are stretchers, headers, soldier, and rowlock.

To estimate the quantity of brick needed for veneer work, multiply the number of square feet to be covered by 6 3/4 (6.75). It takes 675 bricks to cover 100 square feet of wall. Add an allowance for waste to this figure.

REVIEW QUESTIONS

Select the letter preceding the best answer.

1. What kind of brick is used to build piers where appearance is not important?

   a. Roman
   b. Face
   c. Glazed
   d. Common

2. What kind of brick is thinner and longer than most?

   a. Roman
   b. Face
   c. Firebrick
   d. Common

3. What is the name for each of the following positions of a brick?

a.

c.

b.

d.

4. How many common or standard face bricks are needed to lay 100 square feet of brick veneer? Do not allow for waste.

   a. 1,000
   b. 100
   c. 675
   d. 70

5. How many common bricks are needed to lay a chimney with one 8-inch x 8-inch flue, 21 feet high?

   a. 27
   b. 770
   c. 284
   d. 567

6. How many bricks are needed to build steps with a 7 1/2-inch rise and 4 treads, each measuring 10 inches x 36 inches?

   a. 77
   b. 102
   c. 34
   d. 152

unit 14 mortar

OBJECTIVES

After studying this unit the student should be able to

- explain the specifications for mortar.
- list the ingredients of high-quality mortar.
- determine the amount of mortar needed for a specific job.

High-quality mortar is important in masonry construction. Mortar is graded according to the strength it is expected to develop. The same formula is used for both brick and concrete block work. Division 4, Masonry, Section C2 of the specificiations states: *Mortar:* For laying brick and concrete block; one (1) part portland cement, one (1) part lime, and six (6) parts sand. The mortar ingredients shall comply with the following requirements:

a. *Aggregates:* ASTM C-144-52T
b. *Water:* Clean, fresh, free from acid, alkali, sewage, or organic material.
c. *Portland cement:* ASTM C-150-55 Type 1
d. *Masonry cement:* ASTM C-91-55T Type 2
e. *Hydrated lime:* ASTM C-207-49T Type S

The *American Society for Testing Materials* (ASTM) has tested most products used in the building industry and has written specifications to indicate the properties of various materials. Most architects and specification writers use ASTM standards when they specify quality of materials.

The specifications indicate that for laying brick and block, the mason should use one (1) part portland cement, one (1) part lime, and six (6) parts sand. There are 8 shovels of portland cement

the water is added. This makes a strong mortar, which, when dry, will be light gray.

Masonry cement (a premixed blend of portland cement and lime), used in the proportion of 1 part cement to 3 parts sand generally meets all the requirements of a good mortar for any type of work. Masonry cement works better and makes a stronger joint if mixed in the proportions of 1 bag of masonry cement to 20 shovels of sand.

Estimating Mortar

To estimate the quantity of mortar required for brickwork, it is necessary to consider the position in which the bricks will be laid. For brick veneer work with 1/2-inch mortar joints, 8 bags of masonry cement are needed for every 1000 bricks. 70 wall ties should also be included for every 100 square feet of brick veneer.

In estimating mortar for concrete blocks, allow 1 bag of cement for every 28 blocks. This may vary, depending on how individual masons use the material, but it will be fairly close. The sample house in this book has approximately 1400 concrete blocks in the foundation which means approximately 50 bags of masonry cement are needed.

Add this to the material list.

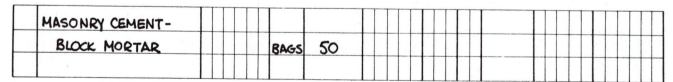

MASONRY CEMENT–																						
BLOCK MORTAR				BAGS	50																	

in a bag. This means that the person mixing the mortar mixes 1 bag of portland cement, 1 bag of mason's lime, and 48 shovels of sand. The ingredients are thoroughly mixed in a mortar box before

In order to make good mortar, it is important to use a good grade of sand for the aggregate. In most localities, good, coarse, washed sand is available. Sand is sold by the cubic yard and is usually

less expensive to buy in truckload lots. Most operators of sand and gravel pits deliver in ten-yard loads. The average house foundation and fireplace require approximately 10 yards of sand. Add this to the material list.

cement, lime, and sand make the strongest mortar. There are eight shovels in a bag of cement. Therefore, if the mix is to be 1-1-6, for every bag of cement, one bag of lime and 48 shovels of sand are needed.

	WASHED SAND-																										
	FOUNDATION & CHIMNEY				YDS.	10																					

SUMMARY

Good mortar is important in masonry construction. Mortar is graded according to the strength it is expected to develop. Portland

Masonry cement mixed in the proportion of 1 part cement to 3 parts sand generally meets all the requirements of the mortar for any type of work.

REVIEW QUESTIONS

A.   Select the letter preceding the best answer.

1. How many shovels of portland cement are there in a bag?

    a. 6   c. 10
    b. 8   d. 12

2. How much sand should be mixed with one bag of masonry cement?

    a. 3 shovels   c. 20 shovels
    b. 6 shovels   d. 48 shovels

3. Approximately how many concrete blocks can be laid with one bag of masonry cement?

    a. 10   c. 100
    b. 48   d. 28

4. How many bags of masonry cement are required to lay 2,000 blocks?

    a. 12   c. 56
    b. 20   d. 72

5. What does ASTM do?

    a. Tests materials and writes specifications for their quality
    b. Writes building construction specifications
    c. Writes formulas for estimating quantities of material
    d. Provides catalogs of materials for architects

6. How much mortar is needed to lay 500 bricks in a veneer wall?

    a. 40 bags   c. 1 bag
    b. 16 bags   d. 4 bags

B.   Briefly answer the following questions.

1. How is mortar for blockwork different from mortar for brickwork?

2. List the three ingredients of mortar and their proportions according to the sample specifications.

unit 15 fireplaces and chimneys

OBJECTIVES

After studying this unit the student will be able to

- discuss the construction of fireplaces.
- locate fireplaces and chimneys on working drawings.
- determine the quantity of the materials necessary to build a fireplace or a chimney.

The outside appearances have varied throughout history. They range from the early-colonial massive brick and hearth style to the more modern two-way corner fireplace, figure 15-1.

The design of the interior portion of fireplaces has remained fairly constant. Through experiences in the past, the proportions necessary to allow a fireplace to *draw* (pull smoke up the chimney) and burn properly have been established. In order for a fireplace to burn properly, it must have a good *draft* (it must draw well) to prevent smoke from coming into the room. The size of the fireplace opening is determined by the damper used, the size of the flue liner, and the height of the chimney. All of these factors influence the draft.

FIREPLACE CONSTRUCTION

Many fireplace forms and fireboxes are prefabricated metal units, figure 15-2. These come in various styles and sizes. The flue size is specified by the manufacturer.

There are standard rules to follow in the construction of a fireplace. If the firebox is to be built of firebrick, the total thickness of the firebox wall should be at least 8 inches.

The base of the firebox is always made with firebrick. The firebricks are set in a bed of fireclay mortar without any side or end joints. These small cracks fill with fine sand and dirt. If mortar joints are used between the firebricks, the joints soon burn out and frequently need repairing.

In the fireplace details for the sample house, a metal ash dump is placed in the floor of the firebox to clean the ashes from the firebox. The ash dump opens into an ash pit which is built into the base of the fireplace. After a period of time they are re-

Fig. 15-1 Corner fireplace

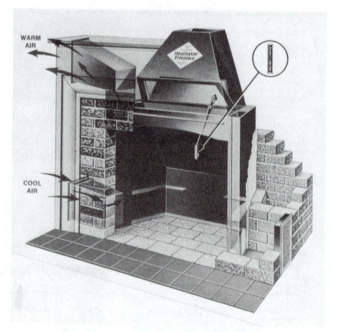

Fig. 15-2 Pre-fabricated metal fireplace

moved through the cleanout door, which is built into the base of the fireplace.

The hearth is made of masonry fireproof construction. This is usually brick, slate, stone, marble, or quarry tile. The hearth should extend at least 16 inches in front of the fireplace opening and at least 8 inches on each side, as shown in the sample plans. Combustible materials should not be used within 8 inches of each side of the fireplace opening.

CHIMNEY CONSTRUCTION

The chimney should have a fireclay flue lining. Standard flue liners are 24 inches long. The most common sizes are 8 1/2 inches x 8 1/2 inches, 8 1/2 inches x 13 inches, 13 inches x 13 inches, and 13 inches by 18 inches. When more than two flues are used in the same chimney, a four-inch masonry division is necessary. The chimney wall should be at least 4 inches thick and should be separated from the wood framing of the building by a 2-inch air space, and from the flooring and sheathing by a 1-inch air space.

The top of the chimney should be at least 3 feet above the ridge of the roof or any portion of the building within 10 feet. The width and depth of the fireplace depend upon its style and its location. A chimney that is exposed along the outside wall of a house takes longer to warm up in cold weather so it takes longer to get a good draft.

The plan for the chimney is found on the basement and foundation plan for the sample house. If a fireplace is included, the chimney may be located either on an outside wall or near the center of the house. If it is a single flue chimney, it is usually located near the center of the building. The purpose of the chimney is to carry the fumes of the heating system out of the house. The heating system is usually placed near the center of the building to save on installation costs and to provide more efficient heat distribution.

The flue for the furnace is usually an 8-inch x 8-inch flue. On the plans, the architect will specify either 8 inch x 8 inch, 8 inch x 12 inch, or 12 inch x 12 inch. The actual flue liner is 8 1/2 inch x 8 1/2 inch, 8 1/2 inch x 13 inch, and 13 inch x 13 inch.

ESTIMATING FIREPLACES AND CHIMNEYS

For a concrete block chimney, find the total height of the chimney in feet from the basement to the top of the chimney cap. To find the number of flue liners needed, divide this height by 2 because standard flue liners are 2 feet long. A chimney is usually built of flue liners and whole chimney blocks up to the roof line. Brick is used where the chimney comes through the roof, figure 15-3.

To build an 8-inch x 8-inch furnace chimney, three whole chimney blocks are needed for each flue liner. Find the height of the chimney from where it comes through the roof. From the low slope side to the top of the chimney (which should be a minimum of three feet above the ridge) figure 27 bricks for every foot of height.

For example, consider a 24 foot high chimney which extends 6 feet through the roof, with a single 8-inch x 8-inch flue. The total height of 24 feet indicates that 12 flue liners are needed. Subtract the 6 feet above the roof from 24 feet. This leaves 18 linear feet of whole chimney blocks. Three blocks are required for every 2 feet of height. Two divided into 18 equals 9 and 9 multiplied by 3 is 27. (Twenty-seven whole chimney blocks are required.) From the roof line to the top of the chimney is 6 feet. 6 times 27 (the number of bricks needed for each foot of height for a single 8 inch x 8 inch flue) equals 162 bricks.

To estimate a fireplace, it is necessary to become more involved with the construction. The floor plans, elevations, and detail drawings must be studied to determine how the fireplace is to be constructed and what materials are required.

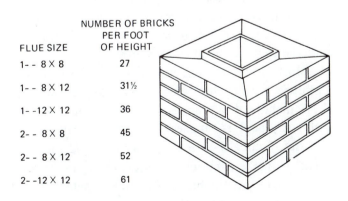

FLUE SIZE	NUMBER OF BRICKS PER FOOT OF HEIGHT
1 -- 8 X 8	27
1 -- 8 X 12	31½
1 --12 X 12	36
2 -- 8 X 8	45
2 -- 8 X 12	52
2 --12 X 12	61

Fig. 15-3 Guide for estimating bricks in chimneys

The base of the fireplace is usually built up with regular concrete blocks. The plans show that the base is 6 feet long and 2 feet 6 inches wide. This makes the perimeter 17 feet. Multiply this perimeter by 3 and divide by 4 to find that 13 concrete blocks are needed to lay one course around the fireplace base. If the base is to be 10 courses high, then 130 blocks are needed for the base. There are 8 inches of masonry between the side of the flue liner and the ash pit. This is usually a concrete block. The base is 10 courses high, so add 10 more blocks for a total of 140.

The ash pit requires a 12-inch x 15-inch cleanout door and the furnace flue requires an 8-inch x 8-inch cleanout door.

The furnace flue liner extends from the basement to the top of the chimney. The firebox flue extends from about one foot above the damper to the top of the chimney. The fireplace is on the first floor, so approximately three less flue liners of the larger size are needed than of the smaller furnace flue liners. The flue from the firebox is usually 8 inches x 13 inches or 13 inches x 13 inches, the size depends on the size of the chimney and the size of the damper.

The top of the fireplace hearth in the sample house is even with the finished floor. (In some cases the hearth may be built up several courses above the finished floor. This is known as a *raised hearth*.) The top of the fireplace base is framed over so that concrete can be poured over the top and extend into the room under the hearth (see sample plans). This concrete should be reinforced with steel rods. A hole is cut into the ash pit so that a metal ash dump can be placed in the base of the firebox. The ash dump is usually the same size as a firebrick. The cement is finished at a depth which leaves room for the firebrick and hearth material to be set in a bed of mortar.

Most fireplaces use approximately 30 regular firebricks for the floor area. Firebricks are usually 2 1/2 inches x 4 1/2 inches x 9 inches. The firebox walls may be 4 1/2 inches thick, with the firebrick laid flat, figure 15-4, or they may be 2 1/2 inches thick, with the firebrick laid on edge, figure 15-5. To estimate the number of firebricks needed, first find the area of the walls in square feet. Multiply this area by 6.4 if the walls are to be 4 1/2 inches

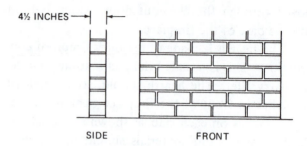

Fig. 15-4 Firebrick laid flat

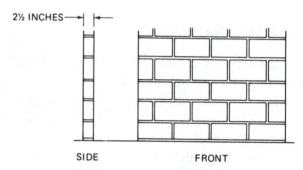

Fig. 15-5 Firebrick laid on edge

thick; if they are to be 2 1/2 inches thick multiply the area by 3.55. When firebricks are used in the firebox wall, they are usually set in fireclay. About 40 pounds of fireclay are required to set 30 firebricks. The firebricks used in the floor of the firebox are usually set without mortar in the end or side joints. If mortar is used, it burns out and has to be replaced frequently. When firebrick is used for the sides of the firebox, a damper must be ordered. Dampers vary in size and design, therefore, the manufacturer, catalog number, width of fireplace opening, and type of control (poker or rotary) must be listed.

A steel angle should be included as a *lintel* to support the masonry work over the fireplace openings. This may not be necessary in all cases; some dampers have a built-in lintel, so a separate piece of angle iron need not be included. Unless it is specified differently on the plans a 3 1/2-inch x 3 1/2-inch steel angle iron should be used. The length of the angle iron should be equal to the fireplace opening plus at least 4 inches of bearing on each end. For example, a 30-inch fireplace opening should have at least a 38-inch angle iron.

Preformed metal fireplace units come with the firebox, damper, and controls all in one unit. These preformed units should be listed by size and manufacturer. The manufacturer furnishes details such

as the necessary flue size for a specific chimney height. Firebrick for the floor of the unit must be ordered separately.

For a fireplace on an inside wall, about 500 common bricks and about 150 4-inch concrete blocks are needed for fills. Approximately 25 bags of masonry cement are required. The hearth and chimney cap require three or four bags of portland cement. If the brick chimney is exposed above the fireplace, allow 6 3/4 bricks for every square foot of finished surface.

Some fireplaces require a single course of bricks stacked one on top of another on each side of the opening and a soldier course across the top. In this case, approximately 50 bricks are required for a 30 to 36-inch opening.

Many fireplaces are laid up on the outside of the house, or have other exposed brickwork. In this case, if the chimney has one 8-inch x 8-inch and one 8-inch x 13-inch flue liner, allow 55 bricks for every foot in height. This figure allows for width of one brick between the flue liners. For solid brick walls and fireplaces of large dimensions, allow 20 bricks for each cubic foot.

The hearth may be finished with tile, marble, brick, flagstone, or other material. These materials are estimated by the square foot.

The following is the material list for the fireplace materials in the sample house.

FLUE LINERS	8" x 8"		17
FLUE LINERS	8" x 13"		14
REGULAR CONCRETE BLOCKS-			
BASE	8" x 8" x 16"		140
ASH DUMP			1
CLEANOUT DOOR	12" x 15"		1
CLEANOUT DOOR	8" x 8"		1
REGULAR FIREBRICKS			30
PORTLAND CEMENT-			
HEARTH & CAP		BGS.	4
PREFABRICATED FIREPLACE			
UNIT - COMPLETE			1
STEEL ANGLE IRON	3½" x 3½" x 40"		1
COMMON BRICKS			350
CONCRETE BLOCKS	4" x 8" x 16"		150
NUMBER 1 COMMON BRICKS			500
RED QUARRY TILE - HEARTH		SQ. FT.	12

The amount of mortar cement that is required to build a fireplace depends partly on the mix that is used. A typical mix would be 1 part cement, 1 part lime, and 6 parts sand. Cement and lime are normally sold in 1-cubic-foot bags.

If 3/8-inch mortar joints are used, then 21 cubic feet of mortar are required to lay approximately 1,000 bricks. At this rate, 1 bag of masonry cement mixed with 1 bag of hydrated lime and 48 shovels of sand will lay approximately 400 bricks.

It has already been determined that the sample house will use 850 bricks and 290 concrete blocks. If 1 bag of masonry cement is enough to lay 28 blocks, then 290 blocks will require 11 bags. 850 bricks will use 5 bags of masonry cement. This is a total of 16 bags. About one-third should be allowed for waste and filling between the blocks and the flue liners. One-third of 16 is approximately 5, so a total of 21 bags of masonry cement should be ordered for the fireplace and chimney.

SUMMARY

Many fireplaces are built around a prefabricated metal unit. These units come in many sizes and styles. The size of the flue liner is specified by the manufacturer of the metal fireplace unit. The size of the fireplace opening is determined by the size of the damper, the size of the flue liner, and the height of the chimney.

All of the materials used in the construction of the fireplace must be fireproof. The floor of the firebox is made with firebrick and the walls are either firebrick or a metal unit. The flue liner is made of fireclay. The hearth is made of a decorative fireplace material, such as slate or quarry tile. No combustible materials should be used within 8 inches of either side of the fireplace opening.

The plan for the chimney is located on the Basement and Foundation Plan. The Floor Plan, Elevations, and Detail Drawings should be consulted for information pertaining to the fireplace.

MASONRY CEMENT -				BAGS	21																
CHIMNEY																					

REVIEW QUESTIONS

A. Select the letter preceding the best answer.

1. If a prefabricated metal fireplace is used, how does the builder determine the flue size?

   a. The manufacturer supplies this information.
   b. From the size of the unit
   c. From the size of the room in which it is installed
   d. From the height of the chimney

2. If the firebox is built of firebrick, how thick should the firebox wall be?

   a. 16 inches   c. 8 inches
   b. 4 inches   d. 6 inches

3. How many flue liners are needed to build a chimney 26 feet high?

   a. 26   c. 13
   b. 78   d. None of these

4. How far should the chimney extend above the ridge of the roof?

   a. 6 feet   c. 1 foot
   b. 4 feet   d. None of these

5. What material is used for the base of the firebox?

   a. Metal   c. Firebrick
   b. Flue liners   d. Quarry tile

6. Where may information be found about the number and size of flues in a chimney?

   a. First floor plan   c. Specifications
   b. Basement plan   d. Elevation

7. Where may information be found about the height of a chimney?

   a. First floor plan   c. Specifications
   b. Basement plan   d. Elevations

8. How many chimney blocks are needed for each flue liner in a single-flue chimney?

   a. 3   c. 4
   b. 2   d. 6

9. How many chimney blocks are required to build a chimney 24 feet high?

   a. 12   c. 18
   b. 24   d. 36

B. 1. List the three things that determine the size of a fireplace opening.

2. Arrange in order the materials needed to build the chimney in figure 15-6. The chimney is to have one 8-inch x 8-inch flue.

   a._____ 8-inch x 8-inch flue liners
   b._____ Chimney blocks
   c._____ Common red bricks
   d._____ Bags of masonry cement for blocks and bricks.
   e._____ Bags of portland cement for cap

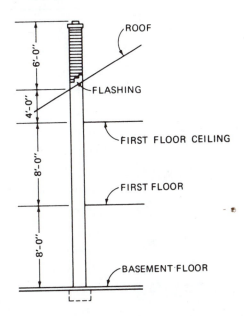

Fig. 15-6

unit 16 metalwork

OBJECTIVES

After studying this unit the student will be able to

- explain the specifications for metals.
- find the number of basement windows, areaways, and steel columns and beams needed for a job from the working drawings.

The specifications for all of the metalwork in the sample house are included in Division 5.

Specifications

DIVISION 5: METALS

A. WORK INCLUDED

The work shall include but shall not be limited by the following:

1. Steel basement columns shall be 4-inch diameter with 3/8-inch x 4-inch x 6-inch welded cap and base plates. Location and length are shown on the plans.
2. *Basement windows* shall be steel 15-inch x 12-inch, two-light, glazed, and with combination storm sash and screens.
3. *Galvanized areaways* shall be 20-gauge galvanized steel, with horizontal corrugations, sizes as shown on drawings. All areaways shall be anchored to the foundation wall. Furnish 6 inches of crushed stone below the areaway to a level of 3 inches below the bottom of the basement windows.

BASEMENT WINDOWS

Most basement windows (commonly called cellar sash) are made of wood, steel, or aluminum, figure 16-1.

Fig. 16-1 Cellar sash

The mason leaves an opening two blocks long and two blocks high when the block foundation wall is built. The cellar windows fit up against the wood sill, with the flanges at the ends imbedded in concrete in the sash blocks. With the sash in place, a concrete sill is formed at the bottom.

Figure 16-2 shows how a cellar window appears on the foundation plan. Notice that the dimensions are given to the center of the opening. A note is included to indicate the size of the window. Although most houses use regular metal basement windows, other types and sizes may be used. On some jobs, the slope of the land is such that full-size windows and exterior doors may be placed in the foundation wall leading out onto a terrace or the backyard. The plans and specifications must be consulted to determine what type of windows are needed.

*NOTE: ALL CELLAR SASH TO BE
15 X 12 TWO LIGHT

Fig. 16-2 Cellar sash on a basement plan.

AREAWAYS

Many homes are designed so that they conform to the ground and become a part of the landscape. These houses may have only nine to twelve inches of foundation wall showing above the finished grade. When a house designed in this manner has basement windows, an areaway is required around each window. Areaways are used to hold back the earth and allow light and air to enter the window or door area, figure 16-3.

Areaways are made of metal or concrete. They are placed so that they will be 2 to 4 inches above the finished grade and go approximately 18 inches below the bottom of the window. The inside of the areaway is filled with crushed stone to within 3 inches of the bottom of the window. This allows water to drain away from the window.

STEEL SUPPORT COLUMNS

Steel posts, sometimes called Lally columns, are used to support the framing members of the building between the foundation walls. The ends of the floor joists rest on a beam or girder. The support posts distribute this weight to the footings. Each post rests on a footing. They are shown on the Basement Plan for the sample house. A detail drawing of the support columns and footings is also included.

Estimating Basement Windows, Areaways, and Basement Support Columns

The number of basement windows is found on the basement and foundation plan. List them according to the size given. If the specifications indicate that the basement sash are to have storms and screens, specify this on the material list.

There are five cellar sash, so five areaways are needed. The specifications indicate that the areaways shall be 20-gauge galvanized steel with horizontal corrugations. If the plans call for concrete areaways, estimate the concrete the same way as for the footings and walls. Multiply the length times the width times the depth (all in feet), then divide by 27 to find the number of cubic yards.

Four-inch steel posts with 3/8-inch x 4-inch x 6-inch welded cap and base plates are specified for

Fig. 16-3 (A) Door areaway

Fig. 16-3 (B) Window areaway

the support columns. Count the number of these shown on the foundation plan and list this on the material list.

needed to provide space around the basement windows. Areaways may be cast of concrete or made of galvanized steel. The inside of the areaway is

CELLAR SASH, GLAZED	15"x 12"	5																						
COMBINATION STORM & SCREEN	15" x 12"	5																						
CORRUGATED, GALVANIZED																								
AREAWAYS	20 GAUGE	5																						
STEEL POSTS w/ WELDED CAP																								
AND BASE PLATES	4"x 7'	5																						

SUMMARY

Most basement windows are 15 inch x 12 inch, two-light sash, made from aluminum or steel. This size fits into an opening two blocks long and two blocks high. If the foundation does not extend far enough above the ground, an areaway may be filled with crushed stone to within 3 inches of the bottom of the window.

Support posts are usually required to support the weight of the building between the foundation walls. These columns rest on concrete footings.

REVIEW QUESTIONS

Select the letter preceding the best answer.

1. What is the thickness of the cap and base plates on the support posts?

   a. 26 gauge
   b. 20 gauge
   c. 1/2 inch
   d. 3/8 inch

2. What is the overall size of a 15-inch x 12-inch, two-light cellar sash?

   a. 32 inches x 13 1/2 inches
   b. 30 inches x 24 inches
   c. 30 inches x 12 inches
   d. 32 inches x 16 inches

3. What is the function of an areaway?

   a. To protect the window
   b. To allow the window to be below grade
   c. To help insulate the basement
   d. None of these

4. What is used to fill the bottom of an areaway?

   a. Concrete
   b. Crushed stone
   c. Earth
   d. Any waterproof material

5. Where is the number of steel support columns found?

   a. Specifications
   b. Elevation
   c. Floor plan
   d. Basement plan

6. Where is the size and number of cellar sash found?

    a. Specifications   c. Floor plan

    b. Elevation   d. Basement plan

7. How are cellar windows fastened into the foundation wall?

    a. They fit into a groove in the sash blocks.

    b. They are fastened with screws and masonry anchors.

    c. They are bolted.

    d. They are nailed with hardened nails.

unit 17 lumber and frame construction

OBJECTIVES

After studying this unit the student will be able to

- tell how lumber is classified.
- identify the abbreviations used in the lumber industry.
- estimate quantities of lumber by the piece, linear foot, and board foot.

LUMBER CLASSIFICATIONS AND GRADES

Lumber is broadly grouped as either hardwood or softwood. Hardwood is not always harder than softwood, but this is generally true. Hardwood comes from *deciduous* (leaf-bearing) trees and softwood comes from *coniferous* (cone-bearing) trees. Approximately three-fourths of the total annual lumber production of the United States is softwood, which is used for all types of building construction. The remainder is hardwood. Some of the most common softwoods are red cedar, cy-press, Douglas fir, eastern hemlock, western hemlock, sitka spruce, eastern spruce, and the pines.

Lumber can be further grouped according to the amount of processing it goes through at the sawmill or the use for which it is intended. The following are some of the classifications (grades) of lumber, figure 17-1.

Grade Stamps

Rough lumber is lumber that has not been planed or milled in any way after it is sawed.

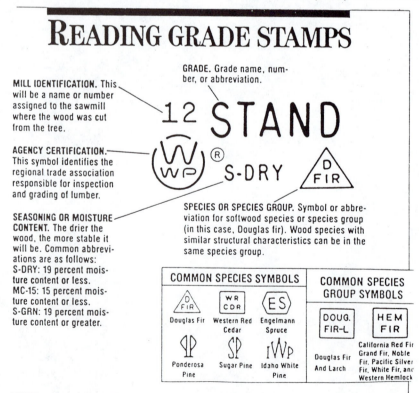

Fig. 17-1 Reading grade stamps

Surfaced lumber has been surfaced on one or more sides in a planer.

Worked lumber has been shaped in a matching machine or molder.

Factory or shop lumber is intended to be cut up for future use. For example, clear pine, which is to be made into window or door parts.

Yard lumber is less than 5 inches thick and is graded for the future use of the entire piece.

Structural lumber is over 2 inches in thickness and width and is graded for strength and the use of the entire piece.

The following abbreviations are in general use throughout the lumber industry. An estimator must be acquainted with them in order to read the plans and specifications.

B1S - Beaded one side	Flg. - Flooring
Bd. ft. - Board foot	Hdwd. - Hardwood
Bev. - Beveled	Lin. ft. - Linear foot
B.M. or b.m. - Board	or 12 inches
(foot) measure	M - 1,000
Btr. - Better	Mldg. - Moulding
Clg. - Ceiling	Qtd. - Quartered
Clr. - Clear	Rfrs.- Roofers
CM - Center matched	Sdg. - Siding
Com. - Common	Sel. - Select
Csg. - Casing	T&G- Tongue and
D&M - Dressed &	

Continued at top of right column.

matched

Dim.- Dimension	groove
S2S - Surfaced two sides	S4S - Surfaced four sides

LUMBER MEASUREMENTS

There are three ways to estimate quantities of lumber. These are:

* the number of pieces required.
* the number of linear feet required.
* the number of board feet required.

Estimate by the number of pieces required if (1) the ends rest on bearings that are a specific distance apart; (2) a number of pieces of the same length are required; or (3) a specific length must be cut, without waste. Estimate by linear feet if the stock can be joined anywhere; for example, the box sill or the sill itself, or the boards for the sides of the footing forms. Fractions of a linear foot are counted as a full foot. Odd lengths are increased to the next even number of feet.

Lumber is commonly sold and listed by the *board foot*. [Board feet = length (ft.) x width (ft.) x thickness (inches).] A board foot is the quantity of lumber equal to one foot by one foot by one inch thick, or 144 cubic inches of wood. To find the board feet contained in a piece of lumber one may convert all dimensions to inches, multiply the width times the length times the thickness, then divide by 144. When calculating board feet, it is

Thickness		Width	
Nominal	Dressed	Nominal	Dressed
1	3/4	2	1 1/2
1 1/4	1	3	2 1/2
1 1/2	1 1/4	4	3 1/2
2	1 1/2	5	4 1/2
2 1/2	2	6	5 1/2
3	2 1/2	7	6 1/2
3 1/2	3	8	7 1/4
4	3 1/2	9	8 1/4
		10	9 1/4
		11	10 1/4
		12	11 1/4

Fig. 17-2 Normal and minimum dressed sizes of lumber

customary to consider a fraction of an inch in thickness to be a full inch. For example, consider a board which measures 3/4″ x 4″ x 6′. After converting all dimensions to inches, the board measures 1″ x 4″ x 72″ (288 cubic inches). Divide by 144 to find that the board contains 2 board feet of wood.

Large areas such as subfloors, roof decks, and wall sheathing are usually covered with 4′ x 8′ panels. These panels are listed by the piece. However, if for some reason the areas are to be covered with boards, list the quantity by board feet.

Nominal size is used to measure board feet. The nominal size of a piece of lumber is the size that it was sawed to at the sawmill, figure 17-2. When the lumber is surfaced, it is further reduced in size. For example, a piece which has a nominal width of six inches is actually only 5 1/2 inches wide after planing.

One hundred board feet of 1 x 6 dressed (planed) lumber will not cover a 100 square foot area. A certain percentage must be added to make up for the loss in dressing, shiplapping, or matching. The amount of waste varies with the kind of material, where it is used, and how it is used.

FRAME CONSTRUCTION

In order to read a set of plans and be able to make a list of materials, the estimator must know what materials are used for each specific purpose, and how to use those materials properly. The one method of framing a house most commonly used is called platform framing.

Platform Framing

This type of framing is considered the fastest and the safest form of construction and is the most widely used. Each floor is framed separately with the subfloor in place before the wall and partition studs of that floor are erected, figure 17-3. Interior and exterior walls are framed in the same manner, assuring balanced shrinkage or settling, if any occurs.

SUMMARY

Lumber is broadly grouped as either hardwood or softwood. Hardwood comes from trees with broad leaves; softwood comes from trees with needle-like leaves. Lumber is further classified as: factory and shop lumber, which is intended to be cut up for further manufacture; yard lumber, which is of all sizes and is intended for general construction purposes; and structural lumber, which is 2 inches or more thick and is intended for use where working stresses are required.

Lumber can be estimated by the piece, linear foot, board foot. Estimate by the piece if the ends rest on bearings that are a specific distance apart, if a number of pieces of the same length are required, or if a specific length is needed from each piece. Estimate by the linear foot if the material can be joined anywhere. Estimate by the board foot when a large area must be covered.

In platform framing, each floor is framed separately, with the subfloor in place before the wall and partition studs of that floor are erected. Interior and exterior walls are framed in the same manner.

REVIEW QUESTIONS

A. Select the letter preceding the best answer.

1. What are the actual dimensions of a 2 x 6?

a. 2 inches x 6 inches

b. 1 3/4 inches x 5 3/4 inches

c. 1 5/8 inches x 5 1/4 inches

d. 1 1/2 inches x 5 1/2 inches

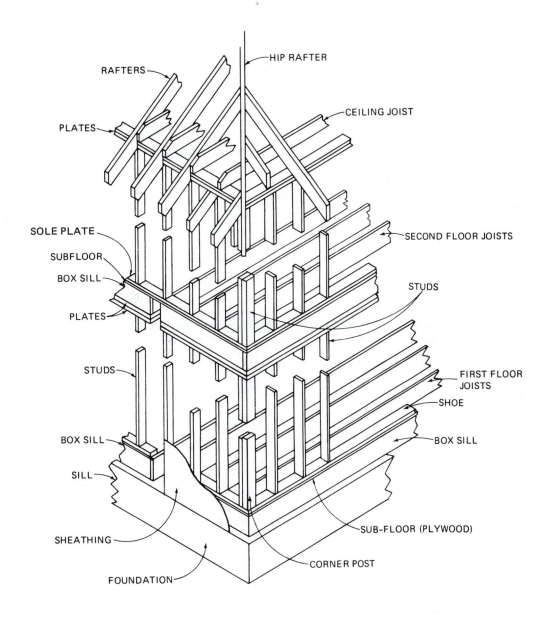

Fig. 17-3 Platform Framing

2. What are the actual dimensions of a 2 x 10?

   a. 2 inches x 10 inches

   b. 1 3/4 inches x 5 3/4 inches

   c. 1 1/2 inches x 9 1/2 inches

   d. 1 1/2 inches x 9 1/4 inches

3. How would the quantity of lumber be estimated for floor joists?

   a. By the piece

   b. By the linear foot

   c. By the board foot

   d. None of these

4. Which of the following is probably the length of the exterior wall studs for platform frame construction?

   a. 8 feet
   b. 12 feet

   c. 16 feet
   d. 20 feet

5. How would the quantity of lumber be estimated for wood siding?

   a. By the piece
   b. By the linear foot

   c. By the board foot
   d. None of these

6. How would the quantity of lumber be estimated for building forms for the footing of a house?

   a. By the piece
   b. By the linear foot

   c. By the board foot
   d. None of these

7. How many board feet does a piece 1 inch x 6 inches x 96 inches contain?

   a. 8 board feet
   b. 16 board feet

   c. 12 board feet
   d. None of these

B. Write the term for which each of the following is an abbreviation.

   Bd. ft.
   B.M.
   Sdg.
   Hdwd.
   Clg.

   T&G
   S2S
   Csg.
   M
   Clr.

unit 18 girders

OBJECTIVES

After studying this unit the student will be able to

- explain the specifications relating to carpentry work.
- tell where to look on working drawings for girders.
- calculate the quantity of material needed for the girder.

After the foundation wall and basement concrete floor has been estimated, the framing estimate begins, figure 18-1. Read the specifications under Division 6, Carpentry and Millwork, and apply

Fig. 18-1 After the foundation and basement floor are in place the framing begins

this information to the working drawings to estimate the framing materials. The specifications for carpentry and millwork in the sample house are as follows:

Specifications

DIVISION 6: CARPENTRY AND MILLWORK

A. MATERIALS

1. All materials are to be the best of their respective kind. Lumber shall bear the mark and grade of the association under whose rules it is produced. Framing lumber shall be thoroughly seasoned with a maximum moisture content of 19 percent. All millwork shall be kiln dried.
2. Properly protect all materials. All lumber shall be kept under cover at the job site. Material shall not be delivered unduly long before it is required for work.
3. Lumber for various uses shall be as follows:
   Framing: No. 2 dimension, Douglas fir or yellow pine.
   Exterior Millwork: No. 1 clear white pine.
   The lumber must be sound, thoroughly seasoned, well manufactured and free from warp which cannot be corrected by bridging or nailing. All woodwork which is exposed to view shall be S4S.

B. INSTALLATION

1. All work shall be done by skilled workers.
2. All work shall be erected plumb, true, square, and in accordance with the drawings.
3. Finish work shall be blind nailed as much as possible, and surface nails shall be set.
4. All work shall be securely nailed to studs, nailing blocks, grounds, furring, and nailing strips.

C. GRADES AND SPECIES OF LUMBER

1. *Framing lumber, except studs and wall plates,* shall be No. 1 Douglas fir.
2. *Studs, shoes, and double wall plates* shall be Douglas fir, utility grade.
3. *Bridging* shall be 1 x 3 spruce.
4. *Joists* shall be 2 x 8's spaced 16" OC except where otherwise indicated. All joints are to be doubled under partitions and around stairways and fireplace openings.
5. *Subflooring* shall be 5/8" plywood APA grade C-D.
6. *Ceiling joists and rafters* shall be 2 x 6's spaced 16" OC. Rafters are to have overhang as noted on drawings. Rafters shall have 1 x 8 collar beams 32" OC.
7. *Wall sheathing and roof sheathing* shall be 1/2" x 4' x 8' plywood APA grade C-D Exterior.
8. *Exterior siding* shall be 1/2" x 8" bevel siding with 6 3/4" exposure.
9. *Cornice material:* Soffit, 1 x 8 clear pine; Fascia, 1" x 6" B or better pine; Moldings are to be clear pine.
10. *Drywall material* shall be 3/8" thick gypsum wallboard.
11. *Insulation:* Ceiling of second floor 6" fiberglass; all side walls 3 1/2" fiberglass or as specified by code.
12. *Interior woodwork* shall be kiln dried clear pine. All interior woodwork is to be machine sanded at the mill and hand sanded on the job.
13. *Hardwood flooring* shall be 1 x 3 select oak.
14. *Underlayment* shall be 5/8" x 4' x 8' plywood APA underlayment.

15. *Cabinet and vanity counter tops* shall be 1/16″ standard grade, laminated plastic surfacing material conforming to NEMA standards. Color is to be selected by the owner.

D. WORKMANSHIP

1. *Framing:* All framing members shall be substantially and accurately fitted together, well secured, braced, and nailed. Plates and sills shall be halved together at all corners and splices. Studs in walls and partitions shall be doubled at all corners and openings. Joists over 8 feet in span shall be bridged with one row of 1 x 3 spruce cross bridging cut on bevel and nailed up tight after the subfloor has been laid.

2. *Gypsum wallboard:* Gypsum wallboard shall be nailed to wood framing in strict accordance with the manufacturer's recommendation. Space nails not more than 7 inches apart on ceilings and not more than 8 inches apart on sidewalls. Dimple the nailheads slightly below the surface of the wallboard, taking care not to break the paper surface.

3. *Insulation:* Fit each batt snugly and securely between joists with the vapor barrier toward the room side. Cut batts to fit irregular spaces, leaving flange on inside for stapling. In removing fiberglass to fit around pipes and other openings, do not cut vapor barrier unless necessary.

4. *Interior trim and millwork:* All exposed millwork shall be machine sanded to a smooth finish, with all joints tight and formed to conceal any shrinkage. Miter exterior angles, butt and cope interior angles and scarf all running joints in moldings.

5. *Hardwood flooring:* All subfloors are to be broom cleaned and covered with deadening felt before the finished floor is laid. Wood flooring, where scheduled, is to be 1 x 3 T&G and end-matched select oak flooring. Flooring is to be laid evenly and blind nailed or stapled every 16″ without tool marks.

6. *Closet rods:* Furnish and install where indicated on drawings. Rods are to be adjustable chrome, fitted, and supported at least every 4′.

E. CLEANUP

Upon completion of work, all surplus and waste materials shall be removed from the building, and the entire structure and involved portions of the site shall be left in a neat, clean, and acceptable condition.

BASEMENT BEAMS OR GIRDERS

These beams support the inner ends of the first-floor joists. The outer ends of the floor joists rest on the sill, on top of the foundation wall. The girder and its supporting posts support the main bearing partitions, as well as part of the weight of the floors and their contents. Girders may be made of either wood or steel. When a built-up wood girder is used, it is usually made from three pieces of wood nailed together. Another type of wood beam which has superior strength and requires a minimum of material is known as box beams. These beams are made up of plywood nailed (sometimes glue is also used) to 2 x 4 or 2 x 6 framing members, figure 18-2. If the girder is made of steel, there are several methods used to frame the wood joists to the steel beam.

One method is to attach a 2 x 6 to the top of the girder and rest the joists on top of the 2 x 6, figure 18-3. Sometimes the beam is flush with the top of the floor joists and the joists are cut to rest on the flange of the beam, figure 18-4.

The size of the built-up wood girder is given on the basement plan. Notes are included to explain how it is to be constructed. For example, a note that the girder is to be 6 x 8 means that

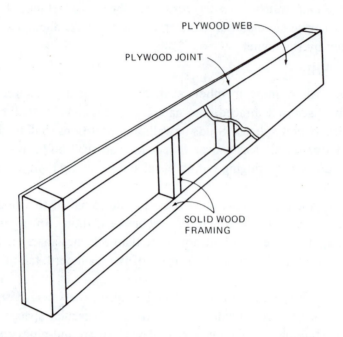

Fig. 18-2 Box beams are strong but use a minimum of material

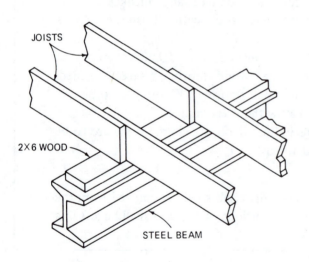

Fig. 18-3 Joists resting on a steel beam

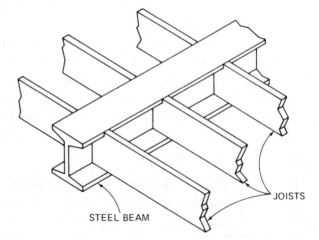

Fig. 18-4 Flush steel beam

three 2 x 8's are nailed together to make the girder. When the girder is built up in this way, the end joints of the 2 x 8's should occur only at the supporting posts, figure 18-5. The joists may be notched to rest on a ledger, as shown in figure 18-6. This method is used where headroom is needed in the basement, such as for a finished room. Another method commonly used is to rest the joists on top of the girder, figure 18-7.

Still another method of attaching joists to a girder is with the use of *joist hangers*. These are specially designed steel devices which are nailed in place on the sides of the girder to form a pocket into which the joist fits. Joist hangers are available in a variety of sizes for different size joists.

Estimating Girders

The Basement and Foundation Plan for the sample house shows the location & size of the girder. Solid lines with arrowheads on each end show the direction of the floor joists. The size of the

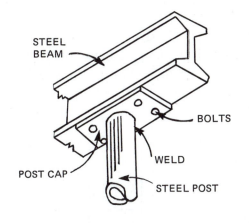

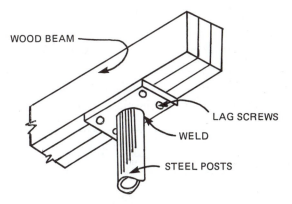

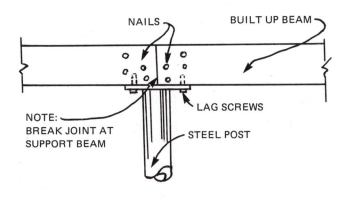

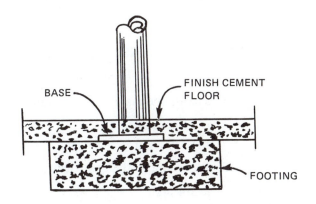

Fig. 18-5

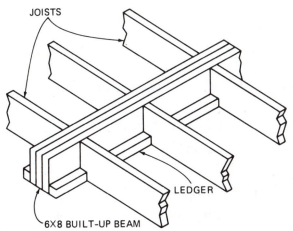

Fig. 18-6 **Joists resting on ledger**

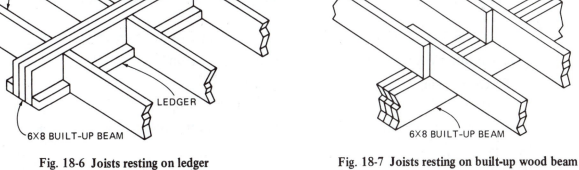

Fig. 18-7 **Joists resting on built-up wood beam**

floor joists and their spacing are noted just above this line. On the plans for the sample house there are three beams. One runs from the front of the house back to the fireplace base. A second one runs from the fireplace base to the rear of the house. A note indicates that the beam is built up of three 2 x 8's.

In figuring the length of the pieces, be sure that all end joints will be at supporting posts. If the girder is not too long, it is better to use pieces that run from wall to wall than to build the girder up from shorter pieces. The sample house requires six 2 x 8 x 8's and three 2 x 8 x 14's. Add these to the material list.

SUMMARY

The girder and its supporting posts support the main bearing partitions, as well as part of the weight of the floors and their contents. Girders may be made of either wood or steel. Wood girders are usu-

ally made of three wood members nailed together. For a built-up wood girder which is flush with the floor joists, a ledger is nailed to the bottom edge to support the joists. The ends of the wood members making up the girder should be joined only at sup- port columns. If a steel girder is used, the joists are cut to fit on the bottom flange of the beam. The size of the beam is shown on the basement and foundation plan.

FIR- BASEMENT BEAMS	2×8×8'			6																										
FIR- BASEMENT BEAMS	2×8×14'			3																										

REVIEW QUESTIONS

A. Select the letter preceding the best answer.

1. What is the maximum allowable moisture content of the framing lumber?

   a. 12 percent
   b. 19 percent
   c. 8 percent
   d. 14 percent

2. According to the specifications, what is the nominal thickness of the fascia?

   a. 3/4 inch
   b. 1 inch
   c. 1 1/4 inch
   d. 2 inches

3. What is indicated by a note that the girder is to be "built-up 6 x 8"?

   a. One solid piece of wood measuring 6 inches x 8 inches
   b. Three 2 x 8's nailed together with the joints spaced equally throughout their length
   c. Three 2 x 8's nailed together with joints centered between the supporting columns
   d. Three 2 x 8's nailed together with the joints directly over the support columns

4. Where is information about the size and type of girder found?

   a. Basement plan
   b. Floor plan
   c. Elevation
   d. Specifications

5. According to the specifications, all framing lumber except studs and wall plates is

   a. Pine.
   b. Spruce.
   c. Fir.
   d. Hemlock.

6. According to the specifications, exterior trim is to be

   a. Pine.
   b. Spruce
   c. Fir.
   d. Hemlock.

7. What size lumber is to be used for the ceiling joists?

   a. 2 x 4
   b. 2 x 6
   c. 2 x 8
   d. 1 x 8

unit 19 floor framing

OBJECTIVES

After studying this unit the student will be able to

- describe the sill seal, sill, box sill, joists, trimmer, and subflooring.

- show where these items can be found on working drawings.

- determine the quantity of material needed for the sill seal, sill, box sill, joists, trimmers, and subflooring.

SILL SEAL

The top of the foundation wall may not be absolutely true and even. To prevent small gaps between the foundation and the sill, a 6 inch wide by 3/4 inch thick fiberglass material called *sill seal* is used, figure 19-1. This is placed on top of the foundation wall and under the sill. This sill seal prevents insects and cold air from entering the basement of the house.

SILL

The sill is a single piece of wood (usually 2 x 6) laid flat on the top of the foundation wall, figure 19-1. The first floor joists and the box sill are nailed to the sill. Usually the sill is fastened to the foundation every six or eight feet by anchor bolts that extend into the foundation.

Estimating the Sill Seal and Sill

Enough sill seal and 2 x 6 (for the sill) must be ordered to cover the perimeter of the building. If a steel supporting beam is used for the main bearing and the joists rest on top of the beam, then enough 2 x 6 to cover the length of the steel beam must be added. If built-up wood girders are used, it is not necessary to cover them with a sill.

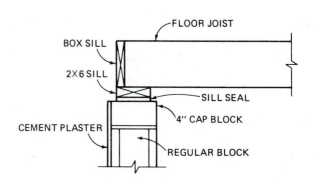

Fig. 19-1 Construction of the box sill

The distance around the outside of the sample house is 108 linear feet, as shown in the Foundation Plan. Sill seal is sold by the linear foot, so add 108 linear feet of sill seal to the estimate. Lumber may be sold by the board foot, but most lumber yards list it by the piece. Standard lengths are from eight to twenty-four feet, in two-foot increments, but pieces over sixteen feet are more expensive. If the sill is 108 linear feet, nine 12-foot pieces are required. (108 ÷ 9 = 12). Best for use with anchor bolts 6'-0" o.c. Add the sill seal and the sill to the material list as follows:

SILL SEAL		6"	LIN. FT.	108	
FIR SEAL	2 X 6 X 12'			9	

BOX SILL

The box sill is the header joist placed at right angles to the ends of the floor joists, figure 19-1. It is the same width as the joists and rests on top of the sill. The box sill is nailed to the other joists and toe-nailed into the sill.

Estimating the Box Sill

The box sill covers the perimeter of the house, so it will also be 108 linear feet. It is the same height as the floor joists and the plan indicates that they are 2 x 8's, so the box sill is 2 x 8's. Again, this is listed either by the linear foot or by the piece.

of the joists and a note just above this line indicates their size and spacing. For example, the Foundation Plan indicates that the floor joists are 2 x 8's spaced 16″ O.C. To find the number of floor joists that are needed, multiply the length of the foundation wall on which they rest by three-fourths, then subtract one. This provides for 3 joists every 4 feet and 1 additional joist at the end of the wall. Also add one joist for each partition that runs in the same direction as the joists.

According to the foundation plan, the joists on the left end of the house run from the porch wall to the fireplace wall and beam. The foundation wall on which these joists rest is twenty feet

FIR - BOX SILL	2x8x12'		9			

JOISTS

The floor joists are the horizontal members that support the floor. They are usually spaced 16″ O.C. For short spans, the joists may extend from one foundation wall to another. Over larger spans, the joists rest with one end on the foundation and the other on the supporting beam or girder. Joist sizes, like beam sizes, are dependent upon the length of the span they have to bridge and the load they have to carry. Joists should overlap for the full width of the beam that supports them, and they should be nailed together, figure 19-2.

Estimating Floor Joists

The size and direction of the floor joists are indicated on the foundation plan. A straight line with arrowheads on each end indicates the direction

long. Using the procedure described above, multiply 20 by 3/4 to get 15, then subtract 1 to get 14 floor joists needed for the left end of the first floor. The span of these joists is 13 feet. However, lumber is sold in even lengths, so these floor joists must be 14 feet long.

The joists for the right side of the first floor area run from the front to the back of the house. They cover an area 17 feet wide, so multiply 17 by 3/4 and get 12 3/4. This is rounded off to 13. Subtract 1 from this and get 12 floor joists for the front of the house and 14 for the back of the house. There are 2 partitions running in the same direction as the floor joists, so 2 joists must be added to the total. These joists must span 12 feet, plus the overlap at the beam, so 14-foot lumber can be ordered.

Add the first-floor joists to the material list.

FIR JOISTS - LIVING ROOM AREA	2x8x14'		14			
FIR JOISTS - FIRST FLOOR AREA	2x8x14'		28			

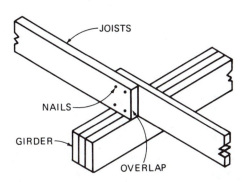

Fig. 19-2 **The floor joists should overlap the full width of the beam and should be nailed together.**

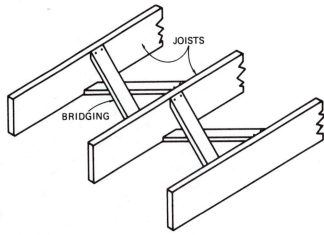

Fig. 19-3 **Diagonal bridging stiffens the floor joists and helps distribute the load over all of the joists.**

BRIDGING

Bridging is the lateral bracing of the floor joists required by code when joists' depth to thickness ratio does not exceed 6. Wood bridging is made of short pieces of 1 x 3 nailed to the top of one joist and the bottom of the next, figure 19-3. Bridging is also made from metal. The purpose of bridging is to keep the joists in alignment and to distribute the load over all the joists.

Estimating Bridging

To determine the amount of wood bridging needed, multiply the number of floor joists by 3. This is the number of linear feet of bridging needed for one row. Spans of 10 to 14 feet may have one row of bridging in the center of the span. If the span is over 14 feet wide, allow for two rows of bridging for each span.

The longest span in the sample house is 13 feet, so only one row of bridging is used for each span of joists. There are 42 floor joists in the first floor so multiply 42 by 3 and get 126 linear feet of bridging needed. If metal bridging is used, allow 2 pieces for every floor joist. Add this to material list (A).

HEADERS AND TRIMMERS

When the joists must be cut to make an opening for a stairway, chimney, or fireplace, headers are used to support the cut ends of the joists, figure 19-4. The joists at the ends of the headers are called *trimmers*. Trimmers should be doubled to support the extra load when more than one joist is cut.

Estimating Headers and Trimmers

The material for headers and trimmers must be the same width as that for the floor joists. The length of the trimmers is the same as the length of the floor joists. Looking at the plan, see that there is a fireplace opening and a stairwell. The fireplace requires a double header for the length of the opening, so two 2 x 8 x 10's are needed. This allows for both sides, because the length of the fire place opening is only seven feet. The stairwell requires two 2 x 8 x 14's.

Add this to material list (B).

A

	SPRUCE BRIDGING	1x3	LIN.FT.	126																		

B

	FIR	2x8x10'	2																	
	FIR	2x8x14'	2																	

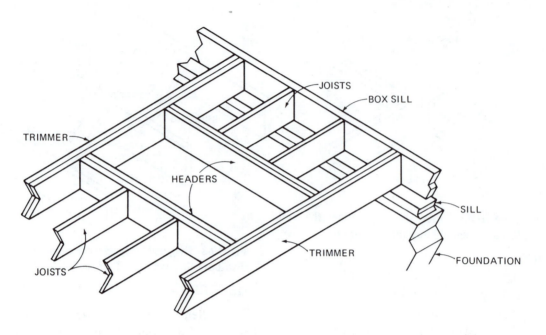

Fig. 19-4 Framing for a stairwell or chimney opening

SUBFLOORING

The subfloor is the first layer of the floor. It is nailed to the floor joists and the finished flooring is applied to it. The subfloor helps keep the joists in alignment, holds the floor level, prevents the movement of dust and dirt between floors, helps deaden sound, acts as insulation, and provides a solid base to receive the finished floor. Most subflooring is done with 4' x 8' sheets of plywood, figure 19-5.

To conserve material and eliminate the need to use underlayment (discussed in a later unit), the American Plywood Association has developed new ways to construct floors using only one application of special plywood materials. Two of these are shown in figure 19-6 and figure 19-7.

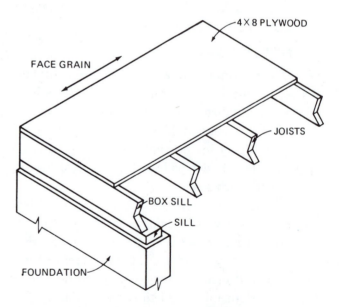

Fig. 19-5 Plywood subflooring

Estimating Subflooring

The area of the first floor is 30 feet by 24 feet or 720 square feet less a 9 foot by 4 foot corner taken out of the living-room end. This makes a total of 684 square feet on the first floor. Each sheet of plywood covers 32 square feet, so divide 684 square feet (the area of the floor) by 32 for a total of 21 12/32 or 22 sheets. Add the subflooring to the material list.

CD PLYWOOD SUBFLOOR								
FIRST FLOOR	5/8"x4'x8'	22						

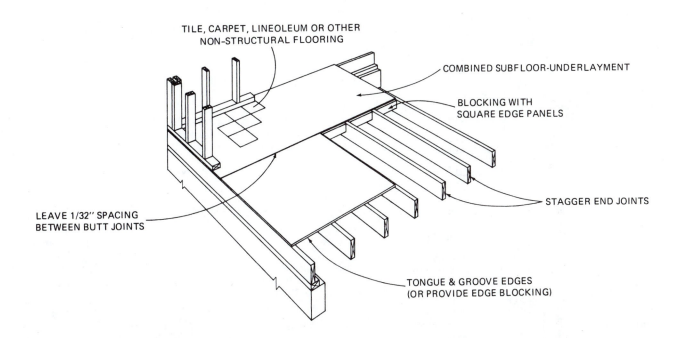

Combined subfloor–underlayment/ For direct application of tile, carpeting, linoleum or other non-structural flooring (Plywood continuous over two or more spans; grain of face plys across supports. Seasoned framing lumber is recommended.)

Plywood Grade[3]	Plywood Species Group	Maximum Support Spacing[1][2]						Nail Spacing (inches)	
		16" o.c.		20" o.c.		24" o.c.		Panel Edges	Intermediate
		Panel Thickness	Deformed Shank Nail Size[4]	Panel Thickness	Deformed Shank Nail Size[4]	Panel Thickness	Deformed Shank Nail Size[4]		
C-C Plugged Exterior	1	1/2"	6d	5/8"	6d	3/4"	6d	6	10
Underlayment with EXT. glue	2 & 3	5/8"	6d	3/4"	6d	7/8"	8d	6	10
Underlayment	4	3/4"	6d	7/8"	8d	1"	8d	6	10

Notes:

(1) Edges shall be tongue and grooved, or supported with framing.

(2) In some non-residential buildings, special conditions may impose heavy concentrated loads and heavy traffic requiring subfloor-underlayment constructions in excess of these minimums.

(3) For certain types of flooring such as wood block or terrazzo, sheathing grades of plywood may be used.

(4) Set nails 1/16" and lightly sand subfloor at joints if resilient flooring is to be applied.

Fig. 19-6 Combined subfloor-underlayment. One of the best ways to save money and time in subfloor construction is to use one layer of plywood as a combination subfloor-underlayment. This gives you all the advantages of plywood in both applications, yet it speeds construction and cuts material costs substantially.

As with plywood underlayment it is important to use only the recommended grades shown in the table above. Satisfactory performance of the finished floor demands good construction practice and reasonable care during installation.

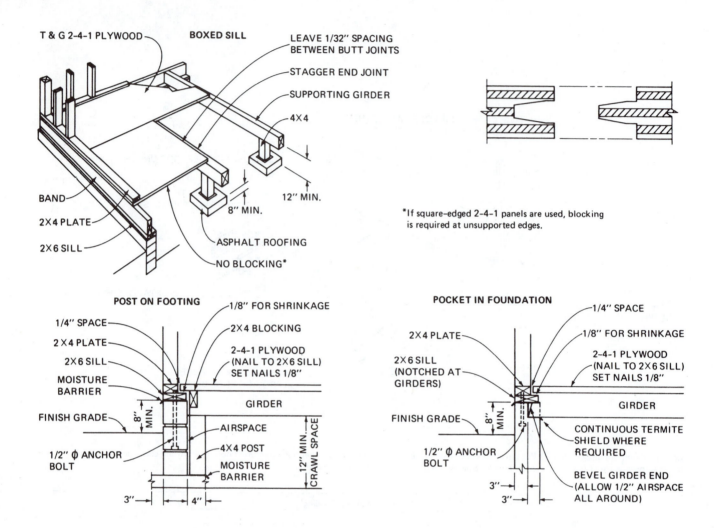

Labels in figure:

T & G 2-4-1 PLYWOOD

BOXED SILL

LEAVE 1/32" SPACING BETWEEN BUTT JOINTS

STAGGER END JOINT

SUPPORTING GIRDER

4 X 4

12" MIN.

8" MIN.

BAND

2 X 4 PLATE

2 X 6 SILL

ASPHALT ROOFING

NO BLOCKING*

*If square-edged 2-4-1 panels are used, blocking is required at unsupported edges.

POST ON FOOTING

1/4" SPACE

2 X 4 PLATE

2 X 6 SILL

MOISTURE BARRIER

FINISH GRADE

8" MIN.

1/2" ø ANCHOR BOLT

3" 4"

1/8" FOR SHRINKAGE

2 X 4 BLOCKING

2-4-1 PLYWOOD (NAIL TO 2 X 6 SILL) SET NAILS 1/8"

GIRDER

AIRSPACE

4 X 4 POST

MOISTURE BARRIER

12" MIN. CRAWL SPACE

POCKET IN FOUNDATION

2 X 4 PLATE

2 X 6 SILL (NOTCHED AT GIRDERS)

FINISH GRADE

8" MIN.

1/2" ø ANCHOR BOLT

3"

3"

1/4" SPACE

1/8" FOR SHRINKAGE

2-4-1 PLYWOOD (NAIL TO 2 X 6 SILL) SET NAILS 1/8"

GIRDER

CONTINUOUS TERMITE SHIELD WHERE REQUIRED

BEVEL GIRDER END (ALLOW 1/2" AIRSPACE ALL AROUND)

2-4-1 Subfloor-underlayment[1]/ For application of tile, carpeting, linoleum or other non-structural flooring; or hardwood flooring (Live loads up to 65 psf—two span continuous; grain of face plys across supports)

Plywood Species Group	2-4-1		Nail Size and Type[2]	Nail Spacing (inches)	
	Plywood Thickness (inches)	Maximum Spacing or Supports c. to c. (inches)		Panel Edges	Intermediate
Groups 1, 2, & 3	1 1/8 only	48[3]	8d ring shank recommended or 10d common smooth shank (if supports are well seasoned)	6	6

Notes:

(1) For additional information, see American Plywood Association Form No. 60-40.

(2) Set nails 1/8" and lightly sand subfloor at joints if resilient flooring is to be applied.

(3) In some non-residential buildings special conditions may impose heavy concentrated loads and heavy traffic, requiring support spacing less than 48".

Fig. 19-7 2-4-1 Subfloor-underlayment. This is one of the fastest, simplest wood floor construction systems ever devised. It consists of a 1 1/8" thick panel that serves as a combination subfloor-underlayment over two spans of 48". These panels are available with a precisely engineered tongue and groove joint that eliminates the need for blocking and provides a smooth surface suitable for any type of finish floor covering—resilient tile, linoleum, wall-to-wall carpeting or hardwood flooring.

SUMMARY

The sill seal is a 3/4 inch thick by 6-inch wide fiberglass piece, which is placed between the foundation and the wood sill. The purpose of the sill seal is to prevent cold air and insects from entering the cellar. Sill seal is sold by the linear foot.

The sill is usually a 2 x 6 piece resting on top of the foundation wall. The rest of the house framing rests on the wood sill. The box sill is the header which is fastened to the ends of the floor joists and rests perpendicularly on the sill. The box sill is the same width as the joists and the same length as the sill.

The floor joists are the framing members which support the floor of the house. They are usually spaced 16 inches O.C. Their width is specified on the foundation plan. Their length is determined by the distance between foundation walls or between the foundation wall and the girder. To find the number of joists needed to cover a particular area, multiply the length of the foundation or girder on which they rest by three-fourths, then deduct one when box

sill has already been estimated. Add one joist for each partition that runs in the same direction as the joists.

Bridging per code helps to distribute the load on a floor and adds stiffness to the floor. It may be made of either wood or metal. To find the number of linear feet of bridging in one row, multiply the number of floor joists by three. If the floor joists span more than 14 feet, two rows of bridging should be used.

Headers and trimmers are used to frame an opening in the floor. These members are the same width as the floor joists. Their lengths are determined by the size of the opening.

Subflooring is usually made of plywood. To find the amount of plywood needed, determine the area of the floor in square feet and divide by thirty-two when floor system is on even module of 4' x 8' panels. If tongue and groove or shiplap are used for the subfloor, it should be nailed diagonally to ensure rigidity. A 25 percent allowance for matching must be added to the total when these materials are used instead of plywood.

REVIEW QUESTIONS

A. Select the letter preceding the best answer.

1. What is the framing member that rests on top of the foundation?

    a. Joist
    b. Header
    c. Sill
    d. Trimmer

2. Where are the size and direction of the first floor joists found?

    a. Floor plan
    b. Basement plan
    c. Elevation
    d. Specifications

3. How should the quantity of material for floor joists be shown on a material list?

    a. Square foot
    b. Board foot
    c. Piece
    d. Linear foot

4. What framing member is used to evenly distribute the load over all of the floor joists?

    a. Box sill
    b. Sill
    c. Girder
    d. Bridging

5. When more than one joist is cut at an opening, how many trimmers should be used?

    a. 1
    b. 2
    c. 3
    d. Trimmers are not used when joists are cut.

B. Using the following given information, construct a material list for the framing up to and including the subfloor for the building in figure 19-8. Framing is to be Douglas fir, bridging is to be spruce, and the subfloor is to be 5/8-inch plywood.

Quantity	Size	Material
a._____	2x8x_____	Fir for girder
b._____	2x_____x_____	Fir for cellar stairs
c._____	2x_____x_____	Fir for cellar stairs
d._____	2x_____x_____	Fir for cellar stairs
e._____lin. ft.	6-inch	Sill seal
f._____	2x_____x_____	Fir for sill
g._____	2x_____x_____	Fir for sill
h._____	2x_____x_____	Fir for sill
i._____lin. ft.	2x8	Fir for box sill
j._____	2x_____x_____	Fir for joists
k._____	2x_____x_____	Fir for joists
l._____	2x_____x_____	Fir for trimmers & headers at stairs
m._____	2x_____x_____	Fir for headers at chimney
n. _____ lin. ft.	1x3	Spruce for bridging
o. _____ sheets	5/8-inch	Plywood for subfloor

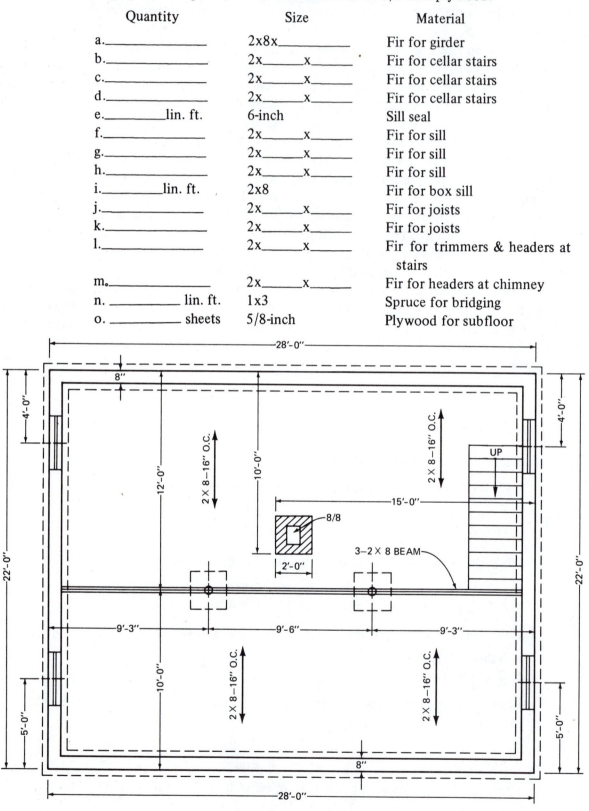

Fig. 19-8 Basement and Foundation Plan

unit 20 wall framing

OBJECTIVES

After studying this unit the student will be able to

- explain what the plate, studs, and headers are.
- tell where plates, studs, and headers are used in framing a house.
- determine the quantity of material for plates, studs, headers and second floor framing.

PLATES

The plates are the horizontal pieces, the same width as the studs (usually 2 x 4's or 2 x 6's) placed at the bottom and the top of a stud wall, figure 20-1. The *sole* plate is a 2 x 4 or 2 x 6 nailed, through the subfloor and into the joists and the box sill for the outside walls, or through the subfloor alone for the inside partitions. Sole plates are used wherever exterior walls and interior partitions are shown on the floor plan. The top 2 x 4's or 2 x 6's, called *top plates*, are doubled. The top

plates are lapped at the corners and where interior partitions and outside walls meet, figure 20-2. The bottom 2 x 4 or 2 x 6 of the top plate is nailed to each stud and corner post. The top 2 x 4 or 2 x 6 is then nailed to the lower one.

Many carpenters lay out the work and frame the wall on the floor before tipping it up into place. The sole plate is nailed to the bottom of the studs and the top plate is nailed to the top of the studs. The wall is then tipped up and set in place and the sole plate is nailed to the subfloor and joists. The top plate is then nailed to the walls and partitions, lapping at corners and where the inside partitions meet the outside walls.

Estimating Plates

Working from the First Floor Plan for the sample house find the distance around the outside of the house where the outside walls are to be erected. This is 108 linear feet for the sample house. If this dimension is found with a scale, be

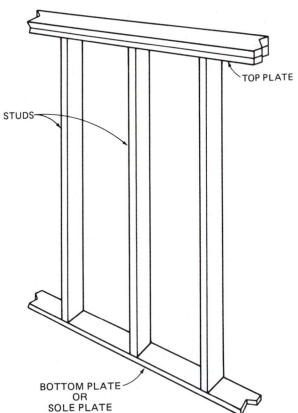

Fig. 20-1 The main parts of a wall frame

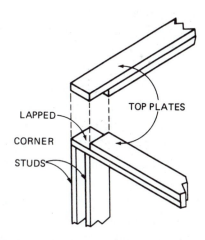

Fig. 20-2 The double top plates are overlapped at the corners to help tie the corner together.

sure the scale used for estimating is the same as the one used for the drawing. The scale of the drawing is given in the title box in the lower right hand corner of the sheet. The scale for most residential floor plans is 1/4" = 1'-0". If a dimension is given, it takes precedence over measurements taken with a scale. Most plans do not show every dimension, so it is necessary to scale dimensions frequently in estimating.

Using an architect's scale, add the length of all the interior partitions. Do not deduct for door or window openings. The sample house has 102 linear feet of interior partitions. Now add the length of all the interior partitions to the length of the exterior walls and get a total of 210 linear feet for all exterior walls and interior partitions on the first floor. All walls and partitions have a single sole plate and a double top plate so the total length of the walls and partitions is multiplied by three to find the quantity of material for the plates. Therefore, 210 linear feet of walls and partitions require 630 linear feet of 2 x 4's or 2 x 6's for plates. Add this to the material list including 10 percent allowance for waste.

There are several methods used for framing corners and partition intersections. Figures 20-3 and 20-4 show methods which are commonly used to provide a nailing surface for both the inside and outside wall covering.

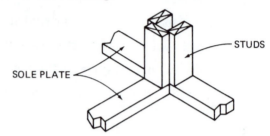

Fig. 20-3

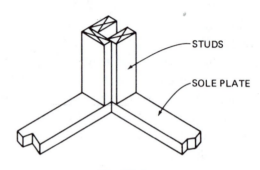

Fig. 20-4

FIR - FIRST FLOOR																							
PLATES	2 x 4	1 IN. FT.	693																				

STUDS

Studs are vertical members, usually 2 x 4's or 2 x 6's spaced 16" O.C., which support the weight of the ceilings, upper floors, and the roof. The studs on the exterior walls are covered with sheathing and siding. On the interior partitions they are covered with the interior wall finish.

There are two methods commonly used for fastening the studs in place. If the sole plate is first nailed to the subfloor and joists, then the studs are toenailed into the sole plate. A more common method is to nail the plates to the ends of the studs before the plates are fastened in place on the subfloor. When this method is used, an entire section of the wall is assembled and then it is tipped up and fastened in place.

Estimating Studs

In estimating for studs spaced 16" O.C., allow 1 stud for every linear foot of interior and exterior walls and partitions. Also add 2 studs for each corner intersection. This provides an allowance for waste and double studs which are needed in some places. From the first floor plan, it can be found that the sample house has 27 corners (either outside corners or places where two walls meet). This means that 54 studs should be allowed for corners. Add this to 210 (the number of linear feet of walls and partitions) for a total of 264 studs required for the first floor. Add this to the material list.

FIR STUDS, FIRST FLOOR	2 X 4 X 8'	264									

HEADERS

Whenever openings for windows or doors are cut out of the walls, *headers* (sometimes called lintels) are installed to carry the vertical loads to other studs. The header is always double and spaced so that its faces are flush with the faces of the studs on each side. Headers are always installed on edge, not flat. Under each header and nailed to the stud is a shorter 2 x 4 or 2 x 6 called a *jack stud*. Figure 20-5 is an illustration of the framing for a door and window opening.

The size of headers varies. They must be of sufficient size to carry the load. If the header sizes are not given on the plan, the following sizes can frequently be used for house construction.

spans up to 4 feet wide. . . . . 2 - 2 x 4's
spans 4 to 5 1/2 feet wide. . . . . . 2 - 2 x 6's
spans 5 1/2 to 7 feet wide. . . . . . 2 - 2 x 8's
spans over 7 feet wide. . . . . . . 2 - 2 x 10's

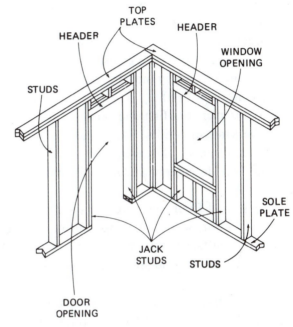

Fig. 20-5

Estimating Headers

To find the length of headers, measure each opening for doors or windows and add 3″ for support. According to the Floor Plan, for the sample house, mullion window in the kitchen is the only opening over three feet wide. There are 16 openings without the mullion window. Allow 3 feet for each opening and multiply 16 x 3 to get 48 feet of opening. Headers are double, so double this figure and get 96 linear feet of 2 x 4 headers. Eight 2 x 4 x 12′ pieces yield 96 linear feet for the headers. The kitchen window has two units, each 2′-4″ wide, with a 2″ mullion between them. An additional 5 inches must be allowed for the rough opening, so the total opening is 5′-2″. Spans between 4 feet and 5 1/2 feet wide need 2 x 6's for the header, so one 2 x 6 x 12′ is required for

this header. Add these to the material list at the bottom of the page.

In some construction, all window and door headers are of the same size material. If this is the case, the estimator may allow twice the width of all openings, plus 3″ to allow for support, plus a 10 percent allowance for waste.

This completes the framing for the first floor. Estimate the second floor the same way as the first. No sill is needed on the second floor, because the top plate of the first floor partitions provides a good nailing surface for the joists. A box sill is required to frame the end of the second floor joists. Studying the First Floor Plan, notice that the perimeter of the building is 108 lin. ft. The second floor box sill is 108 lin. ft. long. According to the First Floor Plan the second floor joists are 2 x 8's,

FIR - WINDOW AND DOOR HEADERS	2 X 4 X 12'	8									
FIR - KITCHEN WINDOW HEADER	2 X 6 X 12'	1									

so the box sill must be of 2 x 8 stock. Add this to the material list (A).

After the length of the box sill has been determined, the second floor joists should be estimated. According to the First Floor Plan, the second floor joists are 2 x 8's spaced 16" O.C. By comparing the Basement Plan and First Floor Plan, it can be observed that the second floor joists run in the same direction as the first floor joists. The estimator should read the plans carefully; the first and second floor joists may not be the same. Add these to material list (B).

On this plan the first and second floor joists run in the same direction. This is not necessarily the case on all plans. The estimator must read all plans carefully to understand how the building is constructed.

The second floor joists are the same as those for the first floor, so the bridging is also the same. There are 42 joists, so multiply by 3 and get a total of 126 linear feet of bridging. Add this to material list (C).

The second floor area of the sample house is the same as that on the first floor, so the subfloor material will be the same. Divide the square-foot area to be covered by 32 when floor system is even module of 4' x 8' panels. The subflooring is plywood which is sold in 4 x 8 sheets covering 32 square feet each. Twenty-two sheets are needed to cover the second floor. Add this to material list (D).

On this house the first floor and the second floor happen to be the same size. This might not always be the case. For example, on a garrison-style house the second floor protrudes beyond the first floor. It is very important to always read the plans and specifications carefully.

The openings for the stairwell and the chimney must be framed at the second floor level. The headers for the stairs are the same as at the first floor but the fireplace chimney is smaller as it goes through the second floor. Refer to the second floor plan for the sample house to determine the length of the headers for the chimney. The opening is 3'-6" long, 2'-0" wide, and the headers are to be double, so a 2 x 8 x 12' is required to cut all of the header pieces. Now add these to material list (E) at the top of the next page.

The order in which materials are estimated for the second floor is the same as for the first floor.

A

FIR - SECOND FLOOR -				
BOX SILL	2 x 8	UN.FT.	108	

B

FIR - JOISTS OVER				
LIVING ROOM	2 x 8 x 14'		14	
FIR - JOISTS - SECOND				
FLOOR	2 x 8 x 14'		28	

C

SPRUCE BRIDGING -				
SECOND FLOOR	1 x 3	UN.FT.	126	

D

CD PLYWOOD - SUBFLOOR,				
SECOND FLOOR	5/8" x 4' x 8'		22	

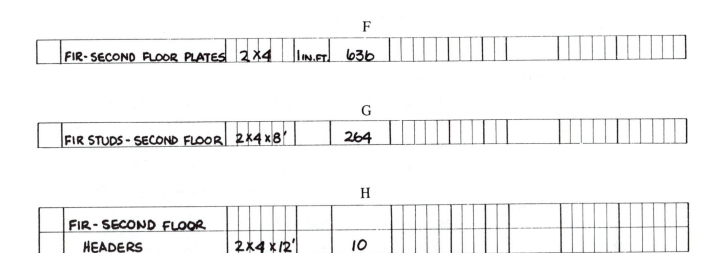

E

FIR-STAIRWELL HEADERS	2x8x10'		2
FIR-CHIMNEY HEADERS	2x8x12'		1
FIR-CHIMNEY AND STAIRWELL HEADERS	2x8x14'		3

F

FIR-SECOND FLOOR PLATES	2x4	LIN.FT.	636

G

FIR STUDS - SECOND FLOOR	2x4x8'		264

H

FIR- SECOND FLOOR HEADERS	2x4x12'		10

The next items are the plates for the walls. However, the Second Floor Plan indicates a different arrangement of rooms and partitions than the First Floor Plan, so the material list will be different.

The outside walls of the second floor are the same as those on the first floor. There is a total of 108 linear feet of outside walls. Add the length of all of the interior partitions on the second floor for a total of 104 linear feet. Add the length of the exterior walls and the interior partitions and get a total of 212 linear feet. As before, the walls will have a single sole plate and a double top plate so multiply 212 by 3 and get a total of 636 linear feet of 2 x 4's for the plates. Add this to material list (F).

The second floor has a total of 212 linear feet of interior and exterior walls and there are 26 corners. Allowing 1 stud for each linear foot of wall and 2 extras for each corner, a total of 264 (52 plus 212 equals 264) studs for the second floor. Add this to material list (G).

This completes the material for the rough framing of the second floor except for the ceiling joists, which are discussed in the next unit, and the door and window headers. To estimate the door and window headers, study the Second Floor Plan to determine the size and number of window and door openings. The windows are 2'-8" wide and the doors average 3'-0" wide. There are nine windows and ten doors for a total of 19 openings. Multiply this by 3 (average width) and get 57 feet of opening. Headers are always doubled, so double this figure and find that 114 linear feet of 2 x 4 is needed for headers. Ten pieces 2 x 4 x 12' should be ordered for headers. Add this to material list (H).

SUMMARY

To estimate material for platform frame construction, each floor of the house is done in essentially the same way. When the first floor or platform is listed, the material for the wall plates is estimated. Find the total length of all walls and partitions and multiply this by 3. This is the number of linear feet of 2 x 4 stock required for the sole plate and double top plate.

The studs are the vertical framing members in the walls. They are usually made of 2 x 4's spaced 16 inches O.C. Allowing 1 stud for each linear foot of wall provides for waste and double pieces to be used around openings. To provide a nailing surface

at corners, 2 extra studs should be allowed for each corner and intersection.

Where openings are framed for windows and doors, a header is installed at the top of the opening to support the weight of the structure above the opening. Headers are double framing members, placed on edge for greater strength and installed with the surfaces flush with the faces of the studs. To estimate the material for headers, find the total width of all openings, then double this figure for the total number of linear feet of header material. Spans of more than four feet require deeper headers.

Material for the second floor of a house is estimated in the same way as for the first floor, except that the second floor does not require a sill. The second floor joists are nailed to the top plate of the first floor walls.

REVIEW QUESTIONS

A. Select the letter preceding the best answer.

1. What is the name of the horizontal framing member at the bottom of a stud wall?
   a. Stud   b. Plate   c. Header   d. Trimmer

2. What is the name of the horizontal framing member at the top of a stud wall?
   a. Stud   b. Plate   c. Header   d. Trimmer

3. What is the name of the vertical members which support the weight of the ceiling, upper floors, and roof?
   a. Stud   b. Plate   c. Header   d. Trimmer

4. What is the name of the framing member which is installed to carry the ends of the cut studs and to distribute the load when an opening is cut for a window or door?
   a. Stud   b. Plate   c. Header   d. Trimmer

5. How many feet of material is required for the plates on 150 feet of walls?
   a. 150 linear feet   b. 300 linear feet   c. 450 linear feet   d. 600 board feet

6. What is the usual spacing for wall studs?
   a. 16" O.C.   b. 12" O.C.   c. 24" O.C.   d. 18" O.C.

7. How many studs are required to frame the corner of a house?
   a. 1   b. 2   c. 3   d. 4

8. What size headers should be used in a 54-inch opening?
   a. One 2 x 4   b. Two 2 x 4's   c. Two 2 x 6's   d. Two 2 x 8's

9. Where is the size of the second floor joists found?
   a. Specifications   c. Second floor plan
   b. Elevation   d. First floor plan

10. Which of the following formulas should be used to estimate studs in a straight wall (no corners)?
   a. One stud every 1 1/3 feet   c. One stud per foot
   b. Two studs per foot   d. One stud every 1 1/2 feet

B. Using the following given information construct a material list. Include the materials required to frame the walls and partitions of the house in figure 20-6.

	Quantity	Size	Material
a.	_____ Lin. ft.	2 x 4	Fir for plates
b.	_____	2 x 4 x _____	Fir for studs
c.	_____	2 x _____ x _____	Fir for picture window header in living room
d.	_____	2 x _____ x _____	Fir for picture window header in dining area

Quantity	Size	Material
e. ____	2 x ____ x ____	Fir for mullion window header
f. ____	2 x ____ x ____	Fir for 4-foot door headers
g. ____	2 x ____ x ____	Fir for small window and door headers

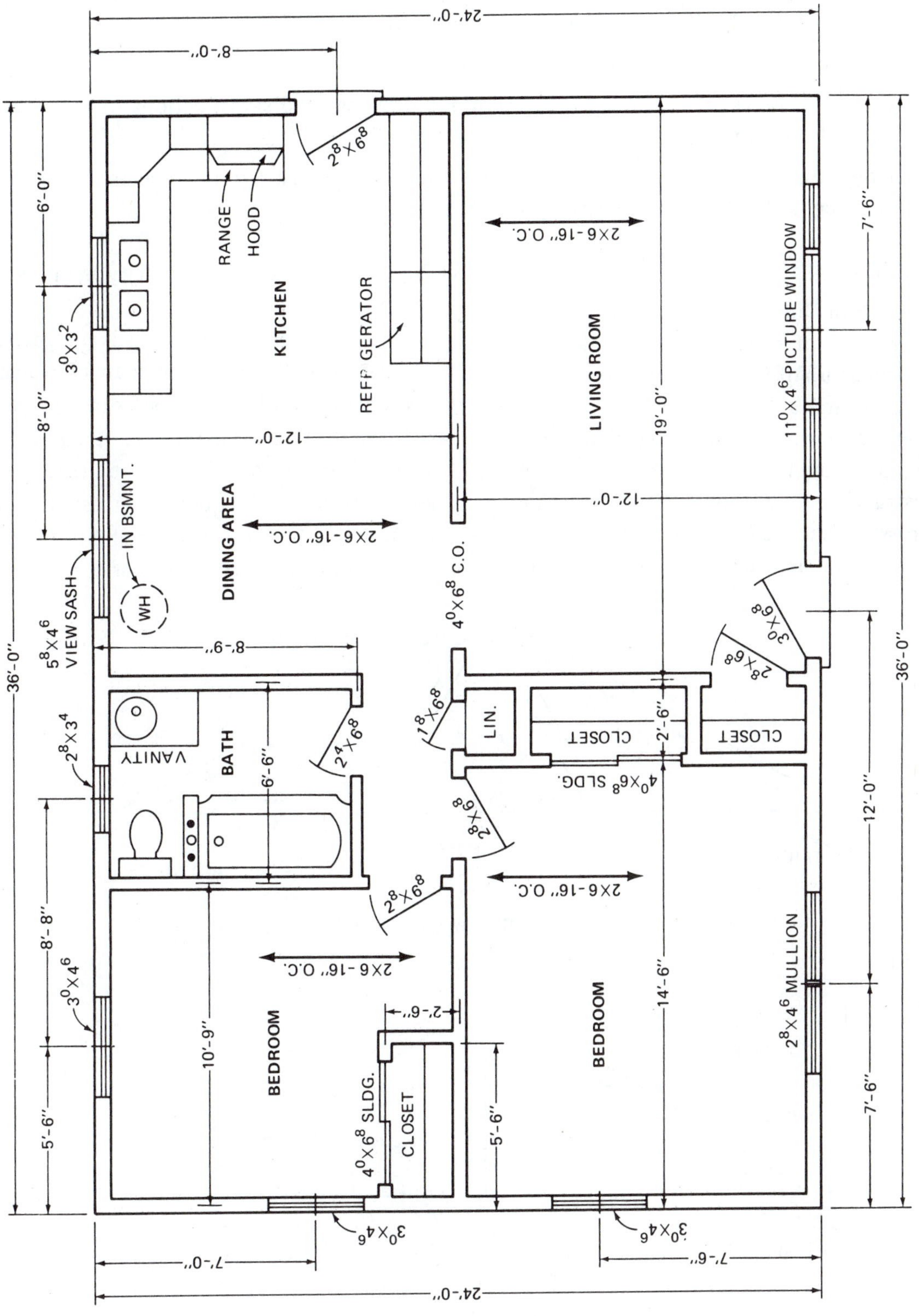

FLOOR PLAN
SCALE 3/16" = 1'-0"

Fig. 20-6

unit 21 roof framing

OBJECTIVES

After studying this unit the student will be able to

- describe the materials used for roof framing.
- locate the necessary information on working drawings and specifications to estimate the roof framing of a house.
- determine the quantity of material needed for ceiling joists, rafters, ridge board, gable studs, collar ties, and roof sheathing on a house.

ROOF TRUSSES

Many modern buildings are designed with roof trusses instead of rafters. A *truss* is a preassembled unit made up of framing members and fasteners. Trusses are available from most major building supply dealers, in a wide range of sizes and styles. The most common styles for roofs on residential construction are W trusses, figure 21-1, and king-post trusses, figure 21-2. The bottom chord of these trusses acts as a ceiling joist. Where no ceiling joist is desired, another type known as scissors trusses is used.

Trusses offer a number of advantages over conventional rafters. Because they are delivered to the job site preassembled, the roof can be constructed faster and with fewer manhours. Being built under controlled conditions in a shop, trusses are frequently stronger. Frequently, trusses are specified for a building because of their ability to span wide areas without supporting posts or partitions.

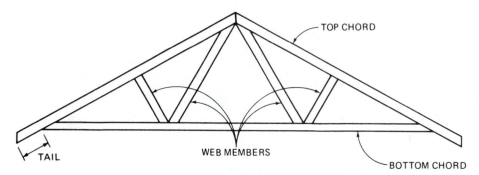

Fig. 21-1 W truss

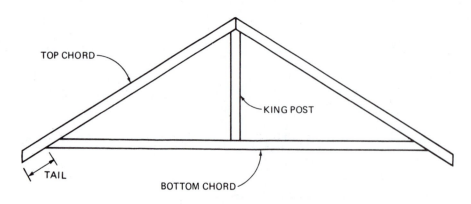

Fig. 21-2 King-post truss

Estimating Roof Trusses

Roof trusses used in residential construction are usually either of the king-post or W-type truss. They are normally spaced 2 feet on centers. To determine the number of trusses required for a straight roof, divide the length of the building in feet by 2 and add 1. If special gable-end trusses are used 2 less king-post or W trusses are needed. When trusses are specified, the style, size of all parts, *pitch* (steepness of the roof), and amount of overhang *tail* must be indicated. It is common practice to include truss detail drawings with the working drawings for the building. Some information is also included in the specifications.

CEILING JOISTS

Ceiling joists support the ceiling of the top floor of the house. They also support the rafters and help secure them to the house. The ceiling joists must span from the outside wall to an interior bearing wall or the opposite exterior wall. If possible, they usually run in the same direction as the rafters. The ceiling joists are shown on the plans in the same manner as the floor joists. They are shown on the uppermost floor plan.

Figure 21-3 shows how ceiling joists are installed. Notice that they tie the rafters and the roof from pushing the outside walls out.

Estimating Ceiling Joists

Find the direction and size of the ceiling joists on the Second Floor Plan for the sample house.

The span over the front bedroom and closets is 13'-3". From the front of the second floor bathroom to the chimney, the span is 9'-3". Over the back bedroom the joists run from the end of the house to the stairway and the span is 12 feet. The span over the stairway is 9 feet. Over the hall the span is 11 feet to the bathroom partition. Over the master bedroom the joists span 12'-9" from the end of the house to the bathroom partition. Figure 21-4 illustrates the layout of the ceiling joists.

Add these joists to the material list as shown on the next page.

RAFTERS

A *common rafter* is one that extends from the ridge to a bearing plate, as in the Second Floor Plan. For a plain roof having just two sides, all of the rafters are common rafters and the roof is referred to as a *gable roof*. Rafters, usually spaced 16 inches O.C., must be strong enough to support the weight of the roof (in some areas the roof must carry a snow load) and to resist wind pressure. The size of the rafter depends on the span and the pitch of the roof. Figure 21-5 shows the locations and names of the various members used to frame a roof. An estimator should be familiar with these.

A *hip rafter* is required where two adjacent slopes meet on a hip roof (one which slopes on all four sides). Hip rafters may be two inches deeper than the jack and common rafters to provide a full

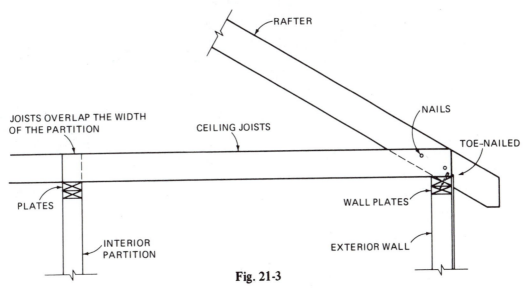

Fig. 21-3

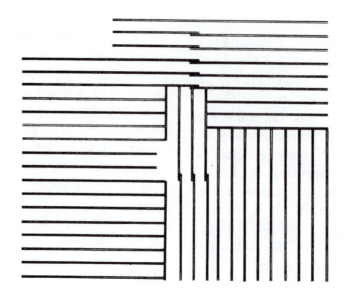

Fig. 21-4 Layout of ceiling joists

	FIR-BATH CEILING JOISTS	2X6X10'	8																								
	FIR-REAR BEDROOM, HALLS, STAIRS, CEILING JOISTS	2X6X12'	18																								
	FIR-FRONT BEDROOM CEILING JOISTS	2X6X14'	10																								
	FIR-MASTER BEDROOM CEILING JOISTS	2X6X14'	16																								

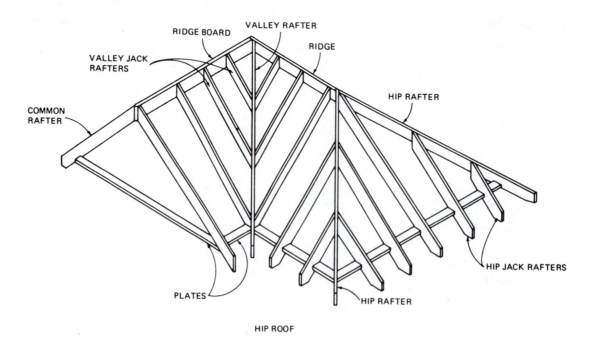

Fig. 21-5 Roof framing parts

nailing surface for the jack and common rafters that are nailed to them.

Jack rafters span the space between the wall plates and the hip rafters, or the ridge and the valley rafter.

A *valley rafter* carries the ends of the jack rafters where two roof surfaces meet to form a valley. The valley rafter, like the hip rafter, may be two inches deeper than the jack rafter to afford a full nailing surface.

The *ridge board* is the top member of the roof framing and runs the length of the roof. The tops of the rafters are nailed to the ridge board. This board is usually a 2 x 8 or a 2 x 10 depending upon the size of the rafters. The ridge board is usually two inches deeper than the rafters, to offer a full nailing surface. One inch nominal thickness stock is sometimes used, but two-inch stock is preferred because the 1 1/2 inch thickness is better for holding the abutting rafters in alignment.

Estimating Rafters

To determine the number of rafters 16 inches O.C. required on each side of a gable roof, multiply the length of the ridge by 3/4 and add 1. Assume jack rafters are the same size and the same length as the common rafter. To find the number of jack rafters needed, take 1 1/2 times the width of the building and add 2. Allow 1 hip rafter for each hip and 1 valley rafter for each valley.

The following procedure can be used to find the true length of hip or valley rafters from the working drawings: (This procedure refers to figure 21-6).

- Work on the elevation drawing showing the hip or valley.

- Draw line CX perpendicular to line CA (the line of the level cornice of the building) passing through point B.

- Mark point D on line CX so that CD is the same length as AB.

- The distance from point A to point D is the true length of the hip or valley. Be sure to use the same scale to measure this as was used in making the drawing.

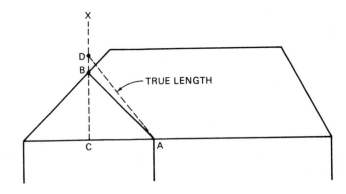

Fig. 21-6 Finding the true length of hip and valley rafters on elevation drawings.

The amount of overhang required by the cornice must be considered when ordering hip or valley rafters. Also remember that if the length of the hip or valley is an odd number, the next higher even number must be used, because lumber is sold in even-foot sizes.

If there is a top view only showing the roof layout, as in figure 21-7, the length of the valley or hip rafters can still be found. To find the true length of AB, draw line AC perpendicular to AB. Draw AC equal to the height of the gable. Then draw CB which is the true length of the valley rafter AB.

The length of the ridge board can be taken from the plans. The sample house plans show that the house is 30 feet long in one direction and has a *tight cornice* (no overhang) at the rakes. The ridge board in this direction is 30 feet. The back gable has a ridge board which is 15 feet. The ridge board is 2 inches wider than the rafters, so 45

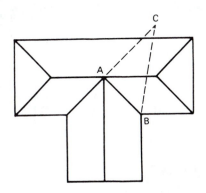

Fig. 21-7 Finding the true length of hip and valley rafters on plan-view drawings.

linear feet of 2 x 8 stock will be used. This can be shown on the material list as three 2 x 8 x 16 pieces.

FIR - RIDGE BOARD	2x8x16'	3												

The sample house has a plain gable roof with rafters spaced 16 inches O.C. To find the number of rafters on one side of the roof, multiply the number of linear feet in the ridge board by 3 and add 1. (30 x 3 = 90; 90 ÷ 4 = 22 1/2 or 23; 23 + 1 = 24 rafters for the front of the house.) On the Half-Section Elevation, figure 21-8, or a side elevation of the building, scale the drawing to determine that the length of the rafter is 13'-0''. For the front of the house, twenty-four 2 x 6 x 14' rafters are needed. The same number of rafters are needed for the back, even though there is another gable cutting into it.

The rear gable extends 5 feet beyond the living-room wing. Use the same formula for this area. (5 x 3 = 15; 15 ÷ 4 = 4; 4 + 1 = 5) 5 rafters are needed for one side of this area. There are 2 sides to the roof so double this figure. This means that a total of fifty-eight 2 x 6 x 14' rafters are required.

The roof has 2 valley rafters and, using the method described for scaling true lengths, their length is found to be 18 feet. The hip and valley rafters are 2 inches deeper than the common rafters, so two 2 x 8 x 18' valley rafters are needed. Add these to material list (A).

or 2 x 6's depending upon the finish required in the area just below them. If the house is a single-story ranch house, the collar beams are probably 1 x 8's; if the attic area of a two-story house is unfinished, the collar beams are probably 1 x 8's; in a house that has an expansion attic or a shed dormer requiring a finished interior next to the rafters, the collar beams are 2 x 4's or 2 x 6's Their size depends on the span and the ceiling load they have to carry.

Estimating Collar Beams

Occasionally one collar beam is used for each pair of rafters, but most buildings have one for every third pair of rafters. Check the plans and specifications carefully to be sure of what is required. If the collar beams are spaced 16 inches O.C., divide the number of rafters by 2 and add 1. If they are spaced 32 inches O.C., divide the number of rafters by 4 and add 1. If spaced 48'' O.C. divide number of rafters by 6 and add 1.

Under Division 6, Section E, number 6, the specifications for the sample house call for 1 x 8 collar beams spaced 32 inches O.C. The house has 58 rafters, so divide this by 4 and add 1 to find that 16 collar beams are required. The length of the collar beams can be scaled from the elevation

A

FIR - RAFTERS	2x6x14'	58											
FIR - VALLEY RAFTERS	2x8x18'	2											

B

PINE - COLLAR BEAMS	1x8x12'	16											

COLLAR BEAMS

Collar beams, figure 21-9, are horizontal members which tie opposing rafters together to prevent sagging. These are usually 1 x 8's, 2 x 4's,

drawing or a section drawing. Find the span of the roof at about half the gable height. On the sample house the collar beams are 12 feet long. Add 16 collar beams to material list (B) above.

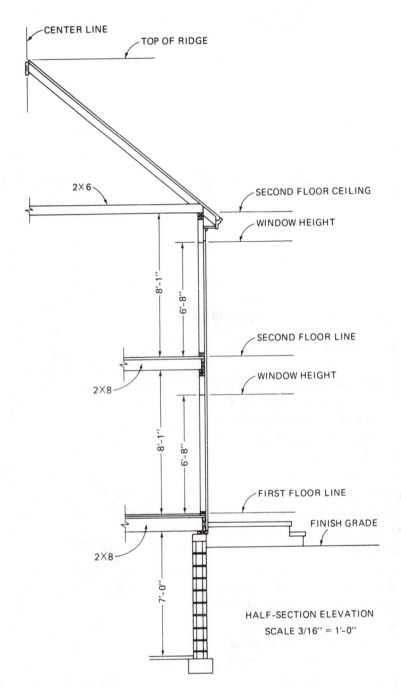

Fig. 21-8

GABLE STUDS

A house with a hip roof does not have gable studs, but when a gable roof is used, the gable end must be framed, figure 21-10. *Gable studs* are vertical studs, the same size as the ones used for the exterior walls and the interior partitions. These are placed between the end rafters of a gable roof and the top plate of each end wall. Gable studs are usually spaced 16 inches O.C. and

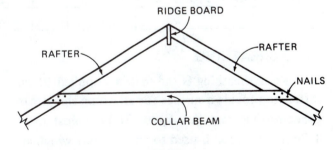

Fig. 21-9

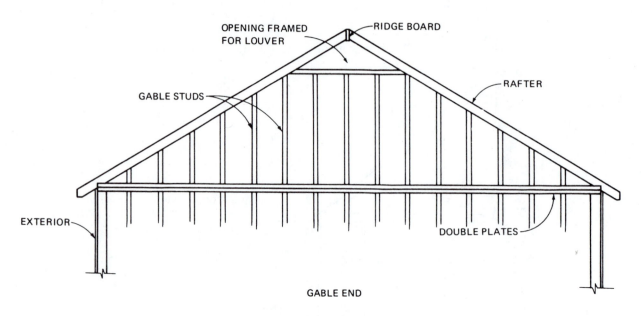

Fig. 21-10 Gable end

are covered with the same type of sheathing as the exterior walls.

On a straight roof with just two gable ends it is only necessary to measure one end of the roof. The height of the gable from the top wall plate to the ridge is the length of the material that must be ordered for gable studs. The number of feet the roof spans is the number of pieces that must be ordered. Although all of the studs in the gable end are not the same length, when a stud is cut for one end of the roof, the remainder of that piece is used for a stud in the opposite gable end.

The sample house has 3 gables, each of which is 7 feet high. The width of each gable is 20 feet. If 1 stud is allowed for every foot of width, 2 gables require 20 studs. However, half this much must be added, again to allow for the third gable. Therefore, the sample house requires thirty 2 x 4 x 8 gable studs. Add this to the material list.

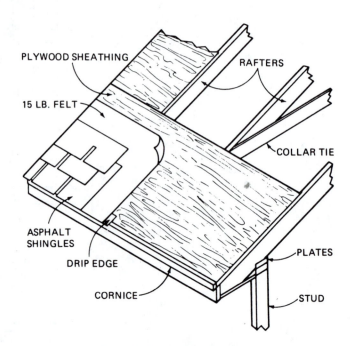

Fig. 21-11 Roof construction

FIR - GABLE STUDS	2×4×8'	30										

ROOF SHEATHING

Roof sheathing is the material applied to the rafters to provide a surface for attaching shingles or other roofing material, figure 21-11. Most often 1/2-inch plywood is used to cover rafters which are spaced 16 inches O.C. On some buildings the sheathing may be 1 x 8 or 1 x 10 shiplap or T & G.

Estimating Roof Sheathing

There are seven different styles of roofs which are commonly constructed on buildings. Figures

21-12 through 21-17 show the types of roofs and the formula for finding the area of each.

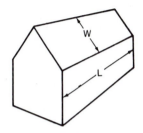

GABLE ROOF
AREA EQUALS 2 × L × W

Fig. 21-12 Gable roof

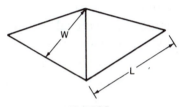

HIP ROOF
AREA EQUALS 2 × L × W

Fig. 21-13 Hip roof

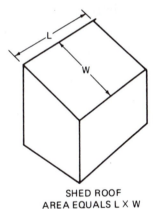

SHED ROOF
AREA EQUALS L × W

Fig. 21-14 Shed roof

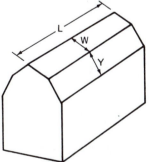

GAMBREL ROOF
TO FIND THE AREA ADD W AND Y.
MULTIPLY THIS SUM BY 2 × L.

Fig. 21-15 Gambrel roof

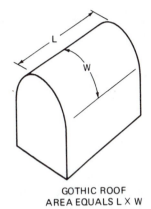

GOTHIC ROOF
AREA EQUALS L × W

Fig. 21-16 Gothic roof

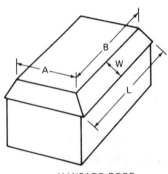

MANSARD ROOF
TO FIND THE AREA
ADD THE FOLLOWING:
A × B
2 × A × W
2 × L × W

Fig. 21-17 Mansard roof

On some buildings with hip or mansard roofs, the pitch may not be the same on the ends and the sides. In this situation, the area of the sides and ends must be calculated separately and then added together.

The sample house has a plain gable roof, so multiply the length of the ridge to a multiple of 32″ or 2.66 ft. times the length of a common rafter to find the area of one side. (48′ x 14′ = 672 sq. ft.) Multiply this by 2 to find the total roof area. (672 x 2 = 1344 sq. ft.) If the roof sheathing is to be shiplap or T & G, add 25 percent to this figure to find the number of board feet needed. The sample house uses plywood, with 32 square feet in a sheet, so divide the total adjusted roof area by 32 to find the number of sheets needed. (1344 ÷ 32 = 42 sheets.) Important: the plywood panels must not be less than 12″ wide or 32″ in length.

If the roof is a complicated one, make sketches and divide the areas into rectangles and triangles, find the area of each, and then add them

together. When the sheathing has been estimated and checked, add it to the material list.

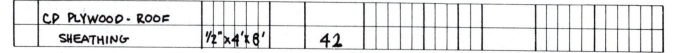

CP PLYWOOD - ROOF SHEATHING	½"x4'x8'	42							

SUMMARY

Ceiling joists support the ceiling loads and tie the rafters to the house. They go from the outside wall to a bearing plate on an interior partition. Ceiling joists usually run in the same direction as the rafters. Their size and direction are shown on the floor plan for the uppermost floor.

Rafters, the framing members which support the roof, are named according to how they are used in the roof. A common rafter extends from the ridge to a bearing plate. On a gable roof, all the rafters are common rafters. A roof that slopes on all four sides is called a hip roof and a hip rafter is required where two adjacent slopes meet. Jack rafters span the space between the wall plates and the hip rafters or between the ridge and the valley rafters. Where two roof surfaces meet to form a valley, the ends of the jack rafters are carried by a valley rafter. The top ends of the rafters are fastened to the ridge board, which runs the full length of the roof.

Frequently the ridge board, hip rafters, and valley rafters are two inches wider than the common rafters. Collar beams are placed about halfway up the rafters to prevent the weight of the roof from pushing the walls outward.

Gable studs are vertical framing members placed between the end rafters on a gable roof and the top plates of the end wall. They are normally the same size as the wall studs and are spaced 16 inches O.C.

The roof sheathing is applied to the top of the rafters. It is usually 1/2-inch plywood. The sheathing strengthens the roof and provides a surface for the shingles or other roofing material.

REVIEW QUESTIONS

A Select the letter preceding the best answer

1. Where is the size of the ceiling joists normally shown?

    a. Specifications
    b. Elevation
    c. Floor plan
    d. None of the above

2. What type of rafter extends from the ridge board to the corner of the building on a roof that slopes down on all four sides?

    a. Hip rafter
    b. Common rafter
    c. Jack rafter
    d. Corner rafter

3. What is the normal width of a ridge board?

    a. The same as the rafters
    b. 2 inches narrower than the rafters
    c. 1 inch wider than the rafters
    d. 2 inches wider than the rafters

4. Which of the following formulas shpuld be used to find the number of rafters required for a gable roof?

    a. Two rafters for every linear foot of ridge board
    b. Multiply the length of the ridge board by 3, then divide by 4 for each side of the roof
    c. Multiply the length of the ridge board by 4, then divide by 3 for each side of the roof
    d. One rafter for every linear foot of ridge board

5. What kind of roof is framed entirely with common rafters?

    a. Gable
    b. Hip
    c. Mansard
    d. Gambrel

6. What kind of rafter is used between the ridge board and a valley?

    a. Common
    b. Hip
    c. Jack
    d. Valley

7. What is the name of the horizontal board running between two opposing rafters about halfway up the roof?

    a. Gable studs
    b. Collar ties
    c. Ridge boards
    d. Joists

8. On a roof with only two sides, what is the name of the framing members used to enclose the ends of the roof?

    a. Collar ties
    b. Joists
    c. Roof headers
    d. Gable studs

B. Using the following given information construct a material list which includes the materials required to frame the roof of the house in figures 21-18, 21-19, 21-20.

Quantity	Size	Materials
a._____	2 x ____x____	Fir for ceiling joists
b._____	2 x 6 x _____	Fir for rafters
c._____	2 x____ x ____	Fir for ridge board
d._____	1 x____x____	Collar ties
e._____	2 x____x____	Fir for gable studs
f._____	1/2" x 4' x 8'	Plywood for sheathing

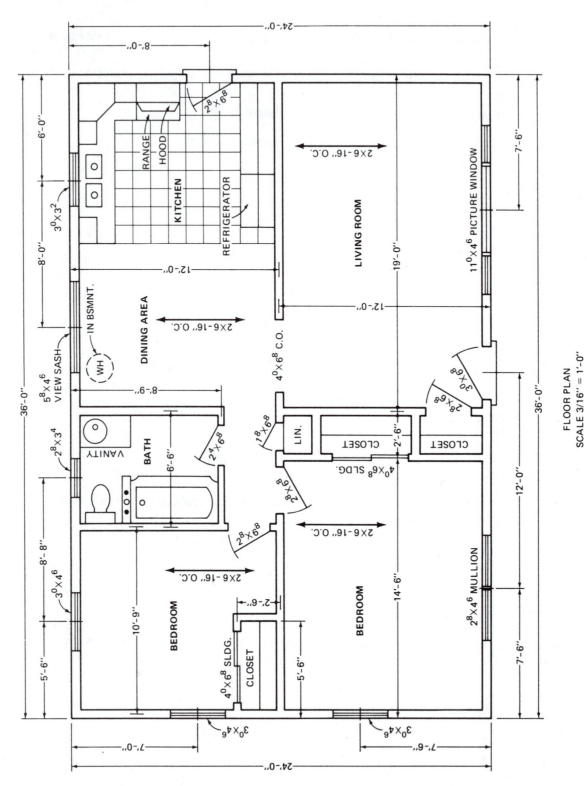

FLOOR PLAN
SCALE 3/16" = 1'-0"

Fig. 21-18

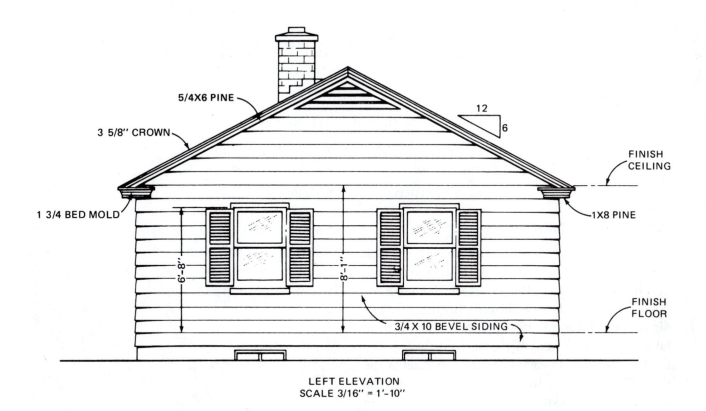

5/4X6 PINE

3 5/8" CROWN

1 3/4 BED MOLD

FINISH CEILING

1X8 PINE

FINISH FLOOR

3/4 X 10 BEVEL SIDING

9'-8"

8'-1"

12 / 6

LEFT ELEVATION
SCALE 3/16" = 1'–10"

Fig. 21-19

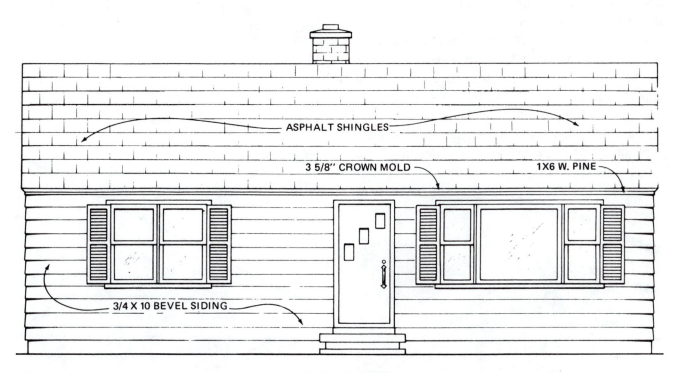

ASPHALT SHINGLES

3 5/8" CROWN MOLD

1X6 W. PINE

3/4 X 10 BEVEL SIDING

FRONT ELEVATION
SCALE 3/16" = 1'-0"

Fig. 21-20

unit 22 cornices

OBJECTIVES

After studying this unit the student will be able to

- describe the various types of cornices.
- identify the parts of a cornice.
- determine the quantity of material needed for a cornice.

CORNICE CONSTRUCTION

A *cornice* is the assembly of boards and moldings used in combination with each other to provide a finish to the ends of the rafters which extend beyond the face of the outside walls, figure 22-1. On a hip roof or a mansard roof, the cornice runs horizontally around the house. When the house has a gable, the cornice runs up each end rafter to meet at the ridge. This is referred to as the *rake,* figure 22-2. The rake cornice can extend beyond the end rafters, in which case it is called a *boxed rake;* or it can be nailed tight against the sheathing and is called a *tight rake,* figure 22-3. With either type, when the house has a gable end, the cornice running up and down the rafters meets the level end of the cornice at the eaves. Where the level cornice at the eaves turns the corner at the end of the house, it is called a *cornice return,* figure 22-4. The length of the cornice return depends upon the style of the cornice and the amount of cornice overhang at the eaves.

Both the material used in the house and the type of architecture influence the design of the cornice. Most cornices include a frieze, bed mold, fascia, soffit, and crown mold. A simple cottage cornice, an eleborate colonial cornice, and a wide

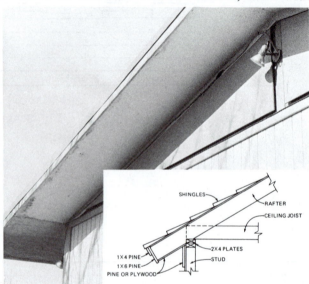

Fig. 22-2 Rake cornice with sloping cornice at the eaves. Inset shows construction at eaves.

Fig. 22-1 Typical box Cornice

Fig. 22-3 Tight rake cornice

Figure 22-4 Cornice return

names of the various styles and their parts. Before estimating the materials, study the working drawings to learn what type of cornice is needed. This information is usually found on a special detail, figure 22-6, or on the Wall Section Detail.

Estimating Cornices

Mouldings used in cornices are running trim which are figured by the linear feet. The sample house drawings show a tight cornice on the rake and a box cornice at the eaves. There are three gables and the rafters are 14'-0" long. The house needs six pieces of 5/4" x 6" x 14' pine for the rake. The bevel siding on the wall butts against the rake at the gable ends, so 5/4-inch stock is used. If 1 x 6 stock were used, the siding would protrude beyond the face of the rake. Most tight cornices use 5/4-inch stock.

sweeping cornice are all made from the same basic parts.

On some houses the rafters bear on a *rafter shoe* which is nailed to the top of the ceiling joists, figure 22-5. This is done to allow more overhang without the cornice being too low. It also makes the cornice high enough to come even with the top of the window casing. Some houses have a wide fascia, requiring two boards to make the required height. In this case, a panel mold covers the crack where the two boards meet.

Study the various styles of cornices, figures 22-2 through 22-5, to become acquainted with the

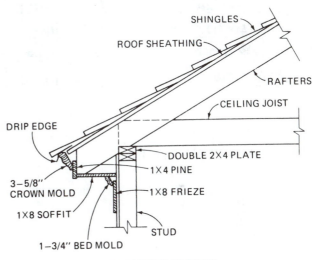

BOX CORNICE AT EAVES

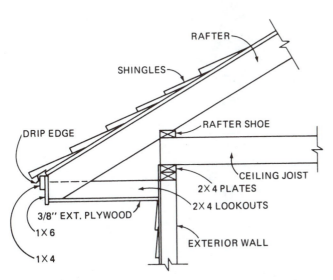

Fig. 22-5 Roof framing with a rafter shoe to allow for wide overhang

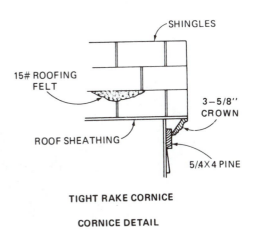

TIGHT RAKE CORNICE

CORNICE DETAIL

Fig. 22-6 Cornice detail

Adding the length of the eaves and the cornice returns, there is a total of 65 linear feet. This means that 65 linear feet of 1 x 6 pine is needed for the fascia. The soffit and the frieze are both 1 x 8, so allow 130 linear feet of 1 x 8 for these parts (65 feet for each). Both the level cornice and the rakes use crown molding, so add the total length of all rakes (84) to the length of the level cornice. This makes 149 linear feet of cornice, using 3 5/8-inch crown molding. Bed molding is used to cover the crack between the frieze and the soffit. Including an allowance for waste, the frieze and soffit are each 65 feet long, so allow 65 feet of 1 3/4-inch bed molding. Add these materials to the material list.

SUMMARY

In general all cornices are made up of finished millwork consisting of boards and moldings used in combination with each other to finish the ends of the rafters which extend beyond the face of the outside walls. The parts of a cornice are named according to their position.

Some types of cornices are the open rafter, the box cornice, the wide overhang cornice, and the colonial cornice. On a box cornice the rake rests on top of the level cornice return. To estimate the material for a cornice, find the sizes of the various parts on the working drawings. List each part of the cornice separately.

PINE - CORNICE RAKE	5/4" x 6" x 14'		6
PINE - FASCIA	1 x 6	LIN.FT.	65
PINE - SOFFIT/FRIEZE	1 x 8	LIN FT.	130
CROWN MOLD	3 5/8"	LIN.FT.	149
BED MOLD	1 3/4"	LIN.FT.	65

REVIEW QUESTIONS

A.   Identify the parts of the cornice in figure 22-7.

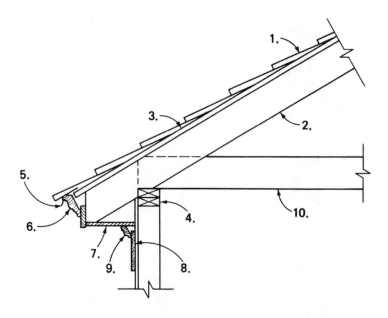

Fig. 22-7

B. Using the following given information, construct a material list which includes the materials required to build the cornices on the house in figures 22-8 and 22-9. The rake is to have a tight cornice and the eaves are to have a box cornice.

	Quantity	Size	Material
a.	_____	_____x_____	Pine for frieze at rake
b.	_____	_____x_____	Pine for frieze at eaves
c.	_____	_____x_____	Pine for cornice return
d.	_____lin. ft.	_____x_____	Pine for fascia at eaves
e.	_____lin. ft.	_____x_____	Pine for soffit
f.	_____lin. ft.	_____	Bed molding
g.	_____lin. ft.	_____	Crown molding

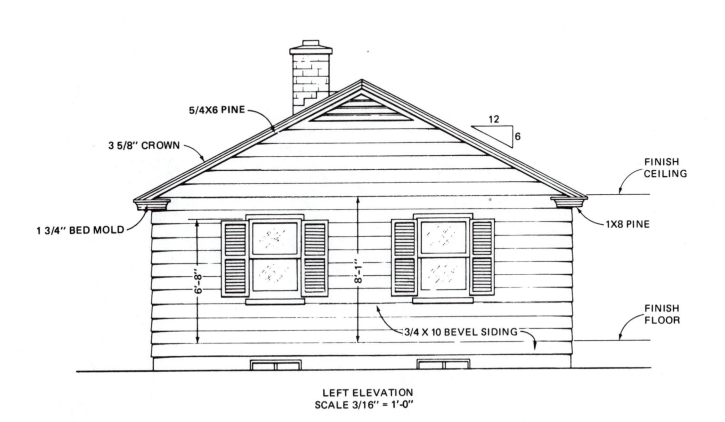

Fig. 22-8

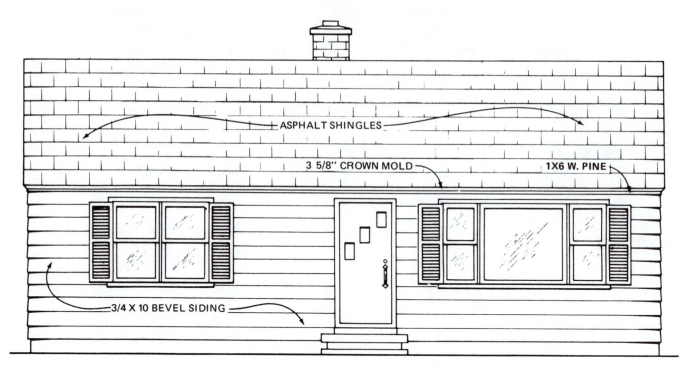

ASPHALT SHINGLES

3 5/8" CROWN MOLD

1X6 W. PINE

3/4 X 10 BEVEL SIDING

FRONT ELEVATION
SCALE 3/16" = 1'-0"

Fig. 22-9

unit 23 sheathing and siding

OBJECTIVES

After studying this unit the student will be able to

- explain specifications for sheathing and siding.
- describe the various types of siding that are available.
- determine the quantity of sheathing and siding needed for a residential building.

Exterior sheathing is the material applied to the outside face of the studs. Sheathing strengthens the frame, acts as insulation, and provides a surface on which the outside finish of the building may be fastened. Sheathing may be plywood, fiberboard, or shiplap, figure 23-1. Fiberboard (except a special type called nail-base) does not provide a strong enough surface for fastening siding. In this case, the siding must be nailed through the fiberboard into the studs.

PLYWOOD

Plywood is commonly used for subflooring, roof sheathing, and wall sheathing. Plywood is a wood material, usually made up of an odd number of thin veneer sheets, although 4 ply is available, cross laminated and bonded together under high pressure with glue. It is available as exterior plywood with waterproof glue and interior plywood with water-resistant glue. It is generally classified according to the quality of wood that is used on the face of the panel. Hardwood plywood is mainly a construction material. It is graded according to the quality and appearance of the face and back veneers. All plywood except sheathing grades are sanded. Panels which are 3/8 inch thick and less have a minimum of three plys. Those that are 1/2 inch to 3/4 inch thick have a minimum of five plys. Most plywood is manufactured in 4-foot by 8-foot sheets, but other sizes may be ordered. Figure 23-2 are guides to engineered and appearance grades of plywood.

Plywood which must be exposed to water should always be made with exterior glue. Several types of plywood are manufactured for marine use, building construction, or with special surfaces for use as siding.

WAFERBOARD

This material is a non-veneer structural panel that is versatile and inexpensive. It is made by using engineered wood wafers longer than they are wide. The multi-layer board is composed of oriented face and back wafers on a randomly oriented core. The wafers are bonded together with phenoic resin under extreme heat and pressure. The result is a rigid flat panel that is dimensionally stable and strong. It is an ideal material for structural and decorative use and may be finished natural, stained, or painted. The physical proper-

Fig. 23-1 Plywood & fiberboard sheathing

Table 1: Guide to Engineered Grades of Plywood

Specific grades and thicknesses may be in locally limited supply. See your dealer for availability before specifying.

Interior Type

Use these terms when you specify plywood	Description and Most Common Uses	Typical Grade-trademarks	Veneer Grade			Most Common Thicknesses (inch) (1)				
			Face	Back	Inner Plies					
C-D INT-APA (2) (3)	For wall and roof sheathing, subflooring, industrial uses such as pallets. Also available with intermediate glue or exterior glue. Specify intermediate glue for moderate construction delays; exterior glue for better durability in somewhat longer construction delays and for treated wood foundations.	C-D 32/16 INTERIOR PS 1/4 000 APA	C	D	D	5/16	3/8	1/2	5/8	3/4
STRUCTURAL I C-D INT-APA and STRUCTURAL II C-D INT-APA	Unsanded structural grades where plywood strength properties are of maximum importance: structural diaphragms, box beams, gusset plates, stressed skin panels, containers, pallet bins. Made only with exterior glue.	STRUCTURAL I C-D 24/0 INTERIOR PS 1/4 000 EXTERIOR GLUE APA	C (6)	D (7)	D (7)	5/16	3/8	1/2	5/8	3/4
UNDERLAYMENT INT-APA (3) (2) (9)	For underlayment or combination subfloor underlayment under resilient floor coverings, carpeting in homes, apartments, mobile homes. Specify exterior glue where moisture may be present, such as bathrooms, utility rooms. Touch-sanded. Also available in tongue-and-groove.	UNDERLAYMENT GROUP 1 INTERIOR PS 1/4 000 APA	C Plugged	D	(8) C & D	1/4		3/8	1/2	5/8	3/4
C-D PLUGGED INT-APA (3) (2) (9)	For built-ins, wall and ceiling tile backing, cable reels, walkways, separator boards. Not a substitute for Underlayment, as it lacks Underlayment's punch-through resistance. Touch-sanded.	C-D PLUGGED GROUP 2 INTERIOR PS 1/4 000 APA	C Plugged	D	D	5/16	3/8	1/2	5/8	3/4
2·4·1 INT-APA (2) (5)	Combination subfloor underlayment. Quality base for resilient floor coverings, carpeting, wood strip flooring. Use 2·4·1 with exterior glue in areas subject to moisture. Unsanded or touch-sanded as specified.	2·4·1 GROUP 1 INTERIOR PS 1/4 000 APA	C Plugged	D	C & D	(available 1-1/8" or 1-1/4")					

Exterior Type

Use these terms when you specify plywood	Description and Most Common Uses	Typical Grade-trademarks	Face	Back	Inner Plies	Most Common Thicknesses (inch) (1)				
C-C EXT-APA (3)	Unsanded grade with waterproof bond for subflooring and roof decking, siding on service and farm buildings, crating, pallets, pallet bins, cable reels.	C-C 42/20 EXTERIOR PS 1/4 000 APA	C	C	C	5/16	3/8	1/2	5/8	3/4
STRUCTURAL I C-C EXT-APA and STRUCTURAL II C-C EXT-APA	For engineered applications in construction and industry where full Exterior type panels are required. Unsanded. See (9) for species group requirements.	STRUCTURAL I C-C 32/16 EXTERIOR PS 1/4 000 APA	C	C	C	5/16	3/8	1/2	5/8	3/4
UNDERLAYMENT C-C Plugged EXT-APA (3) (9) C-C PLUGGED EXT-APA (3) (9)	For Underlayment or combination subfloor underlayment under resilient floor coverings where severe moisture conditions may be present, as in balcony decks. Use for tile backing where severe moisture conditions exist. For refrigerated or controlled atmosphere rooms, pallets, fruit pallet bins, reusable cargo containers, tanks and boxcar and truck floors and linings. Touch-sanded. Also available in tongue-and groove.	UNDERLAYMENT C-C PLUGGED GROUP 2 EXTERIOR PS 1/4 000 APA / C-C PLUGGED GROUP 3 EXTERIOR PS 1/4 000 APA	C Plugged	C	C (8)	1/4		3/8	1/2	5/8	3/4
B-B PLYFORM CLASS I & CLASS II EXT-APA (4)	Concrete form grades with high reuse factor. Sanded both sides. Mill oiled unless otherwise specified. Special restrictions on species. Also available in HDO.	B-B PLYFORM CLASS I EXTERIOR PS 1/4 000 APA	B	B	C					5/8	3/4

(1) Panels are standard 4x8 foot size. Other sizes available.
(2) Also available with exterior or intermediate glue.
(3) Available in Group 1, 2, 3, 4 or 5.
(4) Also available in STRUCTURAL I.
(5) Available in Group 1, 2 or 3 only.
(6) Special improved C grade for structural panels.
(7) Special improved D grade for structural panels.
(8) Ply beneath face a special C grade which limits knotholes to 1 inch or in Interior Underlayment D under Group 1 or 2 faces 1/6 inch thick.
(9) Also available in STRUCTURAL I (all plies limited to Group 1 species) and STRUCTURAL II (all plies limited to Group 1, 2 or 3 species).

Fig. 23-2 Plywood grade-use chart — engineered grades
(Courtesy of American Plywood Association)

Table 2: Guide to Appearance Grades of Plywood[1]

For strength properties of appearance grades, see "Plywood Design Specifications," Y510

Interior Type

Use these terms when you specify plywood (2)	Description and Most Common Uses	Typical Grade-trademarks	Veneer Grade			Most Common Thicknesses (inch) (3)				
			Face	Back	Inner Plies					
N-N, N-A, N-B INT-APA	Cabinet quality. For natural finish furniture, cabinet doors, built-ins, etc. Special order items.	N-N·G-1·INT-APA·PS 1·74	N	N.A. or B	C					3/4
N-D INT-APA	For natural finish paneling. Special order item.	N-D·G-3·INT·APA·PS 1·74	N	D	D	1/4				
A-A INT-APA	For applications with both sides on view. Built-ins, cabinets, furniture and partitions. Smooth face; suitable for painting.	A-A·G-4·INT-APA·PS 1·74	A	A	D	1/4	3/8	1/2	5/8	3/4
A-B INT-APA	Use where appearance of one side is less important but two smooth solid surfaces are necessary.	A-B·G-4·INT·APA·PS 1·74	A	B	D	1/4	3/8	1/2	5/8	3/4
A-D INT-APA	Use where appearance of only one side is important. Paneling, built-ins, shelving, partitions, and flow racks.	A-D GROUP 1 INTERIOR PS 1·74 000 APA	A	D	D	1/4	3/8	1/2	5/8	3/4
B-B INT-APA	Utility panel with two smooth sides. Permits circular plugs.	B-B·G-3·INT·APA·PS 1·74	B	B	D	1/4	3/8	1/2	5/8	3/4
B-D INT-APA	Utility panel with one smooth side. Good for backing, sides of built-ins. Industry: shelving, slip sheets, separator boards and bins.	B-D GROUP 3 INTERIOR PS 1·74 000 APA	B	D	D	1/4	3/8	1/2	5/8	3/4
DECORATIVE PANELS INT-APA	Rough sawn, brushed, grooved, or striated faces. For paneling, interior accent walls, built-ins, counter facing, displays, and exhibits.	DECORATIVE GROUP 1 INTERIOR PS 1·74 000 GROUP 1 FACE APA	C or btr.	D	D	5/16	3/8	1/2	5/8	
PLYRON INT-APA	Hardboard face on both sides. For counter tops, shelving, cabinet doors, flooring. Faces tempered, untempered, smooth, or screened.	PLYRON·INT·APA·PS 1·74			C & D			1/2	5/8	3/4

Exterior Type[7]

A-A EXT-APA (4)	Use where appearance of both sides is important. Fences, built-ins, signs, boats, cabinets, commercial refrigerators, shipping containers, tote boxes, tanks, and ducts.	A-A·G-3·EXT·APA·PS 1·74	A	A	C	1/4	3/8	1/2	5/8	3/4
A-B EXT-APA (4)	Use where the appearance of one side is less important.	A-B·G-1·EXT·APA·PS 1·74	A	B	C	1/4	3/8	1/2	5/8	3/4
A-C EXT-APA (4)	Use where the appearance of only one side is important. Soffits, fences, structural uses, boxcar and truck lining, farm buildings. Tanks, trays, commercial refrigerators.	A-C GROUP 4 EXTERIOR PS 1·74 000 APA	A	C	C	1/4	3/8	1/2	5/8	3/4
B-B EXT-APA (4)	Utility panel with solid faces.	B-B·G-1·EXT·APA·PS 1·74	B	B	C	1/4	3/8	1/2	5/8	3/4
B-C EXT-APA (4)	Utility panel for farm service and work buildings, boxcar and truck lining, containers, tanks, agricultural equipment. Also as base for exterior coatings for walls, roofs.	B-C GROUP 2 EXTERIOR PS 1·74 000 APA	B	C	C	1/4	3/8	1/2	5/8	3/4
HDO EXT-APA (4)	High Density Overlay plywood. Has a hard, semi-opaque resin-fiber overlay both faces. Abrasion resistant. For concrete forms, cabinets, counter tops, signs and tanks.	HDO·AA·G-1·EXT·APA·PS 1·74	A or B	A or B	C or C plgd	5/16	3/8	1/2	5/8	3/4
MDO EXT-APA (4)	Medium Density Overlay with smooth, opaque, resin-fiber overlay one or both panel faces. Highly recommended for siding and other outdoor applications, built-ins, signs, and displays. Ideal base for paint.	MDO·BB·G-4·EXT·APA·PS 1·74	B	B or C	C	5/16	3/8	1/2	5/8	3/4
303 SIDING EXT-APA (6)	Proprietary plywood products for exterior siding, fencing, etc. Special surface treatment such as V groove, channel groove, striated, brushed, rough sawn.	303 SIDING 16 oc GROUP 1 EXTERIOR PS 1·74 000 APA	(5)	C	C		3/8	1/2	5/8	
T 1-11 EXT-APA (6)	Special 303 panel having grooves 1/4" deep, 3/8" wide, spaced 4" or 8" o.c. Other spacing optional. Edges shiplapped. Available unsanded, textured, and MDO.	303 SIDING 16 oc T 1·11 GROUP 1 EXTERIOR PS 1·74 000 APA	C or btr.	C	C				5/8	
PLYRON EXT-APA	Hardboard faces both sides, tempered, smooth or screened.	PLYRON·EXT·APA·PS 1·74			C			1/2	5/8	3/4
MARINE EXT-APA	Ideal for boat hulls. Made only with Douglas fir or western larch. Special solid jointed core construction. Subject to special limitations on core gaps and number of face repairs. Also available with HDO or MDO faces.	MARINE·AA·EXT·APA·PS 1·74	A or B	A or B	B	1/4	3/8	1/2	5/8	3/4

(1) Sanded both sides except where decorative or other surfaces specified.
(2) Available in Group 1, 2, 3, 4 or 5 unless otherwise noted.
(3) Standard 4x8 panel sizes, other sizes available.
(4) Also available in Structural I (all plies limited to Group 1 species)
and Structural II (all plies limited to Group 1, 2 or 3 species).
(5) C or better for 5 plies, C Plugged or better for 3 ply panels.
(6) Stud spacing is shown on grade stamp.
(7) For finishing recommendations, see Form V307.

Fig. 23-3 Plywood grade-use chart — appearance grades
(Courtesy of American Plywood Association)

ties of waferboard make it an excellent choice for a wide variety of applications. The panel is lightweight, easy to handle, is smooth on both sides and is free of knots, core voids, splits, and checks. It is easily fastened by nailing, screwing, gluing, or stapling. It saws cleanly, drills and routs with standard wood working tools. It has become a very popular building material and is used for floors, roofs and sidewalls. It is manufactured in 4 x 8 foot panels and is available in various thicknesses.

Estimating Sheathing

In the specifications for the sample house under Division 6, Carpentry and Millwork, Section C. Number 7 describes the wall sheathing and roof sheathing as follows: Wall sheathing and roof sheathing shall by 1/2″ x 4′ x 8′ plywood APA grade, C–D Interior with Exterior glue. To find adjusted perimeter widths must be a multiple of 16″ or 1.33 feet. Adjusted perimeter equals 109.33.

square feet for the rest of the outside walls and get a total of 2069 square feet.

If the specifications called for shiplap for the sheathing, 25 percent would be allowed for matching, so the total would be 2557 board feet. However, these specifications call for 1/2-inch plywood sheathing, so divide 2069 by 32 to find that 65 sheets of plywood are needed. Add this to the material list.

VINYL SIDING

Vinyl siding is becoming very popular in today's construction of new homes and the residing of older homes because of its premium quality. It is solid vinyl (PVC). It won't rot or attract termites. It won't corrode, chip, or peel; it doesn't conduct heat, cold, or electricity. It needs no painting because the color goes clear through the material. It is virtually maintenance

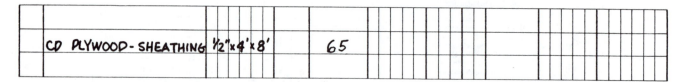

Multiply the perimeter of the house by the height. The height can be scaled from the elevations by measuring the distance from the top of the foundation to the eaves. The sample house is (109.33 adjusted) linear feet around the outside and from the top of the foundation to the eaves is 17 feet. Therefore, there are 1859 square feet of wall space. The house has a gable roof, so the gables must also be covered. Each gable is 20 feet wide and 7 feet high, so there are 140 square feet in two gables. This house has three gables, so add half of this (70 square feet) for the third gable. The total for all three gables is 210 square feet. Add this 210 square feet for the gable ends to the 1859

free for home owners. It's siding that's tough and durable while offering beauty that lasts. It is manufactured in twin 4 or twin 5 horizontal siding, 8″ horizontal siding and vertical siding. As an estimator it is necessary for you to know the styles and accessories used in the application of vinyl siding and how it is installed as well as how to figure the material needed. Vinyl siding is usually installed to eliminate the need for painting so the entire house including the cornice is covered with vinyl. Figure 23-4 shows where the various accessories are used. Also use the estimating guide (figure 23-5) to assist you in determining what is needed.

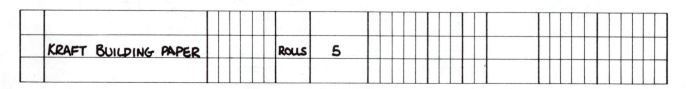

RED CEDAR BEVEL SIDING	½"	x	8"	BD. FT.	2556																
PINE-CORNER BOARDS	5/4"	x	5"	LIN. FT.	170																

WOOD SIDING

Wood siding is available in various patterns and in several widths and thicknesses. The most common sizes and types of siding are: 1/2 x 6, 1/2 x 8, 3/4 x 10 bevel siding, 16-inch shingles, random-width vertical siding, shakes, brick or stone veneer, aluminum siding, and vinyl siding.

Wood shingles and wood shakes may be applied to either *single courses* or *double courses* (a course is a layer). When the double-course method of application is used, first-grade shingles or shakes are applied in each course so that they project about 1/2 inch below the underlayer of third-grade shingles. In many cases, asphalt-impregnated undercover or backer boards are used as the undercourse.

Wood shingles are also remanufactured into processed shakes with square butts and vertical edges finely machined to obtain tight invisible joints. Processed shakes have a face side that has a combed surface making them look like hand split shakes. These shakes are packed so that two bundles cover 100 square feet (one square).

Figure 23-6 shows the wood products that are most commonly used for siding on houses. Some of these may be used in combination with each other, such as hand-split shakes and colonial brick, or stone and vertical board and battens. There is no limit to the combinations that can be used.

A discussion of all of the various types of siding available would be beyond the limits of this textbook. Additional information is available through siding manufacturers' specifications and most lumber yards. It is important for an estimator to be familiar with as many types of siding as possible.

Estimating Siding

The sheathing should be covered with building paper or "housewrap" such as tyvek before the siding is applied. Kraft building paper is sold in 500-square-foot rolls. The sample house has a total wall area of 2046 square feet, so five rolls of building paper are needed. Add this to the material list.

If the material to be used for siding is sold by the square, divide the total square foot area of the walls by 100. This is the number of squares needed to side the house. On the sample house the wall area has been determined to be 2046 square feet, or 20 1/2 squares. Many siding materials are sold by the board foot. The following notes should serve as a guide to estimating siding materials.

- Wood shingles: allow three bundles for each square, if they are to be installed with 12-inch exposure.

- Six-inch beveled siding with 4 3/4-inch exposure: add 30 percent to the wall area to be covered and order by the board foot.

- Eight-inch beveled siding with 6 3/4-inch exposure: add 25 percent to the wall area to be covered and order by the board foot.

- Ten-inch beveled siding with 8 3/4-inch exposure: add 20 percent to the wall area to be covered and order by the board foot.

- Brick veneer: allow 6 3/4 bricks for every square foot of wall area.

- Stone veneer: allow 20 percent for waste and order by the ton.

The sample house specifications indicate that the siding is to be 1/2" x 8" beveled siding. The outside wall area is 2046 square feet. Using 1/2" x 8" beveled siding 25 percent must be added, so 2558 board feet of siding are needed for the sample house.

According to the elevation drawings the outside corners have corner boards. The corner boards will be made of 5/4-inch material. There are 5 outside corners and they are all 17 feet in height with 2 corner boards on each corner. Multiply 17 times

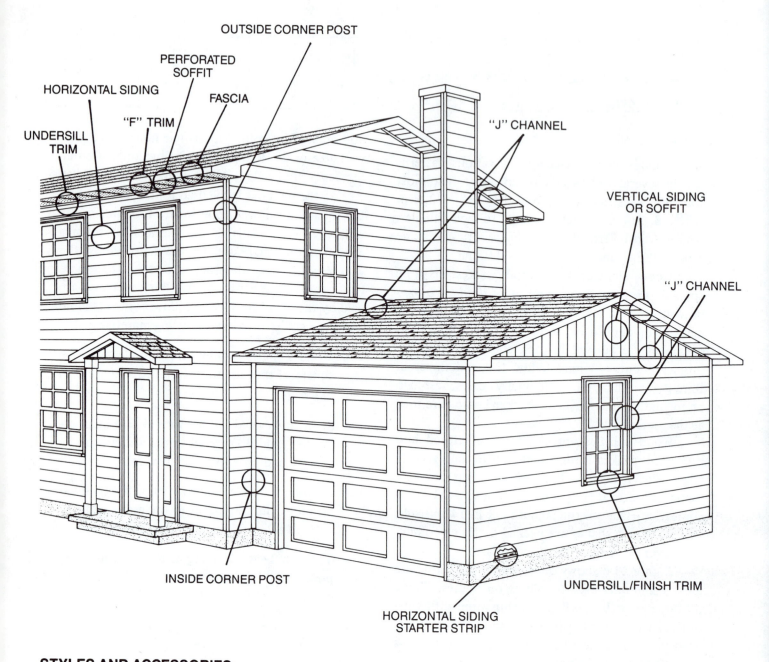

OUTSIDE CORNER POST

PERFORATED
SOFFIT

HORIZONTAL SIDING

FASCIA

"F" TRIM

UNDERSILL
TRIM

"J" CHANNEL

VERTICAL SIDING
OR SOFFIT

"J" CHANNEL

INSIDE CORNER POST

UNDERSILL/FINISH TRIM

HORIZONTAL SIDING
STARTER STRIP

STYLES AND ACCESSORIES

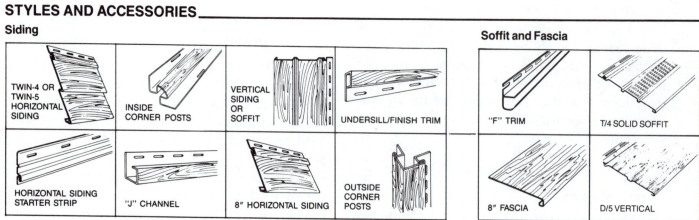

Siding

| TWIN-4 OR TWIN-5 HORIZONTAL SIDING | INSIDE CORNER POSTS | VERTICAL SIDING OR SOFFIT | UNDERSILL/FINISH TRIM |
| HORIZONTAL SIDING STARTER STRIP | "J" CHANNEL | 8" HORIZONTAL SIDING | OUTSIDE CORNER POSTS |

Soffit and Fascia

| "F" TRIM | T/4 SOLID SOFFIT |
| 8" FASCIA | D/5 VERTICAL |

Fig. 23-4 (Chart courtesy of Georgia-Pacific Corporation)

Estimating Guide

1. Surface areas of sides

_____ sq. ft.
_____ sq. ft.
_____ sq. ft.
_____ sq. ft.
_____ sq. ft.
Subtotal _____ sq. ft.

2. Surface areas of dormer, gables, etc.

_____ sq. ft.
_____ sq. ft.
_____ sq. ft.
_____ sq. ft.
Subtotal _____ sq. ft.

3. Surface areas *NOT* TO BE COVERED

Windows _____ sq. ft.

Doors _____ sq. ft.
Chimneys _____ sq. ft.
Other _____ sq. ft.
Subtotal _____ sq. ft.
          (Not to be covered)

4. Add Step 1 and Step 2 together.

_____ sq. ft.

Subtract Step 3 from Total of Step 1 and Step 2.

_____ sq. ft.

5. To determine number of squares needed,
Divide footage by 100
Example: 1,281 sq. ft.
Divided by 100 = 12.8 or 13 squares

6. To determine number of squares of

soffit needed,
Divide by 100
Example: 267 square feet
Divide by 100 = 2.67 or 3 squares
(Remember to use perforated soffit where ventilation is needed.)

7. Measure total linear feet of each accessory needed

Starter strip _____ linear ft.
"F" Trim _____ linear ft.
"J" Channel (siding) _____ linear ft.*
"J" Channel (soffit) _____ linear ft.
Undersill/Finish Trim _____ linear ft.*
Inside Corner Post _____ linear ft.
Outside Corner _____ linear ft.
Fascia _____ linear ft.

*Add 10%

Fig. 23-5 (Illustration courtesy of Georgia-Pacific Corporation)

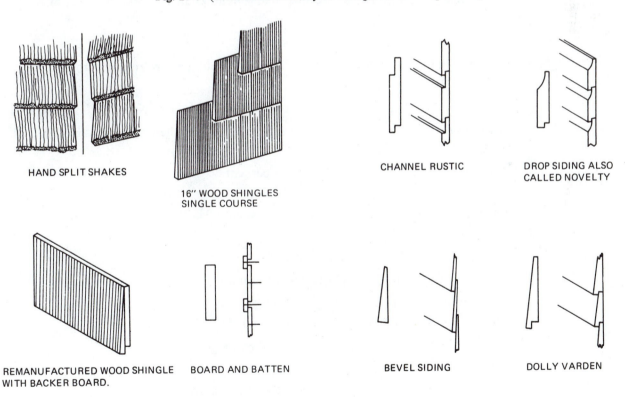

HAND SPLIT SHAKES

16" WOOD SHINGLES SINGLE COURSE

CHANNEL RUSTIC

DROP SIDING ALSO CALLED NOVELTY

REMANUFACTURED WOOD SHINGLE WITH BACKER BOARD.

BOARD AND BATTEN

BEVEL SIDING

DOLLY VARDEN

Fig. 23-6 **Common wood product siding**

5 and get 85. Double this (because there are 2 corner boards at each corner) to get a total of 170 linear feet. This means that 170 linear feet of 5/4 x 5 pine boards are needed for corner boards. Add this to the material list at the bottom of the page.

SUMMARY

Exterior sheathing may be boards applied horizontally or diagonally, or it may be plywood or fiber board. The most common is plywood. To

find the number of sheets of plywood needed for sheathing, divide the total outside adjusted wall area by 32.

Wood siding is available in various patterns and in several widths and thicknesses. The most commonly used types are beveled siding, wood shingles and shakes, random-width vertical siding, aluminum siding, vinyl siding, brick veneer, and stone veneer.

Wood shingles and shakes may be applied in either single courses or double courses with a backer shingle or board. Hand-split shakes take 5 bundles per square. Sixteen-inch wood shingles take 3 bundles per square when used as siding and remanufactured shakes with backer boards are usually packaged 2 bundles per square.

REVIEW QUESTIONS

A. Select the letter preceding the best answer.

1. Which material is used most often for sheathing?

   a. Shiplap
   b. Tongue and groove lumber
   c. Plywood
   d. Wallboard

2. What percentage must be added for matching if 1 x 8 shiplap is used for sheathing?

   a. 10 percent
   b. 20 percent
   c. 25 percent
   d. 30 percent

3. If wood shingles are used for siding, how many bundles are required to cover one square?

   a. 1   b. 2   c. 3   d. 4

4. If hand split shakes are used for siding, how many bundles are required to cover one square?

   a. 1   b. 2   c. 3   d. None of these

5. If 8-inch beveled siding is used, how much must be added to allow for overlap at the edges?

   a. 10 percent
   b. 15 percent
   c. 25 percent
   d. 30 percent

6. A standard roll of building paper is how many square feet?

   a. 100
   b. 250
   c. 400
   d. 500

7. Which kind of plywood might be used for sheathing?

   a. Exterior with interior glue
   b. Interior with exterior glue
   c. Marine
   d. Hardwood

8. Where is the grade of material to be used for sheathing indicated?

   a. Floor plan
   b. Elevation
   c. Specifications
   d. Special detail drawing

B. List five materials commonly used as siding on houses.

C. Using the following given information, construct a material list for the sheathing and siding on the house in figures 23-7 and 23-8. The sheathing is to be 1/2-inch plywood and the siding is to be 3/4-inch by 10-inch cedar. Be sure to include the gable ends.

Quantity	Size	Material
a._____	1/2″ x 4′ x 8′	Plywood for sheathing
b._____bd. ft.	3/4″ x 10″	Red cedar beveled siding
c._____	____x____x____	Pine for corner boards
d._____rolls		Kraft building paper

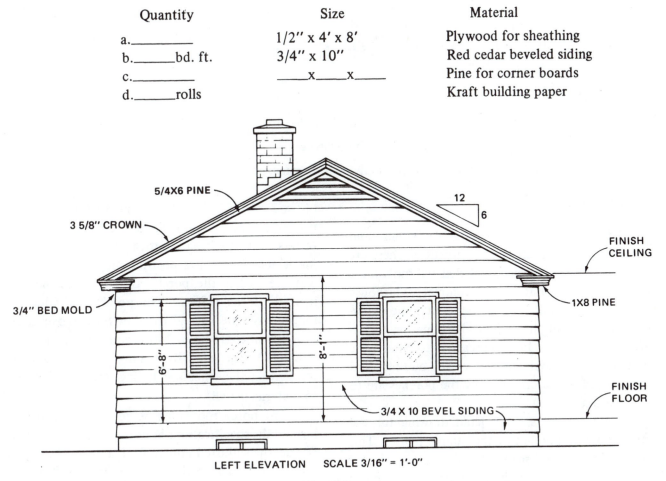

5/4X6 PINE

3 5/8″ CROWN

3/4″ BED MOLD

12 / 6

FINISH CEILING

1X8 PINE

FINISH FLOOR

3/4 X 10 BEVEL SIDING

6′-8″   8′-1″

LEFT ELEVATION   SCALE 3/16″ = 1′-0″

Fig. 23-7

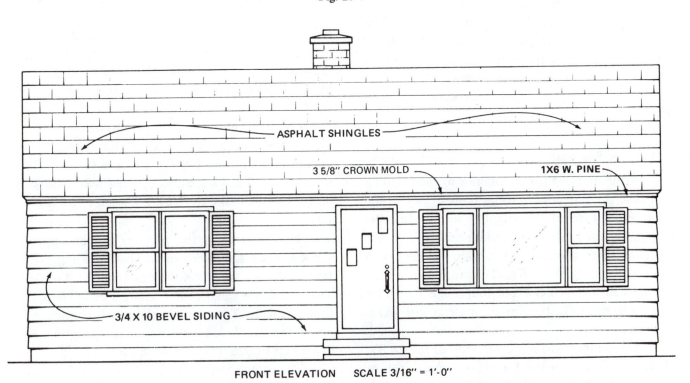

ASPHALT SHINGLES

3 5/8″ CROWN MOLD

1X6 W. PINE

3/4 X 10 BEVEL SIDING

FRONT ELEVATION   SCALE 3/16″ = 1′-0″

Fig. 23-8

unit 24 porches

OBJECTIVES

After studying this unit the student will be able to

- read special detail drawings for porch construction.
- determine the quantity of material needed to construct a porch.

Porch Construction

There are as many possible styles and methods of construction for porches as there are for houses. The porches included on the sample house are appropriate for a colonial style house and are intended to give the student some experience with detail drawings and estimating materials for porches. It is necessary to study the drawings and specifications for each building that is estimated to determine how the porches are to be constructed.

Look at the special detail drawing for the front and side porches of the sample house. This shows the kinds of material and methods of construction. Notice that the 2 x 4 rafters rest on top of the ceiling joists. In this common method of construction, the rafters are said to *barefoot* on the ceiling joists. Some plans indicate that a gable should barefoot on the main roof of a house, to eliminate the use of valley rafters. This should only be done when the main rafters rest on a bearing partition.

Estimating Porches

After studying the detail drawings and finding out how the porches are to be framed, look at the First Floor Plan of the sample house to find the length and quantity of material needed. The side porch is 18 feet long by 8 feet wide and the beams are made up of two 2 x 8's. Therefore, two 2 x 8's are needed for each end and four 2 x 8 x 10's for the long side. List them on material list (A) as follows:

The ceiling joists rest on the top of the 2 x 8's. Three joists are needed for every 4 linear feet, so multiply the length of the porch by 3/4 and add 1.

(18 x 3 = 54; 54 ÷ 4 = 14; and 14 + 1 = 15 joists needed.) The same rule applies to the rafters. The rafters on this job barefoot on the top of the ceiling joists. The joists protrude beyond the beams and are what the porch cornice is fastened to instead of the rafters. The porch is 8' wide, so longer joists are needed to allow for this overhang. Stock lumber is sold in 2-foot multiples, so 10-foot pieces are needed. By scaling the rafters, it is found that they must be 10 feet long. Add these to material list (B).

The porch roof must be covered with sheathing and the specifications indicate 1/2-inch plywood. Divide the adjusted square-foot area of the porch roof by 32 to find the number of 4' x 8' pieces needed. Eighteen feet 8 inches times 10 feet equals 188 square feet. Divide this by 32 to find that 5.89, or 6 pieces are needed. Add this to material list (C).

The next step is to list the finished material needed for the side porch. The roof is supported by 3 full posts and 2 half posts against the side of the house. The posts are made up of 16 pieces of 1 x 6 pine, 8 feet long. The bottoms of the 2 x 8's are covered with pine also, so add three 1 x 4 x 8' pine boards and one 1 x 4 x 10' pine board. The sides of the 2 x 8's are also covered with finished pine. This will take six 1 x 8 x 8's and two 1 x 8 x 10's.

The cornice on the porches is made up of 1 x 4's, 1 x 6's, and crown and bed molding. The fascia will take two 1 x 4 x 8's and four 1 x 4 x 10's. Two 1 x 6 x 8's and four 1 x 6 x 10's are needed for the soffit. To complete the porch cornice, 60 linear feet of 3 5/8-inch crown molding and 34 linear feet of bed molding are rquired.

The porch ceiling is finished with 5/8-inch beaded fir. There are 180 square feet of ceiling

and 25 percent is allowed for matching, so 225 square feet of beaded fir must be ordered. To finish the corner where the fir ceiling meets the boxed-in beams a cove mold is used, so measure this length. The perimeter of the porch ceiling is 52 feet. Add this to material list (D)

A

SIDE PORCH BEAMS	2×8×8'		4	
SIDE PORCH BEAMS	2×8×10'		4	

B

FIR- SIDE PORCH				
CEILING JOISTS & RAFTERS	2×4×10'		30	

C

CD PLYWOOD - SIDE PORCH				
ROOF SHEATHING	½"×4×8'		6	

D

PINE- PORCH POSTS	1×6×8'		16	
PINE- BOTTOM OF BEAMS	1×4×8'		3	
PINE- BOTTOM OF BEAMS	1×4×10'		1	
PINE- SIDES OF BEAMS	1×8×8'		6	
PINE - SIDES OF BEAMS	1×8×10'		2	
PINE - FASCIA	1×4×8'		2	
PINE- FASCIA	1×4×10'		4	
PINE- SOFFIT	1×6×8'		2	
PINE- SOFFIT	1×6×10'		4	
CROWN MOLD	3⅝"	LIN.FT.	60	
BED MOLD	1¾"	LIN.FT.	34	
BEADED FIR - CEILING	⅝"×4"		225	
COVE MOLD	1¾"	LIN.FT.	52	
PANEL MOLD		LIN.FT.	8	

This completes the estimate for the side porch. Now do the same for the front entrance porch. This is framed in the same manner as the side porch. The beams are 2 x 8's, and the ceiling joists and rafters are 2 x 4's.. The finished pine for the posts, the cornice, and boxing the 2 x 8 beams must be figured. The roof sheathing is the same and the roofing material is metal. The porch ceiling is 5/8-inch beaded fir and the same kind of finished moldings are needed. Add this to the material list.

FIR - FRONT PORCH BEAM	2 x 8 x 14'		2
FIR - FRONT PORCH CEILING JOISTS	2 x 4 x 14'		2
FIR - FRONT PORCH RAFTERS	2 x 4 x 8'		1
CD PLYWOOD - ROOF SHEATHING	½" x 4 x 8'		1
PINE - FRONT PORCH BEAMS	1 x 4 x 14'		1
PINE - FRONT PORCH BEAMS	1 x 8 x 14'		2
PINE - FRONT PORCH CORNICE	1 x 6 x 14'		2
PINE - FRONT PORCH CORNICE	1 x 6 x 8'		2
PINE - FRONT PORCH POSTS	1 x 6 x 8'		8
BEADED FIR - CEILING	5/8" x 4"		30
COVE MOLD	¾"	LIN. FT.	20
CROWN MOLD	3 5/8"	LIN. FT.	30
BED MOLD	1¾"	LIN. FT.	14
PANEL MOLD		LIN. FT.	6

SUMMARY

In building the front and the side porch, a different method of framing may be used. The rafters are cut to rest on the top of the ceiling joists and the joists extend beyond the face of the support beam. The cornice is attached to the end and the bottom of the ceiling joists instead of the rafter. This method of framing is referred to as barefoot. In listing the materials needed for the framing of the front and side porches, the details and elevations are referred to, in order to find the method of construction. The first floor plan is used to determine the lengths and the areas of the porches.

REVIEW QUESTIONS

A.   Select the letter preceding the best answer.

1. Which of the following is the best definition of the term barefoot as it applies to rafters?

   a. Rafters which do not have a ridge board
   b. The type of construction in which the rafters form the cornice
   c. The method of construction in which roof sheathing is applied directly to the rafters
   d. The method of construction in which the rafters rest on top of the ceiling joists

2. Where is the width of the porch rafters given?

   a. On elevation drawings
   b. Porch plan drawings
   c. Special detail drawings
   d. Specifications

3. Where is the thickness of the porch ceiling given?

   a. Elevation drawings
   b. Porch plan drawings
   c. Special detail drawings
   d. Specifications

4. How much T&G material should be ordered to cover a 120-square foot porch ceiling?

   a. 120 bd. ft.
   b. 150 bd. ft.
   c. 160 bd. ft.
   d. 120 sq. ft.

5. If the porch in figure 24-1 is 5 feet wide by 8 feet long, how much cove molding is required?

   a. 18 feet
   b. 10 feet
   c. 16 feet
   d. 26 feet

B.   Using the following given information, construct a materials list for the porch in figure 24-2. The porch ceiling and the gables at the ends of the porch are to be covered with 3/8-inch plywood.

Quantity	Size	Material
a._____	2 x____x____	Fir for rafters
b._____	2 x____x____	Fir for ceiling joists
c._____	2 x 8 x _____	Fir for beams
d._____	2 x 8 x _____	Fir for beams

e._____	1 x 6 x _____	Pine for posts
f._____lin. ft.	1 x 6	Pine for trim on posts
g._____lin. ft.	5/4 x 8	Pine for caps on posts
h._____lin. ft.	1 x 4	Pine for bottom of beams
i._____lin. ft.	1 x 8	Pine for sides of beams
j._____lin. ft.	1 x 8	Pine for soffit
k._____lin. ft.	1 x 4	Pine for fascia
l._____	3/8″ x 4′ x 8′	Plywood for ceiling and gables
m._____	1/2″ x 4′ x 8′	Plywood for roof sheathing
n._____lin. ft.	3 5/8″	Crown molding
o._____lin. ft.	1 3/4″	Cove molding

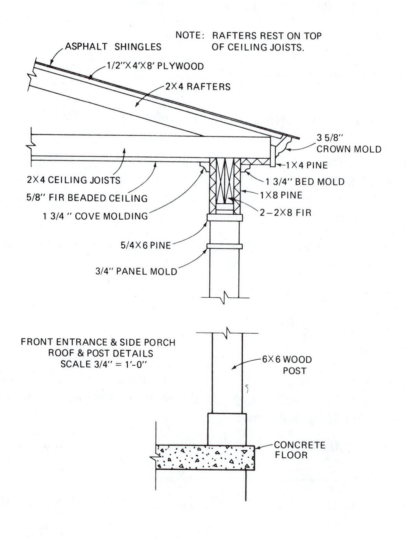

NOTE: RAFTERS REST ON TOP
OF CEILING JOISTS.

ASPHALT SHINGLES
1/2″X4′X8′ PLYWOOD
2X4 RAFTERS
3 5/8″ CROWN MOLD
1X4 PINE
1 3/4″ BED MOLD
1X8 PINE
2—2X8 FIR
2X4 CEILING JOISTS
5/8″ FIR BEADED CEILING
1 3/4″ COVE MOLDING
5/4X6 PINE
3/4″ PANEL MOLD

FRONT ENTRANCE & SIDE PORCH
ROOF & POST DETAILS
SCALE 3/4″ = 1′-0″

6X6 WOOD POST

CONCRETE FLOOR

Fig. 24-1

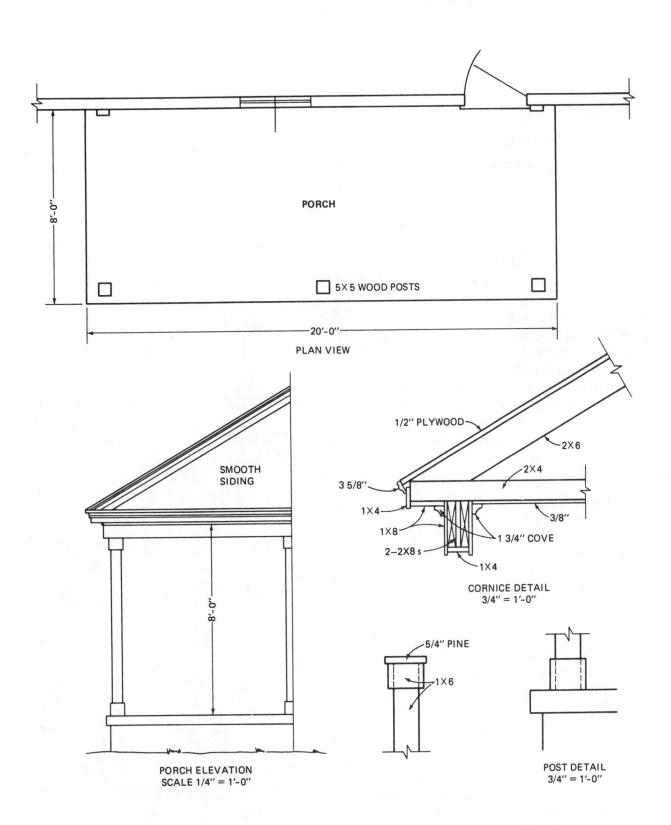

PORCH

5×5 WOOD POSTS

8'-0"

20'-0"

PLAN VIEW

SMOOTH
SIDING

8'-0"

PORCH ELEVATION
SCALE 1/4" = 1'-0"

1/2" PLYWOOD

2×6

2×4

3 5/8"

1×4

3/8"

1×8

1 3/4" COVE

2—2X8 s

1X4

CORNICE DETAIL
3/4" = 1'-0"

5/4" PINE

1X6

POST DETAIL
3/4" = 1'-0"

Fig. 24-2

SECTION 4 INTERIOR AND EXTERIOR FINISHING
unit 25 insulation

OBJECTIVES

After studying this unit the student will be able to

- explain the purpose of insulation in sidewalls and ceilings.
- discuss the specifications pertaining to insulation.
- determine the quantity of material needed to insulate a house

The basic function of insulation is to resist the flow of heat. Insulation is installed in the sidewalls, ceilings, roofs, and floors of buildings to reduce the flow of heat to the outside in the winter and to the inside in the summer.

Insulation may be made from various materials such as wood fibers, plastic foam, rock wool, fiberglass, aluminum foil, and vermiculite. It is sold in loose form, in batts, in rolls, and attached to paper. It is chemically treated to be vermin-proof and flame-resistant. The estimator should be acquainted with the types most often used. See Unit 44 Energy Conservation for additional information.

Moisture Control

Household moisture from daily activities, like cooking and washing, tends to build up in the air. Warm air holds more moisture than cold air; so when warm, humid, inside air comes in contact with cold surfaces, the air releases this moisture in the form of condensation. If the moisture present in the warm air is allowed to pass through the walls, it may condense on or under the cold exterior siding, causing the surface paint to peel. A vapor barrier (material that prevents the movement of water vapor) is needed between the interior of the house and the outer surface of the house walls. Most insulation sold in batts or rolls has a vapor-resistant facing. This vapor barrier, usually a special coated kraft paper or aluminum foil, should always be installed toward the heated side of the wall or ceiling, figure 25-1.

To help prevent condensation, it is important to have adequate ventilation in kitchens, bathrooms, and attics. Allowing sufficient air flow over the

insulation reduces the humidity level before condensation builds up. As a general rule, attics should have one square foot of free-flow ventilation area for every 150 square feet of attic floor space. This ventilation is normally provided for by louvers installed in the gables and soffits, figure 25-2, or by ridge vents. If ridge vents are used only 1 square foot of vent is required for every 300 square feet.

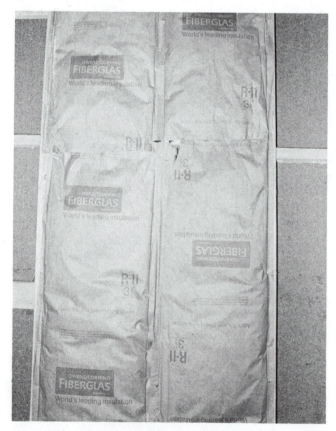

Fig. 25-1 Fiberglass insulation batts installed between wall studs.

Fig. 25-2 Aluminum louver for attic ventilation

Sizes

It pays to use good insulation in both the sidewalls and the ceiling of a house. Ceilings, walls, thermal windows, and insulation all have *thermal resistance values* or *R values*. The higher the R value of a material, the higher is its resistance to the flow of heat. For example, 3 1/2-inch, R-11 fiberglass insulation has the same thermal resistance as a 9-inch wooden wall, a 4 1/2-foot brick wall, or an 11-foot stone wall.

Insulation is manufactured in widths suitable for inserting between the framing members of the building. It is usually 15 inches wide for framing on 16-inch centers and 23 inches wide for 24-inch centers. It has a reinforced paper flange along each edge for nailing or stapling to the studs, joists, or rafters.

Estimating Insulation

Because there are so many types and thicknesses of insulation available, the first step in estimating insulation is to read the specifications and drawings. Division 6, Section C, Number 11 of the sample house specifications describes the insulation.

The ceiling of the second floor is to have 6-inch fiberglass insulation. All sidewalls are to have a 3 1/2-inch fiberglass insulation.

To determine the amount of insulation needed for the ceiling, figure the square foot area to be covered. There are 684 square feet of floor area on the second floor, so the ceiling area should be the same. The 6-inch fiberglass insulation is sold in batt form. The batts are 15″ x 48″ and come 10 pieces to the bag. Each bag covers 50 square feet, so 14 bags of 6-inch fiberglass insulation are listed on material list (A) for the ceiling. (684 sq. ft. ÷ 50 sq. ft. per bag = 13.7 or 14 bags.)

In estimating the sidewall area to be insulated do not include the gable ends. Multiply the distance around the house (108 feet) by the height of the walls (17 feet) to find the total area (1836 square feet). Three and one-half inch thick fiberglass insulation is sold in 70-square foot rolls, so divide the total area by 70. (1836 sq. ft. ÷ 70 square feet per roll = 26.2 or 27 rolls.) Add this to material list (B).

SUMMARY

The basic function of insulation is to resist the flow of heat. Insulation is made from various materials, such as wood fibers, fiberglass, aluminum foil, vermiculite, and plastic foam. It is available in loose form (for pouring), batts, and rolls. It is manufactured in widths which can be inserted between the framing members of the building. Most insulation has a reinforced paper flange along each edge for nailing or stapling to studs, joists, or rafters.

The insulation should include a vapor barrier on the heated side of the wall. This prevents condensation from building up on the outside wall. Moisture condensation on the outside of the wall can cause the surface paint to peel or blister.

A

FIBERGLASS INSULATION-									
CEILING	6″ X 15″	BAGS	14						

B

FIBERGLASS INSULATION-									
SIDEWALLS	3½″ x 15″	ROLLS	27						

REVIEW QUESTIONS

A.   Select the letter preceding the best answer.

1. Which of the following materials is commonly used for insulation?

   a. Plastic foam
   b. Vermiculite
   c. Fiberglass
   d. All of the above

2. What is the purpose of the kraft paper or aluminum foil facing on insulation?

   a. To prevent the flow of heat
   b. To hold the insulating material together
   c. To prevent the flow of water vapor
   d. To reflect heat

3. What is the normal width of fiberglass insulation for stud walls?

   a. 15 inches
   b. 16 inches
   c. 17 inches
   d. 23 inches

4. How is the resistance of insulation to the flow of heat measured?

   a. Thickness
   b. Kind of material
   c. Kind of backing
   d. R value

5. How many square feet will one bag of 15-inch fiberglass batts cover?

   a. 30
   b. 50
   c. 70
   d. 100

6. How many square feet of free-flow ventilation should be provided for a 1,000-square foot attic.

   a. 6.66 or 7 square feet
   b. 10 square feet
   c. 5 1/2 square feet
   d. 9 square feet

7. Which of the following provides the greatest insulating properties?

   a. 6 inch thick, R-19 fiberglass
   b. 6 inch thick, R-21 fiberglass
   c. 3 inch thick, R-23 plastic foam
   d. 3 1/2 inch thick, R-21 plastic foam

B.   Using the following given information, construct a material list to insulate the house in figure 25-3. The walls are to be insulated with R-11, 3 1/2-inch fiberglass and the ceiling is to be insulated with R-19, 6-inch fiberglass.

Quantity.	Material
a._____bags	R-19, 6-inch fiberglass batts for ceilings
b._____sq. ft.	R-11, 31/2-inch fiberglass for walls.

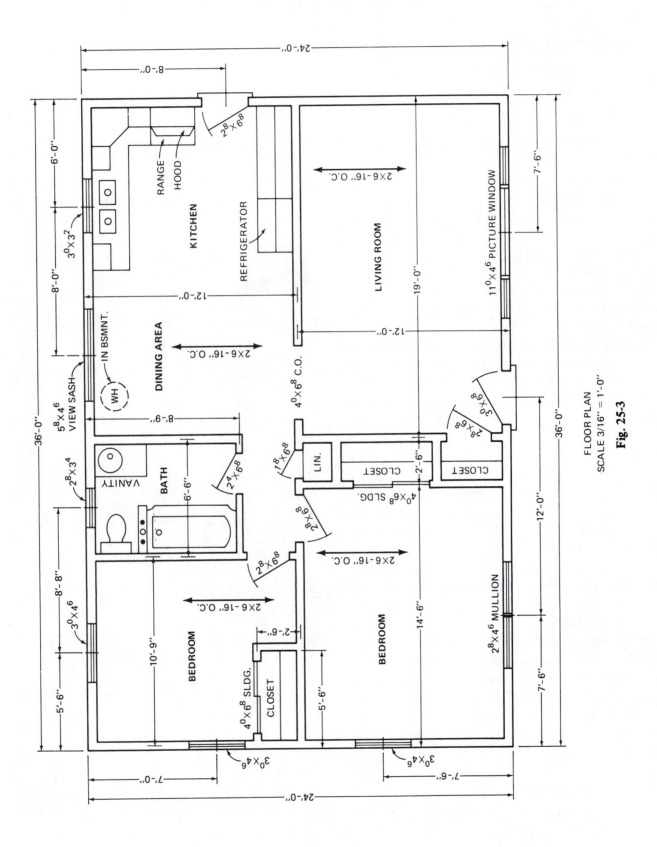

FLOOR PLAN
SCALE 3/16" = 1'-0"

Fig. 25-3

unit 26 wallboard and plaster

OBJECTIVES

After studying this unit the student will be able to

- explain the specifications for wallboard and plaster.
- describe wallboard joint system.
- determine the quantity of material needed for drywall or plaster walls.

Before the interior walls are covered, all work must be completed inside the walls. The rough plumbing, wiring, and heating equipment must be in place and inspected. Then the insulation is installed. With this done, the house is ready for wood paneling, lath and plaster, or gypsum wallboard. The most commonly used wall covering is *gypsum wallboard*, figure 26-1, sometimes referred to as drywall.

Gypsum wallboard is made by encasing a core made of gypsum rock and other ingredients between two sheets of paper. It is manufactured in widths of four feet, and lengths from six to sixteen feet. The long edges are reinforced and tapered. Its most popular thicknesses are 3/8 inch and 1/2 inch; fire rated 5/8 inch is used as required by code. The tapered edge allows the joints to be treated in such a way that they are entirely concealed. This treatment consists of using a special cement, reinforced with a perforated tape which fits into the recess formed at the joints by the boards, figure 26-2.

Gypsum wallboard is usually applied directly to the wood framing. The ceilings are applied first, then the sidewalls. The boards should be accurately cut and positioned, but not forced together. Horizontal application, with the long edges at right angles to the framing members is preferred because it minimizes joints and strengthens the wall or ceiling. Outside corners that are not protected by millwork are reinforced with steel corner bead. Annular-ring nails or screws are recommended for wallboard application.

Estimating Gypsum Wallboard

The first step in estimating interior walls is to read the specifications to determine what kind of

material is to be used. This is found under Division 6, Section C Number 10. The sample house is to have 3/8 inch thick gypsum wallboard.

Fig. 26-1 Gypsum wallboard being nailed into place

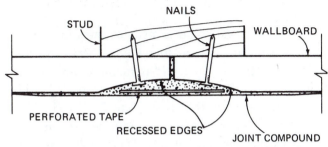

Fig. 26-2 Joints between pieces of gypsum wallboard are concealed with special joint compound and reinforcing paper tape

The quantity needed can be found by two methods. One method is to lay out each room by making a rough sketch of each wall and ceiling, showing the dimensions. From these sketches, the number of pieces and their lengths can be determined. Because some contractors apply the wallboard horizontally and other contractors apply the wallboard vertically, using a second method by square foot measurement may be more practical. This allows the contractor to order the needed lengths when the wallboard is to be installed. The price will be the same whether it is listed by the piece or by the square foot.

the second floor has 104 linear feet of partitions. Add these together for a total of 206 linear feet of interior partitions. The walls are eight feet high, so multiply 206 by 8 and get 1648 square feet. The interior partitions have to be covered on both sides, so double this to find that 3296 square feet of wallboard is needed for the interior partitions. Add the ceiling areas of 1368, the outside wall area of 1836, and the interior partition area of 3296 to find that 6500 square feet of 3/8-inch gypsum wallboard is needed.

Joint compound is a pre-mixed, vinyl-based cement that may be used directly from the container. It is used to apply the paper tape to the wallboard joints and for concealing nails *(spotting)*.

The tape that goes over the seams in the wallboard comes in 250 and 500-foot rolls. Joint compound comes in 5 and 1-gallon cans. Allow one 250-foot roll of tape and six gallons of compound for every 1,000 square feet of wallboard. The sample house has 6,500 square feet of wallboard, so seven rolls of tape and eight 5-gallon cans of joint compound are needed. Add these to the material list. Corner beads of metal are required for all outside corners. 8 foot lengths are standard.

GYPSUM WALLBOARD –															
CEILINGS AND WALLS	3/8"	SQ. FT.	6,500												
PERFORATED TAPE –															
WALLBOARD	250'	ROLLS	7												
JOINT COMPOUND –															
WALLBOARD	5 GAL.	CANS	8												

NOTE: cathedral or sloped ceilings will require an area equal to length of slope ceiling x length. If no cathedral or sloped ceiling then the amount of wallboard needed for the ceiling is the same as the area of the floor. The area of the floors has already been found to be 684 square feet at each level, so multiply this by two (for two floors) and find that 1368 square feet of wallboard is needed for the ceilings.

In figuring wallboard, no deduction is made for openings, unless the opening is larger than 32 square feet, so take the length of the outside walls times the height (108 x 17) and get a total of 1836 square feet.

The length of the interior partitions was found when the framing material was estimated. The first floor has 102 linear feet of partitions and

PLASTER

Plastered walls are not as common as gypsum wallboard, because of the easy application of wallboard and the long drying time involved with the use of plaster. Plaster contains a lot of moisture and the building must be allowed to dry out before finished woodwork is taken into the building. However, some buildings do have plastered walls, so an estimator must be familiar with this material.

Gypsum Lath

Gypsum lath has a gypsum core encased between two sheets of paper. It is manufactured in

two sizes; 16 inches x 32 inches and 16 inches x 48 inches; and in two thicknesses; 3/8 inch, which is standard, and 1/2 inch, which is recommended for use where the studs are more than 16 inches O.C. but not more than 24 inches O.C. It is manufactured either plain or perforated. The perforated type is the same as the plain except that it has one 3/4-inch hole for every 16 square inches of surface. This provides a mechanical key for the plaster coat. The amount of gypsum lath needed for a building is found in the same way as the amount of wallboard is found.

To find the amount of plaster required for a house, first find the number of square yards to be covered. To do this divide the number of square feet of lath by nine. For every 100 square yards to be plastered the following materials are required:

Neat plaster for brown coat..... 850 lbs.

Gauging plaster .... 125 lbs.

Hydrated lime .... 250 lbs.

Note: Plaster is sold is 100 lb bags and hydrated lime is sold in 50 lb. bags.

The sample house has 6500 square feet or 723 square yards of wall area to be plastered, so base the estimate on 750 square yards. Multiplying the quantities for 100 square yards by 7 1/2 (7 1/2 x 100 = 750) gives the following quantities required for the sample house:

Sixty-four 100-lb. bags of neat plaster

Ten bags of gauging plaster

Fifty 50-lb. bags of hydrated lime

SUMMARY

Gypsum wallboard is made by encasing a core of gypsum and other ingredients between two sheets of strong paper. It is manufactured in 4' widths and in lengths ranging from 6' to 16'. The two most popular thicknesses are 3/8" and 1/2".

The tapered edge allows a joint treatment that conceals the joint completely. This treatment consists of applying special cement, reinforced with a perforated tape, into the recess formed at the joint by the boards.

Gypsum wallboard is usually applied directly to the wood framing members. The ceiling is covered first, then the sidewalls. Annular ring nails or screws are recommended for applying wallboard to the framing members.

Joint compound is a vinyl-base cement that may be used directly from the container. It is designed for tape application and complete joint finishing.

Gypsum wallboard can be figured in two ways. It can be figured by the square-foot area to be covered, or by the piece. To estimate by the piece, make a rough sketch of each room of the house showing the dimensions. Then figure the number of pieces needed and their lengths.

If the specifications call for plastered walls instead of gypsum wallboard, it is necessary to find the number of square feet of lath, the number of bags of neat plaster for the brown coat, the amount of gauging plaster, and the amount of hydrated lime that is needed. Plaster is estimated by the square yard. To find the number of square yards divide the square-foot area by nine. For every 100 square yards of wall, 8 1/2 bags of neat plaster, 1 1/4 bags of gauging plaster, and five 50-lb. bags of hydrated lime are needed.

REVIEW QUESTIONS

A.   Select the letter preceding the best answer.

1. How wide is gypsum wallboard?

a. 18 inches
b. 2 feet
c. 3 feet
d. 4 feet

2. Which of the following thicknesses is most common for gypsum wallboard?

a. 1/4 inch
b. 1/2 inch
c. 7/8 inch
d. 1 inch

3. How are the nails or screws concealed when applying gypsum wallboard?

    a. They are hammered flat and painted over.
    b. Finishing nails are used.
    c. They are covered with joint compound.
    d. They are covered with plaster.

4. If a house is to be plastered, why must all plastering be done before the finished woodwork is taken into the house?

    a. The plaster contains a large amount of moisture, which will be absorbed by the woodwork.
    b. Spilled plaster may stain or damage the woodwork.
    c. Stored woodwork will be in the way of plasterers.
    d. The plaster may be damaged by workers handling the stored woodwork.

5. What is the proper thickness for gypsum lath which is to be applied to studs that are 16 inches O.C.?

    a. 1/4 inch
    b. 3/8 inch
    c. 1/2 inch
    d. 5/8 inch

6. How much perforated tape is required to finish the joints on 1,000 square feet of wallboard?

    a. 750 feet
    b. 500 feet
    c. 300 feet
    d. 250 feet

7. How much joint compound is required to finish the joints on 1,000 square feet of wallboard?

    a. 10 gallons
    b. 6 gallons
    c. 5 gallons
    d. 3 gallons

B. Using the following given information, construct a material list for covering all of the walls and ceilings of the house in figure 26-3 with gypsum wallboard. The walls are to be covered with 1/2-inch wallboard and the ceilings are to be covered with 3/8-inch wallboard.

Quantity	Material
a._____sq. ft.	3/8-inch gypsum wallboard for ceilings
b._____sq. ft.	1/2-inch gypsum wallboard for walls
c._____rolls	Perforated tape for wallboard joints
d._____gallons	Joint compound

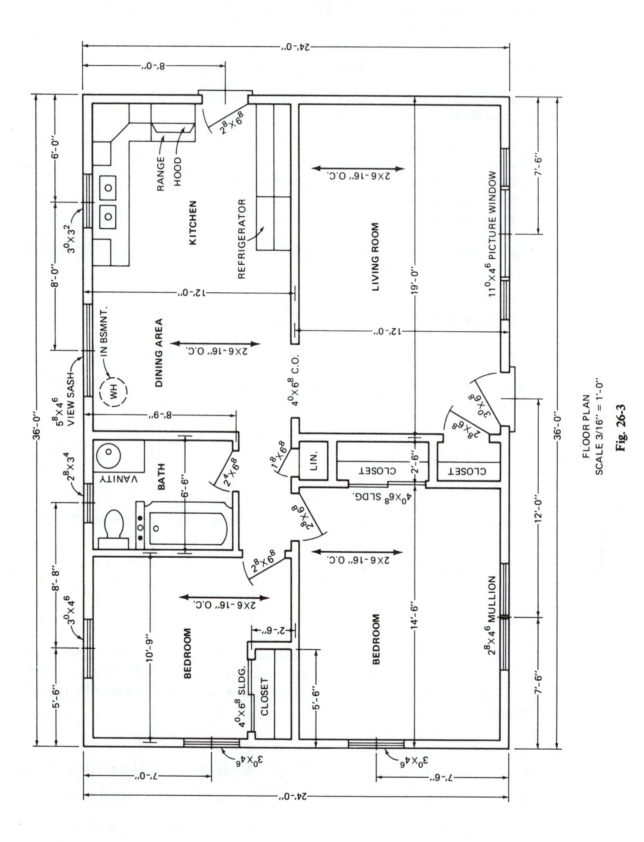

FLOOR PLAN
SCALE 3/16" = 1'-0"

Fig. 26-3

unit 27 moldings

OBJECTIVES

After studying this unit the student will be able to

- identify several types of molding.
- locate the molding in the working drawings and specifications for a house.
- determine the quantity of molding needed for a job.

The interior millwork or woodwork includes all of the finished trim and moldings. This may vary with each individual building, so it is important for the estimator to check the working drawings and specifications carefully before beginning the estimate.

The inside of each window and each side of the door jambs are trimmed with *casing. Stop molding* is the small piece that stops the door from swinging through the opening as it is closed. Windows also use stop molding to hold the sash in place. In addition, the bottom of the window is trimmed with the *stool,* which forms the finished windowsill, and the *apron,* which is placed against the wall directly below the stool. These door and window parts are all considered interior millwork, but they are estimated with the doors and windows. Several common molding shapes are shown in figure 27-1. There are two categories of trim. One is "running trim," figured in linear feet for baseboard, crown, mold, and chair rails. The second is "standing trim," figured for a specific length for door and window casings and stool (which forms the finished window sill).

COVE MOLDING

The corner where the sidewalls and the ceiling meet is often covered with a decorative molding. This corner is called a cove and therefore is most often covered with *cove molding,* but *crown molding* and *bed molding* are also used here. In some cases combinations of two or more kinds of moldings may be used.

To find out how the corner next to the ceiling is to be finished, look at the drawings. This information is found on the drawing of the half section view, the cornice detail, or a special detail of the cove. The wall section detail is usually drawn to the scale of 3/4" = 1'-0". It shows the construction of the house from the bottom of the footings up to the rafters at the eaves. The wall section detail for the sample house calls for 1 3/4-inch cove molding.

Estimating Cove Molding

Molding is estimated by the linear foot. The distance around the outside of the house is known. There are two floors and the perimeter is the same for each. (108 plus 108 equals 216 linear feet) The total length of interior partitions on both floors has already been found to be 206 linear feet. The cove mold goes on each side of the interior walls so 412 feet of molding is needed for 206 feet of partitions. Add this to the length of the exterior walls and get a total of 628 linear feet. Add 10 percent for waste.

BASEBOARD AND CARPET STRIP

These are the finished moldings at the bottom of a wall where the wallboard or plaster and the finished floor meet. The *carpet strip* (sometimes called a shoe mold) is nailed to the bottom of the baseboard and covers the crack formed where the base meets the finished floor. This allows for any movement between the hardwood floor and the wall.

| COVE MOLDING - CEILINGS | 1¾" | LIN. FT. | 691 | | | | | | | | | | | | | | | | |
| BASEBOARD - CARPET STRIP | 3" | LIN. FT. | 594 | | | | | | | | | | | | | | | | |

147

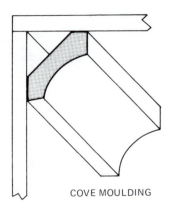

COVE MOULDING

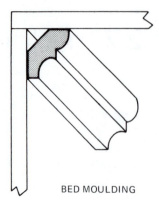

BED MOULDING

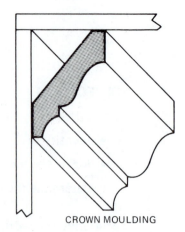

CROWN MOULDING

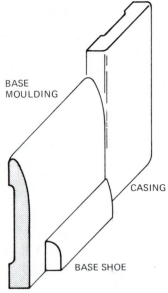

BASE MOULDING

CASING

BASE SHOE

BASE: Applied where floor and walls meet, forming a visual foundation. Protects walls from kicks and bumps, furniture and cleaning tools. Base may be referred to as one, two or three member. The base shoe and base cap are used to conceal uneven floor and wall junctions.

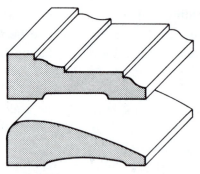

CASING: Used to trim inside and outside door and window openings.

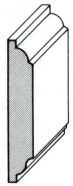

MULLION CASING: The strip which is applied over the window jambs in a multiple opening window. Sometimes called a panel strip, used for decorative wall treatments.

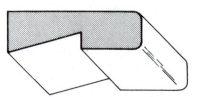

STOOL: A moulded interior trim member serving as a sash or window frame sill cap.

CHAIR RAIL: An interior moulding usually applied about one third the distance from the floor, paralleling the base moulding and encircling the perimeter of a room. Originally used to prevent chairs from marring walls. Use today is as a decorative element or a divider between different wall covering such as wallpaper and paint or wainscoting. A key decorative detail in traditional and Colonial design.

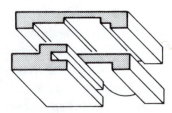

JAMB: The top (header) and two sides (legs) of a door or window frame which contacts the door or sash. Flat jambs are of fixed width while split jambs are adjustable.

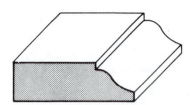

STOP: In door trim, stop is nailed to the faces of the door frame to prevent the door from swinging through. As window trim, stop holds the bottom sash of a double-hung window in place.

Fig. 27-1 Some common types of wood molding

Estimating Baseboard and Carpet Strip

There are two ways that the length of the baseboard can be found. One is to find the perimeter of each room and deduct the width of the door openings. Another is to deduct twice the width of the interior door openings and the width of exterior door openings from the total length of ceiling cove moldings.

There are 628 linear feet of 1 3/4-inch cove molding. Add twice the number of interior doors and the number of exterior doors on both floors and get a total of 35. Multiply this by 2 1/2 (average door width in feet) and get 87 1/2 feet. Subtract this 87 1/2 feet from 628 and get a total of 540 1/2 feet of baseboard and carpet strip. Add 10 percent for waste Add this to the material list.

SUMMARY

The interior millwork or woodwork is made up of finished trim and moldings. Cove molding is used where the sidewalls and the ceiling meet in a corner. This is usually either 1 3/4-inch or 2 1/4-inch cove molding.

The baseboard is the molding attached to the bottom of a wall, where it meets the floor. A small molding called a carpet strip or base shoe is nailed to the bottom of the baseboard to cover the crack formed where the base and the finished floor meet.

Molding is estimated by the linear foot. There are two ways to find the quantity of cove molding needed. One way is to measure each room separately. The other is to add the perimeter of the outside walls and twice the length of the inside partitions. The amount of baseboard and carpet strip needed can be found by subtracting the total width of all doors from the amount of ceiling cove molding.

REVIEW QUESTIONS

A. Select the letter preceding the best answer.

1. What kind of molding is used in the corner between the walls and the ceiling?
   a. Casing
   b. Apron
   c. Cove
   d. Base

2. What kind of molding is used to cover the edge of the floor covering at the wall?
   a. Casing
   b. Apron
   c. Cove
   d. Shoe

3. What unit of measurement is normally used for listing moldings?
   a. Square feet
   b. Linear feet
   c. Board feet
   d. Piece

4. On which of the following is the type of molding to be used at the top and bottom of the walls normally indicated?
   a. Specifications
   b. Wall section
   c. Elevation
   d. Floor plans

5. Once the quantity of cove molding is known how can the quantity of base be determined?
   a. The quantity of base is the same as the quantity of cove molding.
   b. Add the width of all doors to the quantity of cove molding.
   c. Subtract twice the width of all interior doors and the width of all exterior doors from the quantity of cove moldings.
   d. Subtract the width of all doors and windows from the quantity of cove moldings.

B. Using the following given information, construct a material list which includes the quantity of cove molding, base and base shoe required for the house in figure 27-2.

The cove molding is to be 1 3/4 inch in all rooms except the living room. The living room cove molding is to be 2 1/4 inch.

	Quantity			Material	
a. _____ lin. ft.		c. _____ lin. ft.		1 3/4-inch cove molding	Base
b. _____ lin. ft.		d. _____ lin. ft.		2 1/4-inch cove molding	Base shoe

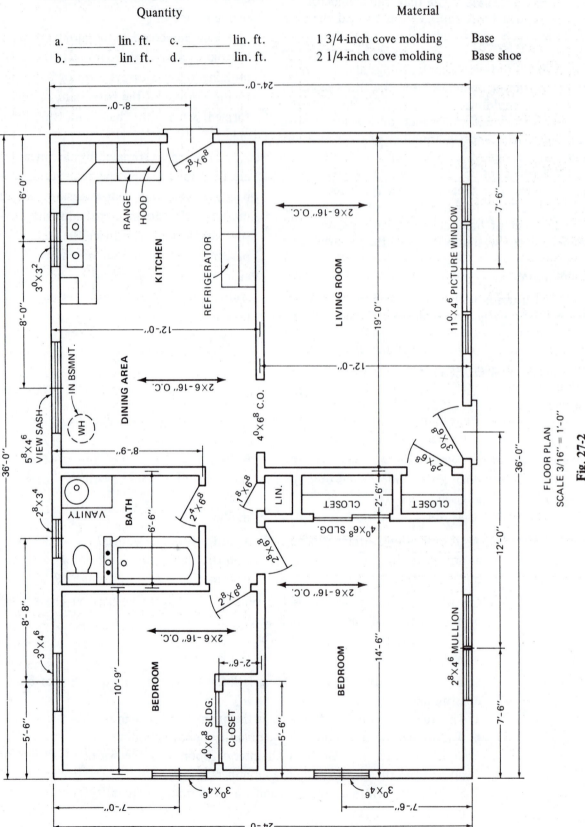

FLOOR PLAN
SCALE 3/16" = 1'-0"

Fig. 27-2

unit 28 flooring and underlayment

OBJECTIVES

After studying this unit the student will be able to

- describe hardwood flooring and underlayment.
- explain the specifications for hardwood flooring and underlayment.
- determine the quantity of hardwood flooring and underlayment required for a building.

Most homes have a different kind of floor covering on the kitchen, bathroom, and entrance floors than on the other floors of the house. Kitchen floors are usually tile or vinyl sheet material. Bathroom floors are usually vinyl sheet, vinyl tile, or ceramic tile. Entrances are usually of a masonry material such as ceramic tile, quarry tile, slate, brick, or marble. Check the plans and specifications carefully to find out what is called for in these areas.

All of the floors should be laid on the same level to eliminate the possibility of tripping on their edges. Different materials are manufactured in various thicknesses. For example, hardwood flooring is 25/32 of an inch thick, but vinyl is 1/8 of an inch thick. To overcome this difference, underlayment is nailed down on top of the subfloor before the finished floor material is installed.

Underlayment can be any material that will give a smooth surface to receive the finish flooring. Often plywood that has a smooth sanded surface on one side is used. Other materials such as hardboard and tempered particle board are also used. The thickness of the underlayment varies according to the thickness of the material used for the finished floor. The thickness of the underlayment plus that of the finished flooring material should be the same as the floor it will butt against.

In some entrance and bathroom floors where masonry material is set in a bed of mortar to form the finished floor, the subflooring is cut to rest flush with the top of the floor joists instead of on top of the joists, figure 28-1. This allows for a thicker bed of mortar. However, because of the high strength of modern adhesives, most ceramic floors are cemented in place with adhesives instead of mortar.

Where a masonry floor comes next to finished wood floors, a marble threshold is placed in the doorway, figure 28-2. In entrance halls, the common practice is to have the masonry installed, and then the finished carpet or wood floor fitted against it.

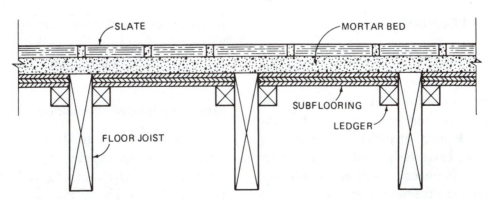

Fig. 28-1 Where slate or ceramic tile is to be set in mortar, the subfloor may be installed on ledgers between the floor joists.

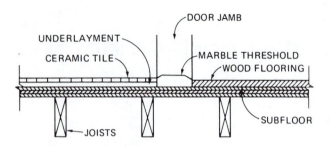

Fig. 28-2. A marble threshold is installed in a doorway between a wood floor and a masonry floor.

Estimating Underlayment

The sample house specifications call for 5/8-inch plywood for the underlayment. To find the quantity needed, take the square-foot area of the kitchen, bathrooms, and entrance hall. Divide this total area by 32 to find the number of pieces of underlayment needed.

The front entrance and closet are 6 1/2 feet x 5 1/2 feet or 35 3/4 square feet. The first floor bathroom is 3 1/2 feet x 7 feet or 24 1/2 square feet. The kitchen is 9 feet x 12 feet or 108 square feet. The cellar stairs and closet are 3 feet x 12 feet and 1 1/2 feet x 6 feet or a total of 45 square feet. The second floor bath is 7 feet x 9 feet or 63 square feet. Add all of these together for a total of 276 1/4 square feet. Divide 32 into 276 1/4 to find that 8 5/8 or 9 sheets of underlayment are needed. Add this to the material list.

The principle American producers of hardwood flooring have adopted uniform grading rules and regulations. Every bundle of flooring produced by a member of either the National Oak Flooring Manufacturers' Association or the Maple Flooring Manufacturers' Association is identified as to grade. Usually the manufacturer's name and mill mark or identification are found on each bundle.

Standard oak flooring grades have been established. Quarter sawed is clear or select, and plain sawed is clear, select, No. 1 common or No. 2 common. Hardwood flooring comes in random lengths.

Strip flooring is usually nailed through the building paper directly into the subflooring and joists. The nails are driven through the base of the tongue of the flooring at a 45-degree angle, figure 28-3, page 153, so that the nail heads don't show. Staples which are made for use with a flooring hammer, may be used instead of nails.

When wide-plank flooring is installed, the width of the boards make blind nailing impractical. In this case, the flooring is drilled and counterbored so that the flooring can be fastened to the subflooring with screws. The heads of the screws are then covered with wood plugs and finished the same as the rest of the floor. Where it is necessary

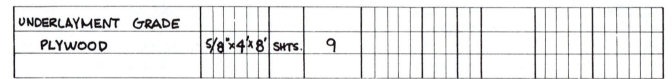

UNDERLAYMENT GRADE																
PLYWOOD	5/8"x4'x8' SHTS.	9														

HARDWOOD FLOORING

The hardwoods most commonly used for flooring are oak, maple, beech, birch, and ash. Oak is the most plentiful and is, by far, the most used. Hardwood flooring may be classified as strip flooring, plank flooring, or parquet. Strip flooring is the type most extensively used. It is tongued and grooved on the ends as well as the edges so that each piece joins the next when it is laid. Strip flooring is manufactured in a variety of sizes. The most popular is 25/32 inches thick and 2 1/4 inches wide.

to face nail the flooring near the walls, the nail heads are set and the holes filled with wood putty and sanded the same as the floor.

Estimating Hardwood Flooring

Division 6, Section D, Number 5 of the specifications for the sample house pertain to wood flooring: *HARDWOOD FLOORING:* All subfloors are to be broom cleaned and covered with deadening felt before the finished flooring is laid. Wood flooring, where scheduled, is to be 1 x 3 T&G and end matched, select oak flooring.

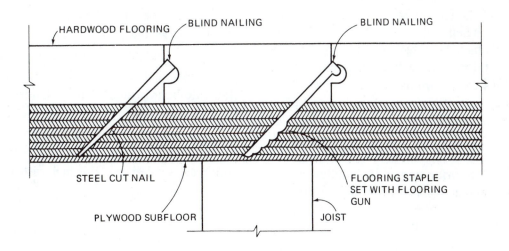

Fig. 28-3 Hardwood flooring is blind nailed.

Flooring is to be laid evenly and blind nailed every 16 inches without tool marks.

To figure the quantity needed, find the area to be covered. The plans show a total of 684 square feet on each floor or a total area of 1368 square feet. Deduct 276 1/2 square feet for the entrance, baths, and kitchen. This leaves 1091 1/2 square feet to be covered with hardwood flooring. Allow one third of this (364) for matching. Add this to the material list.

SUMMARY

Underlayment is the material that is installed over the subfloor and to which the finished flooring material is attached. It is used as a base for such material as vinyl tile, sheet vinyl, slate, quarry tile, and ceramic tile. The thickness of the underlayment depends upon the thickness of the material that is to be applied on top of it. The purpose of underlayment is to bring the finished floor to the same level as the hardwood flooring and to provide a smooth,

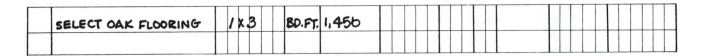

	SELECT OAK FLOORING		1 X 3		BD. FT.	1,456															

The specifications require that the floors be covered with deadening felt before the finished floor is laid. The deadening felt is also laid between the subfloor and the underlayment. To find the amount of deadening felt required, find the area of both floors. The total floor area is 684 square feet for each floor or a total of 1368. A roll of felt covers 500 square feet, so divide 500 into 1368 to find that 3 rolls of deadening felt are needed. Add this to the material list.

level base for the material. Plywood, with the good side toward the finished floor, is commonly used for underlayment, but it is not the only material used. In many cases, tempered particle board or hardboard are used as underlayment.

The hardwoods most commonly used for flooring are oak, maple, beech, birch, and ash. Oak is the most frequently used and the most popular size is 25/32 inches x 2 1/4 inches (1 x 3 nominal). Strip flooring is tongued and grooved and

	DEADENING FELT-FLOORS		500'		ROLLS	3												

end matched at the factory so that each piece joins the next snugly when laid.

Standards have been set so that the producers of hardwood flooring have uniform grading rules and regulations. Standard oak flooring grades have two general classifications; quarter sawed and plain sawed. Quarter-sawed oak is graded as clear or select. Plain-sawed oak is graded as clear, select, No. 1 common, or No. 2 common.

Hardwood flooring comes in bundles of random lengths, and is usually blind nailed through the building paper and directly into the subfloor at a 45 degree angle.

If wide-plank flooring is used, the width of the boards makes blind nailing unsuitable because the wide boards might have a tendency to buckle. In this case, the flooring is counterbored and fastened to the subflooring with screws. The heads of the screws are then covered with wood plugs which are sanded and finished with the rest of the floor.

REVIEW QUESTIONS

A . Select the letter preceding the best answer.

1. Which of the following materials may be used as underlayment?
   a. Hardboard
   b. Particle board
   c. Plywood
   d. All of these

2. Why are various thicknesses of underlayment used?
   a. To accomodate varying thicknesses of flooring
   b. To obtain the required strength
   c. To use the least expensive material needed
   d. None of the above

3. Which of the following is not commonly used for wood flooring?
   a. Pine
   b. Spruce
   c. Fir
   d. Oak

4. How is strip flooring fastened down?
   a. Nailed through the face of the flooring
   b. Nailed through the tongue of the flooring
   c. Through the face with the screw heads covered
   d. With special adhesive

5. What is applied to the subfloor before the hardwood flooring is applied?
   a. Glue
   b. Insulation
   c. Aluminum felt
   d. Deadening felt

6. How is wide-plank flooring fastened down?
   a. Nailed through the face of the flooring
   b. Nailed through the tongue of the flooring
   c. Screwed through the face of the flooring with the screw heads covered
   d. With special adhesive

7. What percentage of the area of a floor must be allowed for matching strip flooring?
   a. 20 percent
   b. 25 percent
   c. 33 1/3 percent
   d. 40 percent

B. Using the following given information, construct a material list to include what is required for underlayment and hardwood flooring in the house in figure 28-4, page 155. The kitchen and bathroom are to have underlayment for vinyl tile. The remainder of the house is to have strip flooring. All subfloors are to be covered with deadening felt.

Quantity		Material
a.	pieces	5/8″ x 4′ x 8′ underlayment plywood
b.	bd. ft.	1 x 3 select oak flooring
c.	rolls	Deadening felt

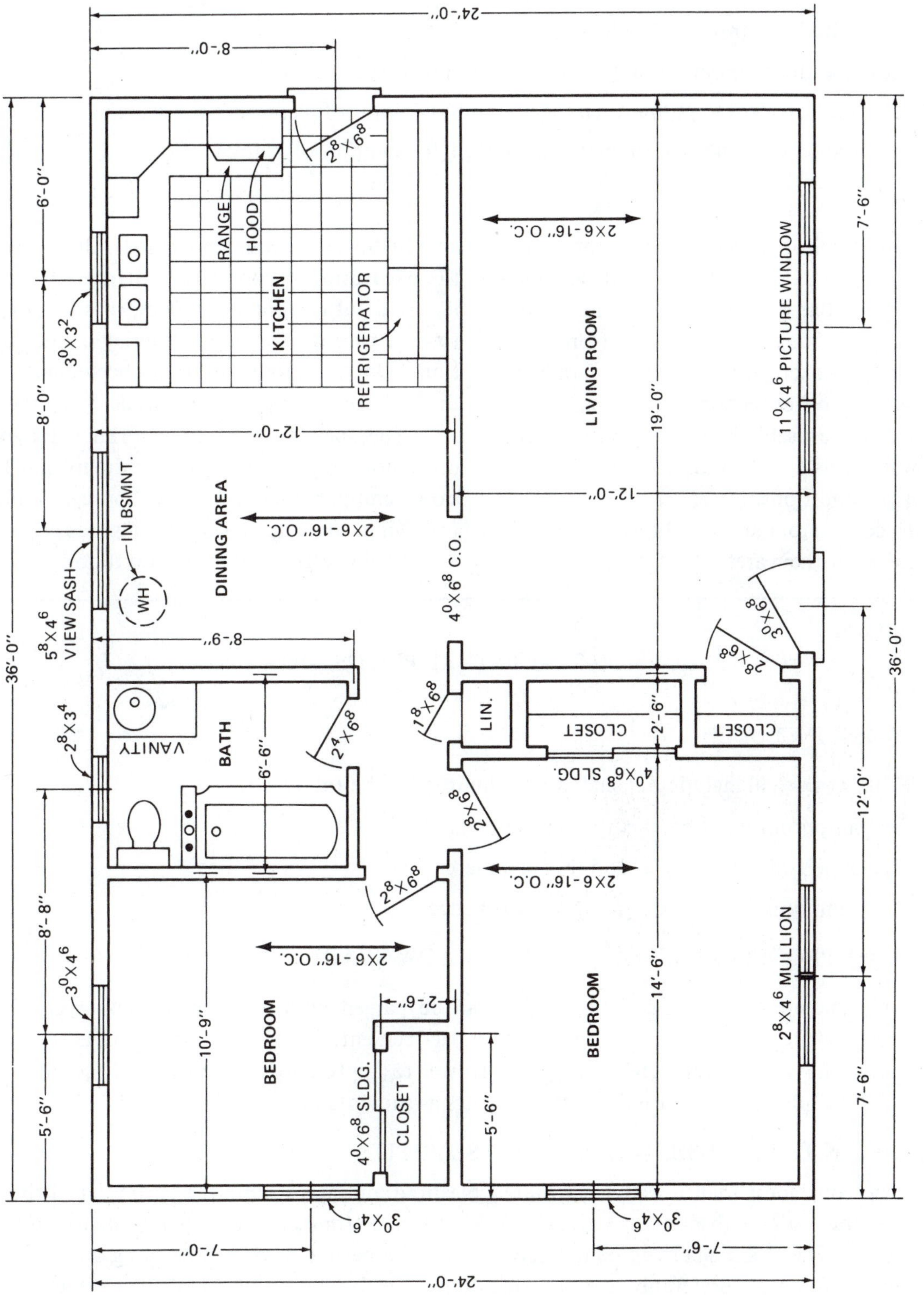

FLOOR PLAN
SCALE 3/16″ = 1′-0″

Fig. 28-4

unit 29 foundation dampproofing

OBJECTIVES

After studying this unit the student will be able to

- describe the materials usually used for dampproofing.
- explain how perimeter drains and foundation dampproofing prevent moisture in a cellar.
- determine the quantity of material needed to dampproof a foundation and install a perimeter drain.

Due to the level of the *water table,* (the natural level of water in the ground), it is not possible to have a usable cellar in many parts of the country. In these areas most homes are built on concrete slabs or have a crawl space underneath the first floor joists. In other sections of the country it is possible to have a usable basement, even where subsurface water exists. The building must be designed for these conditions.

Building contractors are usually aware of the water conditions in their area. However, unfavorable water conditions may not be discovered until the excavation is completed. If there is a possibility of a future water problem, it is less expensive to take care of it at the beginning of the construction. Once the foundation has been backfilled it is difficult to waterproof the walls from the inside.

The specifications and working drawings for most houses require perimeter drains and foundation dampproofing when the house is built. The following are the specifications for the sample house being estimated in this textbook.

Specifications

DIVISION 7: MOISTURE PROTECTION

A. **WORK INCLUDED**

This work shall include but shall not be limited by the following:

1. Dampproofing of basement walls below grade
2. Asphalt shingle roofing, 235 lb. per square
3. Aluminum flashing and galvanized drip edge

B. **DAMPPROOFING BASEMENT WALLS BELOW GRADE**

All concrete block walls below grade shall be parged with a 1/2-inch thick layer of portland cement plaster. Use a mix of 1 part cement, 1 part lime, and 6 parts sand. After the wall is completely cured, use trowel to apply two coats of asphalt coating. Cant the plaster over the edge of the footings as shown on the drawings.

C. **DRAIN TILE, CRUSHED STONE, AND SUMP PIT**

The contractor shall install a perforated plastic drain pipe around the perimeter of the footings fully imbedded in a trench of #2 crushed stone and sloping 1/8 inch to the foot. The drain shall run completely around the perimeter of the building and then into a sump pit preferably to an outside drain. The drain pipe and crushed stone are to be installed completely free of any loose sand or earth.

PERIMETER DRAIN

The most practical way to prevent a wet cellar is to use a *perimeter drain* (an underground drain pipe around the footings) and crushed stone to carry the water away from the foundation. If the ground has a natural slope, the pipe can be run around the foundation wall and away with the slope of the ground. In areas where there is no natural drainage, the drain is run to a sump on the inside of the foundation. A sump pump is installed to take the water out of the area, either to a storm sewer or dry well away from the building.

The perimeter drain pipe is normally perforated plastic pipe, figure 29-1. Plastic drain pipe is manufactured in 10-foot lengths and 250-foot rolls and may be either perforated (with a series of holes to allow water to pass) or solid. Rigid plastic pipe is joined with special couplings. An assortment of plastic fittings is available, such as tees, elbows, and a variety of angles. An advantage of plastic pipe is that the longer lengths and tighter fitting joints eliminate the possibility of soil filtering into the joints and clogging the pipe.

A trench the size of the footing is dug just outside the footing, figure 29-2. The drain is imbedded in crushed stone in this trench. The drain pipe is pitched 1/8 inch per foot, so that the water will run toward the sump or drain opening.

Estimating the Perimeter Drain

To figure the amount of plastic pipe needed, find the distance around the house and divide by 10. This is the number of 10 foot pieces of plastic pipe needed. A 90-degree elbow is required at every corner; and where the drain goes under the footing to a sump, a tee is required. Crushed stone is sold by the ton. The perimeter drain requires approximately one ton for every 18 feet of pipe.

The sample house has a perimeter of 108 feet, so eleven pieces of perforated plastic pipe are needed. There are six corners, so, six 90-degree elbows are needed. If the type that comes in 250-

Fig. 29-1 Perforated plastic drain pipe may be either flexible or rigid

foot rolls is used it is not necessary to include elbows. One tee must be used where the pipe goes under the footing and into the sump pit. Allowing one ton of number 2 crushed stone for every 18 feet of pipe, the drain will use 6 tons of stone. Add these to the material list.

PARGING

The concrete block foundation wall is to be plastered with a 1/2-inch coat of cement and lime below grade. When the plaster is dry, it is to be coated with two coats of asphalt, figure 29-3. The mason trowels on the plaster, smoothing it as much as possible. Where the foundation rests on top of the footings, it is plastered from the top of the footing up. The plaster is *canted* so that it forms a rounded surface from the wall out to the edge of the footing, figure 29-4. This plastering is called *parging*.

Portland cement, when mixed with sand, does not work well and does not hold together to make a good plaster. This is the reason that the mason's

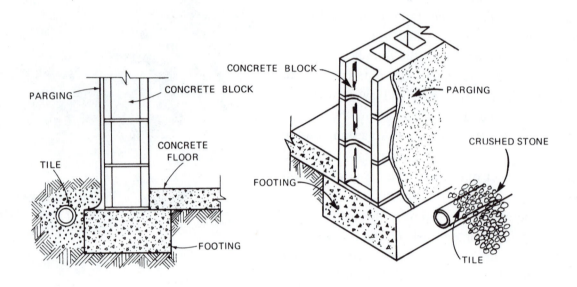

Fig. 29-2

PERFORATED PLASTIC DRAIN																	
PIPE - RIGID	10'		PCS.	11													
90-DEGREE ELBOWS- DRAIN			PCS.	6													
TEE - DRAIN PIPE			PCS.	1													
#2 CRUSHED STONE -																	
PERIMETER DRAIN			TON	6													

lime is added. The lime acts as a binding and carrying agent for the sand and cement. The cement and lime mix makes a strong mortar and a good dampproofing plaster. This is because the lime fills any voids that might exist between the particles of sand and cement.

Estimating Dampproofing

To estimate the quantity of materials for foundation dampproofing, first find the area to be covered. The perimeter of the building is 108 feet and the foundation is approximately 6 feet below

Fig. 29-3 The foundation wall is coated with asphalt to prevent water from passing through into the basement.

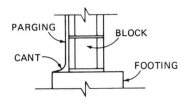

Fig. 29-4

grade. This means there is an area of 648 square feet to be parged. One bag of portland cement will make enough plaster to cover about 40 square feet, so divide the area to be covered by 40. (648 ÷ 40 = 16 bags of cement.) Add this to material list (A) shown below.

The specifications also call for two coats of asphalt dampproofing material to be applied over the parging. The rate of coverage depends upon the manufacturer and the type of coating used. Asphalt emulsion is available in brush or trowel consistency. The brush type is the most widely used. It will cover about 30 to 35 square feet of wall per gallon. Divide 648 (the area to be covered) by 30 and get approximately 22 gallons. The second coat will only require about half as much as the first, so add eleven more gallons for a total of

33 gallons. Asphalt is sold in 5-gallon cans, therefore, approximately seven 5-gallon cans of coating material will complete the job. Add this to material list (B) shown below.

SUMMARY

If subsurface water is present in an area where a house is being constructed, it is much less expensive to guard against the situation as the house is built than later. A drain pipe is imbedded in crushed stone around the perimeter of the footings to carry away excess water. This perimeter drain may carry the water to a storm drain, a natural slope, or to a sump inside the foundation of the building. If a sump is located in the building, a sump pump removes the water to a suitable drain.

Plastic pipe comes in 10-foot lengths and 250-foot rolls and is joined by special couplings. Pipe is laid with a 1/8-inch per foot pitch.

In addition to the drain, which carries the water away, most houses also have some type of dampproofing on the foundation walls. This consists of a 1/2-inch coat of cement plaster called parging and one or two coats of asphalt foundation coating. The dampproofing prevents the water from seeping through the foundation and makes it filter down through the soil to the perimeter drain.

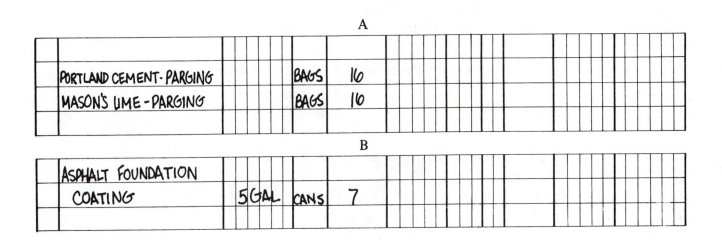

	A			
PORTLAND CEMENT - PARGING		BAGS	16	
MASON'S LIME - PARGING		BAGS	16	

	B			
ASPHALT FOUNDATION COATING	5 GAL	CANS	7	

REVIEW QUESTIONS

A. Select the letter preceding the best answer.

1. What is the length of rigid plastic drain pipe?

   a. 4 feet   c. 1 foot
   b. 2 feet   d. 10 feet

2. What is used to join the sections of plastic drain pipe?

   a. Cement
   b. Special couplings
   c. The joints are wrapped with building paper.
   d. The sections slip together.

3. How much should a perimeter drain be pitched?

   a. 1/8 inch per foot   c. 1/2 inch per foot
   b. 1/4 inch per foot   d. 1 inch per foot

4. How much crushed stone is required for 144 linear feet of perimeter drain?

   a. 28 tons   c. 12 tons
   b. 18 tons   d. 8 tons

5. What is parging?

   a. Waterproofing applied to a foundation wall
   b. A type of drain pipe
   c. Plaster applied to a foundation wall
   d. The rounded shape formed by the plaster at the corner between the foundation wall and the footing

6. Why is mason's lime added to portland cement to make plaster for a foundation?

   a. To make it waterproof
   b. To make it stronger
   c. To neutralize the acid in the soil
   d. To help bind the particles of sand and cement

7. What is applied over the plaster to dampproof a foundation?

   a. Asphalt   c. Building paper
   b. Plastic sheeting   d. Parging

8. Where can the water from a perimeter drain be carried to?
   a. A storm sewer   c. A dry well
   b. A sump   d. All of these are correct

B. Using the following given information, construct a material list for the dampproofing and perimeter drain for a building 24 feet wide by 42 feet long. The footings are 5 feet below the finished grade. All parging is to be sand, portland cement, and lime.

Quantity	Size	Material
a._____tons	#2	Crushed stone
b._____pieces		10-foot plastic pipe
c._____		90-degree elbows for plastic pipe
d._____		Tee's plastic pipe
e._____bags		Portland cement
f._____bags		Hydrated lime
g._____gallons		Asphalt foundation coating

unit 30 roofing

OBJECTIVES

After studying this unit the student will be able to

- list several common roofing materials.
- explain specifications for roofing.
- determine the quantity of material needed for a roof.

Many factors affect the selection of material to be used as roofing. It must provide protection against sun, snow, ice, wind, and sleet. It must also show consideration for the ease of application, the appearance in relation to the design of the house, permanence, and the investment.

The most common roofing material for either new construction or the replacement of older roofs is asphalt or fiberglass roofing. Asphalt or fiberglass roofing material is made from a base of dry felt, which is processed from rag, wood, or other cellulose fibers and then saturated with asphalt or fiberglass. Additional coats of asphalt may be applied to the surface to produce a smooth-surfaced roll roofing, or mineral granules of selected colors may be rolled into the coated surface to produce mineral-surfaced roll roofing or shingles. Fiberglass roofing usually has a better fire rating.

Saturated felt, to which no coating of asphalt has been added, is used as a building paper between the exterior sheathing and the finished siding, as underlayment between the roof sheathing and the finished roof, and between the subfloor and the finished floor. It is also used to form layers of built-up roofing.

Roofing material is available in different weights. The weight referred to is the weight of enough roofing material to cover one square (100 square feet of roof area). Mineral-surfaced roll roofing comes in a 90-pound weight. Smooth-surfaced roll roofing comes in 15, 30, 45, 55, and 65-pound weights. Double-coverage roll roofing weighs 140 pounds per square. Double-coverage roofing is nailed and cemented down in such a way that only a 19-inch edge is exposed and all nails are covered.

tabs pointing upward. Another course is applied directly over this with the tabs pointing downward.

Asphalt roofing shingles, figure 30-1, are made in a wide variety of colors and in straight or blended shades. They are manufactured so that they have a maximum of four square or three hexagonal tabs to the strip. Square-edge strip shingles are 36″ long and 12″ wide. They are applied with either a 4-inch or 5-inch exposure. They are packaged in bundles which cover one-third of a square each.

The second most widely used product for roofing is wood shingles. Western red cedar is the most widely used wood. Wood roofing is manufactured in two basic products: shingles and hand-split shakes, figure 30-2. When a roof is covered with shingles or shakes, a double course (2 layer) is applied at the eaves. Extra shingles must also be allowed for ridges, hips, and starter courses. The first *course* (row) of shingles is applied with the

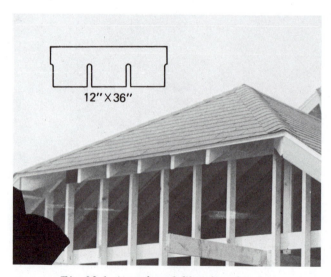

Fig. 30-1 Asphalt and fiberglass shingles

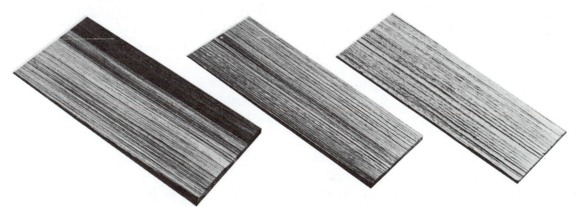

#1 Handsplit-resawn shakes. Tapered. One side heavily textured, one side sawn smooth. Thicker and heavier through the butt than other shakes.

#1 Tapersplit shakes. Medium texture and thickness. Produced by reversing the cedar block and handsplitting to achieve a natural taper.

#1 Straight-split shakes. (Also called Barn Shakes.) Medium textured. Hand-split without reserving the block, so a generally uniform thickness is achieved.

#1 Cedar Shingles. Shingles are pro-duced by sawing both faces to produce two smooth sides. #1 grade is 100% clear, heartwood, edge-grained.

#1 & #2 Rebutted-Rejointed shingles. Trimmed with straight edges, right-angle corners, for close tolerance application on exterior walls. Also available with smooth-sanded faces.

Grooved sidewall shakes. A #1 rebutted-rejointed shingle with grain-like grooves. Available natural or in a variety of factory-applied tones.

#2 shingle, flatgrain, limited sapwood; #3 utility shingle, for economy applications; #4 undercoursing, also effective on interior walls as accent covering.

Fig. 30-2 Wood shingles

It is used primarily on low-pitched roofs. Asphalt shingles weigh 210 to 315 pounds per square.

The following notes apply to the coverage of common roofing materials:

- One square of shingles covers 100 square feet when applied with the following exposure:

  Asphalt . . . . 5 inches
  24-inch wood . . . . 7 1/2 inches
  18-inch wood . . . . 5 1/2 inches
  16-inch wood . . . . 5 inches
  24-inch shakes. . . . . 10 inches

- Wood shingles provide 240 feet of starter course per square.
- Wood shakes provide 120 feet of starter course per square.
- One bundle of shingles or shakes covers 16 2/3 feet of ridge or hip.

Flat roofs commonly have what is known as a built-up roof. A built-up roof consists of several layers of felt laid one on top of another. Each layer is nailed and cemented with either hot tar or cold-process roofing cement. After the last layer is applied, the roof is mopped with hot tar or cold-process cement and then covered with granular stone, figure 30-3. Hot tar built-up roofs are usually bonded, meaning that the contractor guarantees the roof for a number of years.

OTHER MATERIAL USED FOR FINISHED ROOFING

As an estimator you should be familiar with other products used in roofing and how to estimate the quantity.

Slate

Slate roofing is the most expensive and is installed one slate at a time. Its advantage is its durability. Slate is manufactured in various sizes, thicknesses, as well as colors. Standard slate is the quarry run of 3/16 inch thickness with considerable variations above and below 3/16 inch. In estimating slate the usual lap or cover of the lowest course of slate by the upper two is three inches, so a square of roofing slate means a sufficient number of slate of any size to cover 100 square feet with a

3 inch lap. To determine the exposed length, deduct the lap from the length of the slate and divide by 2. For example, if a 14″ x 20″ slate is used, subtract 3 from 20 which is 17 and divide by 2 for 8 1/2 inches exposed area. Next multiply 14 by 8 1/2 to get a total of 119 square inches. There are 14,400 square inches to a square so divide 119 into 14,400 to get 121 slate for each square using a 14″ x 20″ slate with an 8 1/2 inch exposure.

METAL ROOFING

The metals used for roofing material are tin, copper, galvanized iron, zinc, and aluminum. Tin is manufactured in two thicknesses; IC or 29 gauge, which weighs approximately 8 ounces to a square foot; and IX or 27 gauge, which weighs approximately 10 ounces to a square foot. Using standing joints, a 14″ x 20″ sheet will cover 235 square inches: one box (112 sheets) will cover 182 square feet. With a flat lock seam, allowing 3/8 inch all around for joints, one box will cover 198 square feet. Depending upon the shape of the roof, you will have to add for waste and cutting and fitting for any corners, protrusions, etc.

Copper, galvanized iron, zinc, and aluminum are manufactured in sheets. You should contact the manufacturer or dealer to find out about size and coverage.

Fig. 30-3 Built-up roof

The specifications for roofing on the sample house are included under Division 7, Moisture Protection.

  D. Roofing shall be of saturated mineral surface 235 pound strip shingles, applied in accordance with the manufacturer's instructions.

  E. Sheet Metal Flashing shall be 24 gauge, best commercial grade, aluminum 28 inches wide, and shall be used for all valleys, chimneys, and general flashing.

Estimating Roofing

Roofing material is measured by the square. A square is enough material to cover 100 square feet when applied according to the manufacturer's instructions. Divide the square-foot area of the roof by 100 to find the number of squares required. If asphalt or fiberglass strip shingles are being used, add 1 1/2 square feet for every linear foot of ridge, eaves, hip, and valley.

The sample house has 1,444 square feet of roof area including the porches. This is approximately 14.5 squares. There are 45 feet of ridge, 36 feet of valley and 75 feet of eaves. Add this together for a total of 146 linear feet. One and a half times 146 is 219 or 2.2 squares. Add the 2.2 to 14.5 to get a total of 16.7 squares of roofing.

Allow 10 percent for waste. Asphalt or fiberglass shingles are sold in 1/3-square bundles so 18 2/3 squares must be listed on material list (A).

When asphalt or fiberglass shingles are used, building felt is placed between the roof sheathing and the finished roof. This is not required for wood shingles or a metal roof. Building felt is sold in 432 square foot rolls, which covers 400 square feet allowing for overlap. The area of the roof is 1260 square feet divided by 400 square feet = 3 1/2 or 4 rolls. Add this to material list (B).

The valleys and chimney require 28-inch aluminum flashing. There are two valleys and each valley rafter is 18 feet long, so there is a total of 36 feet of valleys. The perimeter of the chimney is 10 feet. This makes a total of 46 feet of flashing. Aluminum flashing comes in 50-foot rolls, so one roll of flashing is added to material list (C).

When asphalt shingles are used on a roof, a metal drip edge is installed along the eaves and the rake of the house. Drip edge is manufactured in 10-foot lengths. This is nailed over the top of the roofing felt along the eaves and up the rake of the cornice. The asphalt shingles go over the top of the edge of the drip edge. Measure the length of the eaves and the rakes. On the sample house this is 133 feet. Fourteen 10-foot pieces or 140 feet of drip edge must be added to material list (D).

SUMMARY

The most commonly used roofing material is

A

235-LB. ASPHALT SHINGLES		SQ.	18 2/3

B

15-LB. ASPHALT FELT ROOF	432 sq. ft.	ROLLS	4

C

ALUMINUM FLASHING	28" x 50'	ROLL	1

D

GALVANIZED DRIP EDGE	10'	PCS.	14

asphalt and fiberglass shingles. They come in various sizes, weights, and colors. The most popular are square edge strip shingles that are 36″ long and 12″ wide. They are made from saturated felt with a mineral surface. They are sold by the square and are packaged three bundles to the square.

Western red cedar is the wood used most often for wood shingles. Wood shingles are manufactured as three basic products: shingles, hand-split shakes, and grooved sidewall shakes.

Built-up roofing is made up of a series of layers of felt and hot tar. The final layer is mopped with a coating of tar and then coarse gravel is poured over the top. A bonded roof, such as most built-up roofs, is guaranteed by the contractor to last for a definite period of time without any defects or leaks.

All roofing material is sold by the square. A square is the amount of material that will cover 100 square feet when applied according to the manufacturer's instructions.

REVIEW QUESTIONS

A. Select the letter preceding the best answer.

1. What is the primary difference between asphalt roofing material and saturated felt which is used as building paper?

a. Saturated felt has no asphalt on the surface.

b. Saturated felt contains no asphalt at all.

c. Roofing material is not saturated.

d. The base materials are different, but they are both saturated with asphalt.

2. What does the weight indicate when asphalt shingles are listed as 240-pound shingles?

a. 240 pounds per bundle

b. 240 pounds per 10 bundles

c. 240 pounds per square

d. They contain 240 pounds of asphalt per square.

3. How many bundles of asphalt strip shingles are required to cover 100 square feet of roof?

a. 2   c. 4

b. 3   d. 5

4. If 16-inch wood shingles are used on a roof, how much exposure will allow one square to cover 100 square feet?

a. 4 inches   c. 10 inches

b. 7 1/2 inches   d. 5 inches

5. What is a square of roofing?

a. Enough material to cover 1,000 square feet of roof

b. Enough material to cover 100 square feet of roof

c. One bundle of shingles or one roll of roll roofing

d. 100 pounds of roofing material

6. What is the size of a roll of saturated felt?
    a. 3 feet x 100 feet
    b. 4 feet x 250 feet
    c. 100 square feet
    d. 500 square feet

7. How long is a roll of aluminum flashing?
    a. 25 feet
    b. 40 feet
    c. 50 feet
    d. 100 feet

8. How long is one length of galvanized drip edge?
    a. 6 feet
    b. 8 feet
    c. 10 feet
    d. 12 feet

9. What is a built-up roof?
    a. One which is made of plywood, saturated felt, and shingles
    b. One in which the rafters rest on top of the ceiling joists
    c. One which is made up of a new roof built over an old roof
    d. One which is made of several layers of roll roofing and tar

B. Using the following given information, construct a material list for the roof of the house in figures 30-4 and 30-5.

Quantity	Material
a._____bundles	240-pound asphalt shingles
b._____rolls	15-pound saturated felt
c._____ linear feet	Galvanized drip edge

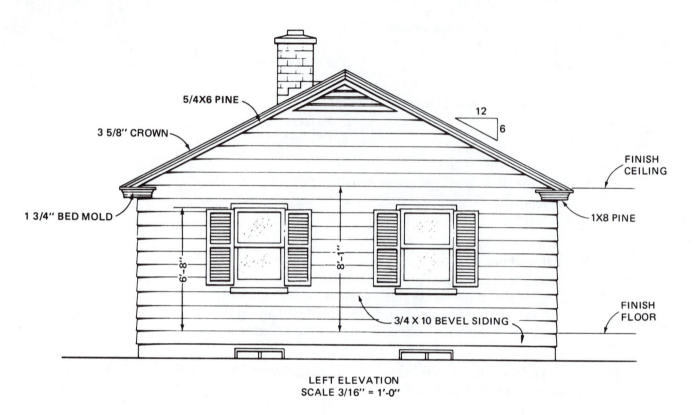

LEFT ELEVATION
SCALE 3/16″ = 1'-0″

Fig. 30-4

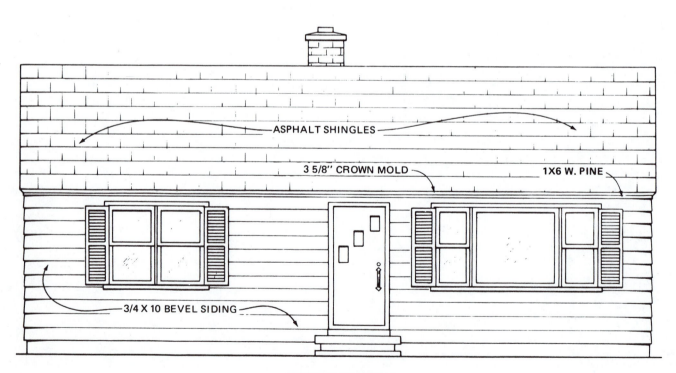

FRONT ELEVATION
SCALE 3/16″ = 1'-0″

Fig. 30-5

unit 31 exterior doors and interior doors

OBJECTIVES

After studying this unit the student will be able to

- describe the difference between exterior and interior doors.
- explain where the information for exterior and interior doors is found in the specifications and the working drawings.
- determine the quantity, size, and styles of exterior and interior doors required for a house.

Doors are classified as either exterior or interior, depending on whether they are hung in exterior walls or interior partitions. They can also be classified as either *flush* (with a smooth surface) or *panel* doors (with decorative panels).

EXTERIOR DOOR

Exterior doors are usually 1 3/4 inches thick, solid pine, fir, or metal. Flush exterior wooden doors usually have a hardwood veneer on the surface and have a solid wood center. A great assortment of styles of exterior solid wood panel doors is available. A colonial door, for example, is constructed with various size panels and moldings. Some have glass panels and others have all solid wood panels. A colonial door is constructed so that the four top panels form a cross. All exterior doors are glued with a waterproof glue.

Gaining in popularity for use as exterior doors are metal doors. They have none of the drawbacks of wooden doors, which tend to swell in hot weather, shrink in cold, and warp with water damage. The metal door's growing attractiveness of design since its relatively recent inception is a good selling point too. Figure 31-1 shows a fine example of such a door, with Crystal Etch™ Collection glass in place of the upper panels.

Doors and windows are always designated with the width first, the height second, and the thickness last. Most main entrance doors in resi-

Fig. 31-1 Pease metal exterior door (Courtesy of Pease Industries Inc.)

dential construction are 3'-0" x 6'-8" x 1 3/4". This means that they are 3 feet wide by 6 feet 8 inches high by 1 3/4 inches thick. Other entrance doors, such as a back door, are usually 2'-8" x 6'-8" x 1 3/4". These sizes may be varied, but they have almost become standard for residential construction throughout the building industry.

INTERIOR DOORS

Interior doors are manufactured in a variety of styles and sizes. They may be 1 3/4 inches, 1 3/8 inches, or 1 1/8 inches thick. Most interior doors are 1 3/8 inches thick, however, narrow louvered doors are 1 1/8 inches thick.

Flush interior doors are usually hollow-core doors which are covered with hardwood veneer. The top, bottom, and side rails are made of solid wood, as is the area where the hardware is installed. The remainder of the inside of the door is filled with a lightweight material, such as honeycombed paper. This makes a lightweight door that is attrac-

tive and serviceable. Flush doors are usually finished with a light stain or a clear finish. This is why manufacturers use select veneers for the finished surface.

Flush doors are manufactured in 6'-0", 6'-4", 6'-8", and 7'-0" heights. They are manufactured in widths from 9 inches to 3 feet.

Panel doors are manufactured in various styles from two to eight panels, figure 31-2. The most popular style panel door is a 6-panel colonial door which has two small panels at the top and four larger panels directly below.

Louver doors, figure 31-3, are popular in such places as the laundry, family room, closets, and kitchens. These doors are manufactured in various widths heights, and styles.

Door heights may vary according to the available headroom where they are to be installed. It is common practice to have door heads and window heads the same height from the finished floor. The most common height for both interior and

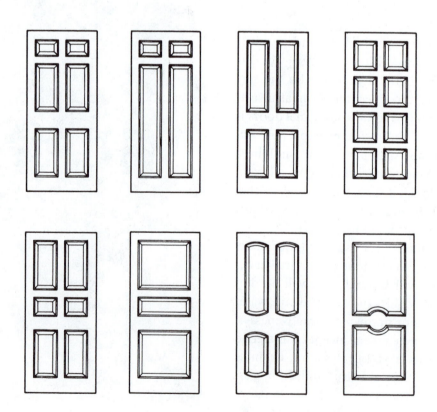

Fig. 31-2 Panel doors

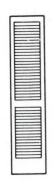

**Fig. 31-3
Louver door**

exterior doors is 6'-8''. This allows the casing for the top of the doors and windows to be at the same height, keeping the appearance of the interior of the room in balance.

PRE-HUNG DOORS

A pre-hung door is a complete unit with the door hung on the jamb at the factory. These units are assembled at the factory with the door, jambs, stop, and casing all in place. The door jamb slides apart so that the two halves can be slid into the opening. With the unit in place, the casings are nailed through the wall and into the studs. To order a pre-hung door, the size, style of the door, style of the casing and hardware, and the swing of the door must all be listed. To determine the *swing of a door* stand on the side toward which the door opens. If the knob is on the right side, it is a right-hand door; if the knob is on the left side, it is a left-hand door. Division 8 of the specifications provides information about the sizes and styles of the doors.

Specifications

DIVISION 8: DOORS, WINDOWS, AND GLASS

A. **WORK INCLUDED**

This work shall include but shall not be limited by the following:

1. Wood doors and wood door frames
2. Hanging and fitting of doors
3. All window units
4. All window and door trim
5. Mirrors in the bathroom over the vanity

B. **WOOD DOORS**

Exterior doors: Doors shall be as follows:

Front door — 3'-0'' x 6'-8'' x 1 3/4''
Porch door — 2'-8'' x 6'-8'' x 1 3/4''
Rear door — 2'-8'' x 6'-8'' x 1 3/4''

Interior doors: All interior doors are to be 6 panel except where shown on plans.
Sizes and thickness as shown on the drawings.

C. **WOOD DOOR FRAMES**

All wood door frames shall be clear pine, dadoed together at the head with joints set with white glue. All frames shall be accurately set, plumb, level, and true. Hinge locations shall be solidly shimmed.

D. **HANGING AND FITTING OF DOORS**

All doors shall be accurately cut, trimmed, and fitted to their frames. All doors shall operate freely without binding, and all hardware shall be adjusted properly. Exterior doors shall be fitted with weather stripping.

E. **INTERIOR DOOR AND WINDOW TRIM**

All interior door and window trim shall be clear pine, machine sanded. Casing to be 11/16'' x 2 1/4'' base to be 9/16'' x 3''. Door and window trim shall be colonial style.

F. WOOD WINDOWS

All window units shall be double hung with removable grilles. Sizes and styles shall be shown on the drawings. Wood combination storm and screen units will be furnished with each unit.

Estimating Doors

List the exterior doors first. The specifications list the size and style of the exterior doors for the sample house. When this information is not included in the specifications it can be found on the drawings or a special schedule of doors.

When job-hung doors are used instead of pre-hung, the parts must be listed separately. Exterior door frames may be assembled by a local lumber dealer. In this case the frame is delivered to the job site with the jambs and the outside casing installed. The oak threshold is sent separately and the frame and sill are installed at the site.

except where listed differently on the plans. The linen closet door on the second floor is to be a louver door. All doors on this job are to be job-hung so list the parts separately.

Look at the first floor plan to see that two sets of jambs 4'-0" x 6'-8" are needed — one leading from the front entrance to the living room and one from the entrance hall to the dining room. The other door sizes shown on the first floor plan are one 2'-6" x 6'-8" between the dining room and the kitchen, one 2'-8" x 6'-8" between the kitchen and the cellarway, and five 2'-0" x 6'-8" doors in the closets and half bath. The quantity of door casing is found by multiplying the number of sides

EXT. DOOR FRAME W/ SILL	3^0 x 6^8		1
EXT. DOOR FRAMES W/ SILL	2^8 x 6^8		2
EXT. DOOR - FRONT	3^0 x 6^8 x $1\frac{3}{4}$		1
EXT. DOORS - SIDE & REAR	2^8 x 6^8 x $1\frac{3}{4}$		2

Add the exterior doors and frames to the material list shown below.

The door stop and the interior door casing are listed separately and are in the category of "standing trim." Estimate the casing on the interior side only, since the exterior casing is installed when the frame is assembled. It takes 2 pieces 7'-0" length and one 3'-6" piece of casing for each side of a door. There are three outside doors, so multiply quantities by 3 and get 6 pieces 7'-0" length and 3 pieces 3'-6". To prevent the door from swinging through the doorway, molding called door stop is installed around the inside of the jambs. The door closes against the door stop.

Each door requires 2 pieces 7'-0" and 1 piece 3'-0" feet of stop, so multiply quantities by 3 to get a total of 6 pieces 7'-0" and 3 pieces 3'-0" of door stop. Add these to material list (A).

List the interior doors next. According to the specifications all interior doors are to be 6 panel,

of doors to be cased by 17. The amount of stop is found by multiplying the number of doors by 17. List the jambs, doors, stop, and casing to material list (B).

Estimate the doors for the second floor in the same manner as those for the first floor. Remember that one door is a louver door. Add these to material list (C).

This is also an appropriate point in the estimate to list the closet rods. There are six closets, so six adjustable closet rods are required. Add these to material list (D).

SUMMARY

Doors can be classified as either exterior or interior, and either panel or flush. Exterior doors are generally 1 3/4 inches thick and are assembled with waterproof glue. Interior doors are usually, but not always, 1 3/8 inches thick. The most common height for doors in residential construction is 6'-8",

but other heights are available. The size of a door is always listed with the width first, the height second, and the thickness last.

To determine the swing of a door, stand on the side toward which the door opens. If the knob is on the left, it is a left-hand door, if the knob is on the right, it is a right-hand door.

A pre-hung door is a complete unit, ready to be installed in the building. It includes the door, jambs, stop, casing, and hardware. The jambs are split, so that they may be taken apart to slide the unit into the wall. Once the pre-hung door is in place, the casing is nailed to the wall studs.

If job-hung doors are specified, each part of the doorway must be listed separately. If pre-hung doors are specified, the size, swing, style of door, style of hardware, and style of trim must be listed.

A

	COLONIAL STYLE CASING-				
	EXT. DOORS	6 pcs. 7'-0" 3 pcs. 3'-6"	LIN. FT.		
	DOOR STOP MOLDING -	6 pcs. 7'-0"			
	EXT. DOORS	3 pcs. 3'-0"	LIN. FT.		

B

	DOOR JAMBS	4⁰ x 6⁸		2
	DOOR JAMBS	2⁶ x 6⁸		1
	DOOR JAMBS	2⁸ x 6⁸		1
	DOOR JAMBS	2⁰ x 6⁸		5
	6-PANEL, INT. DOORS	2⁰ x 6⁸ x 1⅜		5
	COLONIAL STYLE			
	CASING DOORS		LIN. FT.	223
	DOOR STOP MOLDING		LIN. FT.	85

C

	DOOR JAMBS	2⁶ x 6⁸		1
	DOOR JAMBS	2⁴ x 6⁸		1
	DOOR JAMBS	3⁰ x 6⁸		2
	DOOR JAMBS	2⁰ x 6⁸		3
	DOOR JAMBS	1⁰ x 6⁸		1
	6-PANEL, INT. DOORS	2⁶ x 6⁸ x 1⅜		3
	6-PANEL, INT. DOORS	2⁴ x 6⁸ x 1⅜		1
	6-PANEL, INT. DOORS	3⁰ x 6⁸ x 1⅜		2
	6-PANEL, INT. DOORS	2⁰ x 6⁸ x 1⅜		3
	LOUVER DOOR	1⁰ x 6⁸ x 1⅛		1
	COLONIAL STYLE			
	CASING - DOORS		LIN. FT.	340
	DOOR STOP MOLDING		LIN. FT.	170

		D														
CHROME CLOSET RODS—																
ADJUSTABLE			6													

REVIEW QUESTIONS

A.   Select the letter preceding the best answer.

1. What is the most common thickness for exterior doors?

   a. 1 3/8 inches   c. 1 3/4 inches
   b. 1 1/2 inches   d. 2 inches

2. What is the most common thickness for interior doors?

   a. 1 3/8 inches   c. 1 3/4 inches
   b. 1 1/2 inches   d. 2 inches

3. Where is the manufacturer of the doors for a house indicated?

   a. Floor plan   c. Specifications
   b. Elevations   d. None of these

4. Approximately how many feet of casing are required for an interior doorway?

   a. 15   c. 21
   b. 17   d. 34

5. Approximately how many feet of stop are required for a doorway?

   a. 15   c. 21
   b. 17   d. 34

6. How many panels does a colonial door have?

   a. 2   c. 6
   b. 4   d. 8

7. Which of the following is the proper order for listing the dimensions of a door?

   a. Thickness, width, height   c. Height, thickness, width
   b. Width, thickness, height   d. Width, height, thickness

B.   1. What is the swing of the door shown in figure 31-4?

   2. What is the width of the door shown in figure 31-4?

   3. What is the thickness of the door shown in figure 31-4?

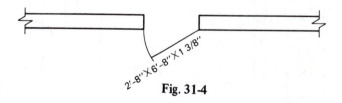

2'-8"X6'-8"X1 3/8"

Fig. 31-4

C. Using the following given information, construct a material list for the doors in figure 31-5.   All doors are to be flush with birch veneer, except closet sliding doors. Closet sliding doors are to be louvered. Interior doors are to be pre-hung and exterior doors are to be job-hung. Remember to list the swing of the doors, where necessary. It is not necessary to list the style or size of casing and stop for this problem. to list the style or size of casing and stop for this problem.

Quantity	Size (and swing)	Material
a._____	_____	Exterior door frame w/sill for front door
b._____	_____	Exterior door frame w/sill for side door
c._____	_____	Exterior birch flush door for front door
d._____	_____	Exterior birch flush door for side door
e._____	_____	Set of door jambs for cased opening between living room and dining area
f._____	_____	Pre-hung birch door for hall closet
g._____	_____	Pre-hung birch doors for bedrooms
h._____	_____	Pre-hung birch door for linen closet
i._____	_____	Pre-hung birch door for bathroom
j._____	_____	Pre-hung louver sliding door units for bedroom closets
k._____lin. ft.		Casing
l._____lin. ft.		Stop

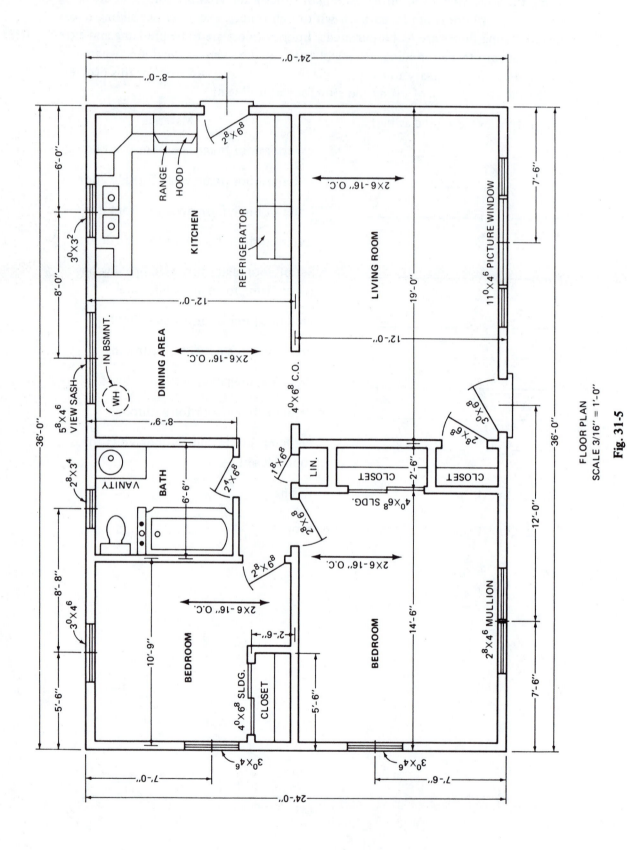

FLOOR PLAN
SCALE 3/16" = 1'-0"

Fig. 31-5

unit 32 windows

OBJECTIVES

After studying this unit the student will be able to

- describe four styles of windows and the combinations in which they are commonly used.

- determine the sizes and styles of windows from working drawings and specifications.

- determine the quantity of trim required for the windows in a house.

TYPES OF WINDOWS

There are too many sizes and styles of windows available to list them all in this textbook. The following is a brief description of the windows used most commonly in house construction.

Double-hung windows, figure 32-1, consist of two *sash* (the frame holding the glass) which slide up and down in the *jambs* (the side members of the window unit). The cross member where the two sash meet is called the *meeting rail* or *check rail.* The sash may be locked in the closed position by a *sash lock* installed on the check rail. Double-hung windows have *spring balances* which help relieve the weight of the sash for ease in opening the window.

Sliding windows, figure 32-2, are similar to double-hung windows, except that they slide from side to side instead of up and down. Because there is no weight to overcome in opening sliding windows, they do not normally have balance mechanisms.

Awning windows, figure 32-3, open out from the bottom of the sash. They may be operated by either a push bar or cranking mechanism.

Casement windows, figure 32-4, are similar to awning windows except that they open from the side instead of the bottom. Some manufacturers produce units which can be installed as either casement or awning windows.

The various types of movable sash are often used in combination with fixed sash (those which cannot be opened) to produce picture windows, bow windows, bay windows, and mullion windows. Picture windows, figure 32-5, have a large fixed

Fig. 32-1 Double-hung window

Fig. 32-2 Sliding window

sash in the center and usually some type of movable sash on the sides for ventilation. Bow windows, figure 32-6, curve out from the face of the house with a series of small fixed or awning windows or casement windows. Bay windows, figure 32-7, are similar to bow windows except that they do not form as uniform a curve. They consist of three windows, usually double-hung, at 30-degree or 45-degree angles with each other. A mullion is the vertical piece between two windows when they are mounted side-by-side. Windows with mullions are called mullion windows, figure 32-8.

Windows come from the manufacturer with weather stripping and sash installed. Normally double hung and sliding windows are supplied with sash locks installed as well. The glass may be either *standard glazing* (a single piece of glass) or *insulated glazing* (two or three pieces of glass with space between them), figure 32-9. Insulated windows eliminate the need for a storm sash.

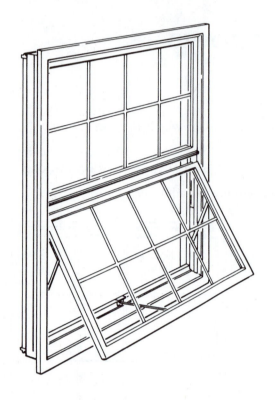

Fig. 32-3 Awning window

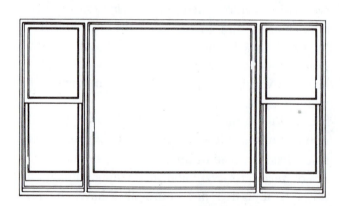

Fig. 32-5 Picture window

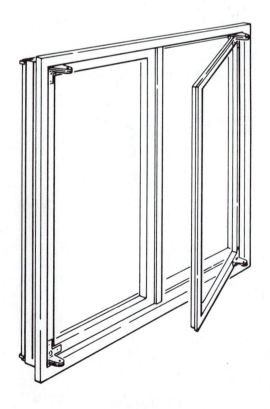

Fig. 32-4 Casement window

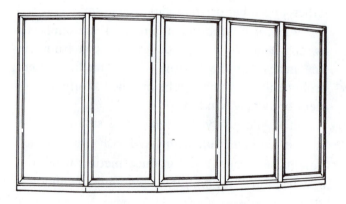

Fig. 32-6 Bow window

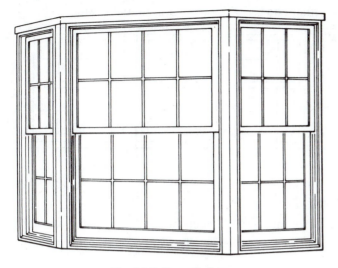

Fig. 32-7 Bay window

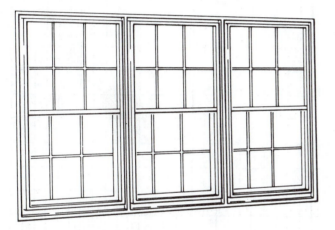

Fig. 32-8   Windows that are separated by a strip of wood (mullion) are called mullion windows.

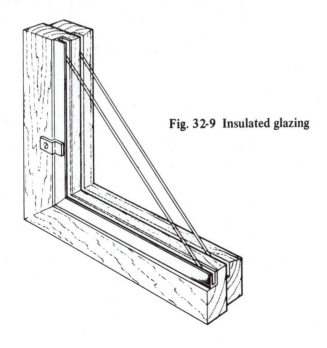

Fig. 32-9 Insulated glazing

WINDOW SIZES

Windows and doors are always listed with the first dimension shown being the width and the second dimension being the height. For example, a 3'-0" x 4'-6" window is three feet wide and four feet-six inches high. The symbol used on the drawing indicates whether it is a single, mullion, or triple unit.

Usually the size of a window is referred to by its nominal size. For example, double-hung windows may be available in sizes ranging from 1'-8" x 2'-6" to 3'-8" x 4'-6". The sizes increase in multiples of two inches in width and four inches in height.

Sometimes the size of a window is referred to by its glass size. To convert from glass size to nominal size; double the glass size (for a double-hung window) and add six inches to the height. For example, if the glass size is shown as 24/24 the window is 24 plus 4 or 28 inches wide. The height is 6 plus two times 24 or 54 inches. Therefore, a 24/24 glass size window is 2'-4" x 4'-6". These figures are intended only as a guide. Each manufacturer may vary slightly. The manufacturer's catalog should always be checked for sizes and styles of windows.

In addition to size and type of window, the estimator must know the number of lights (panes of glass) to be installed in each window. On double-hung windows this is given as a fraction with the top number being the number of lights in the top sash and the bottom number being the number of lights in the bottom sash, figure 32-10. Some windows have removable grilles which simulate small lights, but have the advantage of being easier to clean.

Estimating Windows

The number and size of the windows are shown on the first and second floor plans. The specifications indicate that they are to be double-hung windows with removable grilles. Wood combination storm sash and screens are to be furnished with each window also. Looking at the floor plans

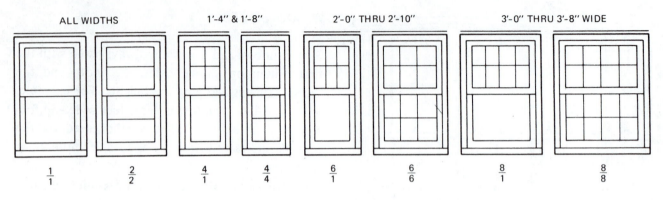

ALL WIDTHS | 1'-4'' & 1'-8'' | 2'-0'' THRU 2'-10'' | 3'-0'' THRU 3'-8'' WIDE

$\frac{1}{1}$ $\frac{2}{2}$ $\frac{4}{1}$ $\frac{4}{4}$ $\frac{6}{1}$ $\frac{6}{6}$ $\frac{8}{1}$ $\frac{8}{8}$

Fig. 32-10

for the sample house, the following windows are listed:

D.H. WINDOW w/ REMOVABLE GRILLES	$3^0 \times 4^6$	5
D.H. WINDOW w/ REMOVABLE GRILLES	$1^8 \times 3^{10}$	1
D.H. MULLION WINDOW w/ REMOVABLE GRILLES	$4^2 \times 3^2$	1
D.H. WINDOW w/ REMOVABLE GRILLES	$2^8 \times 4^6$	9
COMBINATION STORM & SCREEN	$3^0 \times 4^6$	5
COMBINATION STORM & SCREEN	$1^8 \times 3^{10}$	1
COMBINATION STORM & SCREEN	$2^0 \times 3^2$	2
COMBINATION STORM & SCREEN	$2^8 \times 4^6$	9

Notice that the list for combination storm sash and screens is different from the list of windows. This is because the mullion window includes two double-hung units, so an extra combination unit must be ordered.

Estimating Window Trim

Window units come complete, except for the interior trim and are in category of "standing trim." After the window is installed the stool, apron, casing, stop and mullion casing must be

approximately 1/2 inch beyond the casing at each side of the window. To find the length of stool installed, figure 32-11. The window stool extends needed, add 6 inches to the width of windows. Another two inches must be added for each mullion window. On the sample house there will be 50'-10'' of window stool listed by specific sizes.

The window apron is only slightly shorter than the stool, so include the same specific size of apron on the estimate. The casing is applied to both sides and the head of the window. The amount of casing required for the heads of the windows is the same as the amount of stool or apron. Casing size is 6'' longer than window.

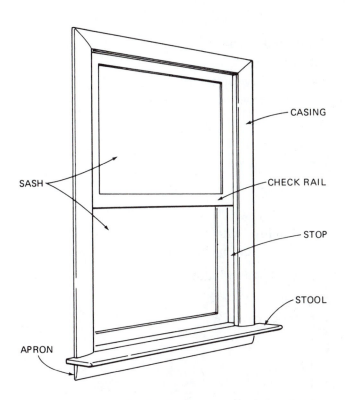

SASH

CASING

CHECK RAIL

STOP

STOOL

APRON

Fig. 32-11 Parts of a double-hung window and adjoining molding

Where the two units of the mullion window meet, a piece of mullion casing is used for trim. This is a flat piece of wood that is 1/2 inch thick by 1 7/8 inches wide. It extends from the top of

the stool to the underside of the head casing. A piece 3'-6" long is sufficient for the kitchen mullion window. Add the window trim to material list (A).

The front elevation shows shutters on the windows. The front windows are the only ones with shutters. Add these to material list (B).

SUMMARY

The basic types of windows named according to the way they open are double hung, sliding, awning, casement, and fixed. These may be used in combinations to make up window walls, picture windows, bow windows, and bay windows. Windows are supplied by the manufacturer completely assembled with outside casing, weather stripping, and sash locks. In addition they may be ordered with screens and storm sash, removable grilles, and insulated glass.

The size and type of windows to be used can be found on the floor plan or in a separate window schedule. The width of the window is always the first dimension shown and the height is the second.

After the window is installed the interior trim is applied. This consists of stool, apron, stop, casing, and mullion casing (on mullion windows). These materials are sold by the linear foot.

A

WINDOW STOOL				LIN.FT.	56										
WINDOW APRON				LIN.FT.	56										
COLONIAL STYLE CASING				LIN.FT.	210										
MULLION CASING				LIN.FT.	3½										

B

WOOD SHUTTERS		4'-6" LONG	PR.	5										

REVIEW QUESTIONS

A. Select the letter preceding the best answer.

1. What is the first dimension listed for the size of a window?

    a. Height   c. Width

    b. Width   d. Area

2. What is a double-hung window?

    a. One which is fastened from both the inside and outside
    b. One with two sash which slide up and down
    c. One with two thicknesses of glass and a space between them
    d. One which may be installed in either one of two ways

3. What is the nominal size of a double-hung window with a glass size of 24/28?

a. 2'-0" x 4'-0"	c. 2'-4" x 4'-8"
b. 2'-4" x 5'-2"	d. 2'-4" x 2'-8"

4. What does $\frac{2}{2}$ indicate about a window?

a. A double-hung window	c. A window which is 2'-0" x 2'-0"
b. Insulated glass	d. A window with two lights on top and two lights on the bottom

5. Which of the following parts normally comes with the window?

a. Interior casing	c. Apron
b. Stool	d. Exterior casing

6. How much casing does a 2'-6" x 4'-6" window require?

a. 12 feet	c. 14 feet
b. 8 feet	d. 15 feet

7. How much apron does a 2'-8" x 4'-6" window require?

a. 3'-2"	c. 4'-6"
b. 2'-8"	d. 5'-0"

8. How much stool does a 2'-8" x 4'-6" window require?

a. 10'-0"	c. 3'-2"
b. 2'-8"	d. 4'-6"

9. Where is information about the windows found?

a. Specifications	c. Elevations
b. Floor plans	d. All of these

B.   Using the following given information, construct a material list for the windows in the house shown in figure 32-12.

Quantity	Size	Material
a._____	____x____x____	Picture window unit — side windows are D.H. $\frac{2}{2}$; center sash in insulated glass
b._____	____x____x____	Insulated glass view sash
c._____	____x____x____	Mullion D.H. window unit $\frac{2}{2}$
d._____	____x____x____	D.H. window for kitchen
e._____	____x____x____	D.H. window for bathroom
f._____	____x____x____	D.H. windows for bedrooms
g.____lin. ft.		Casing
h.____lin. ft.		Stool
i.____lin. ft.		Apron

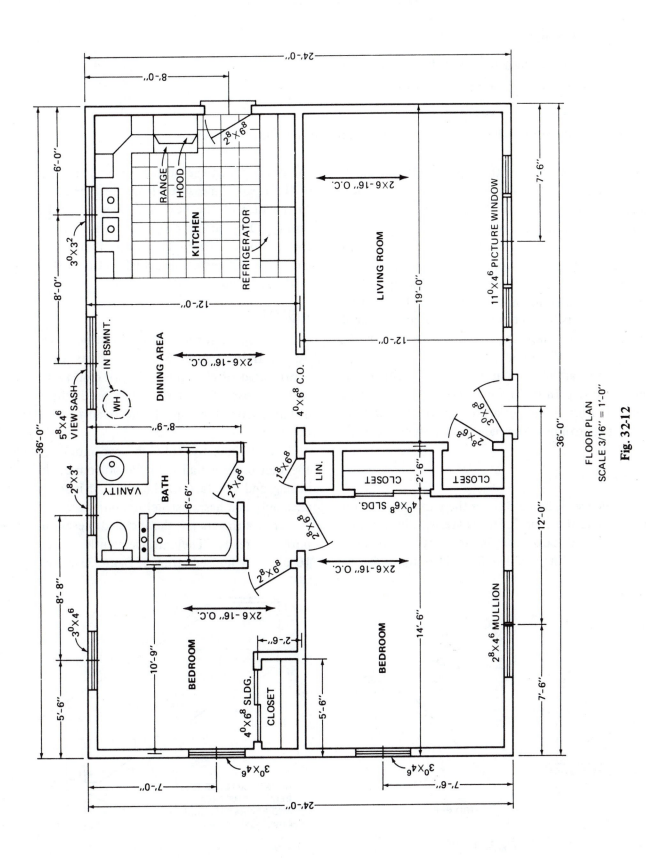

FLOOR PLAN
SCALE 3/16" = 1'-0"

Fig. 32-12

unit 33 stairs

OBJECTIVES

After studying this unit the student will be able to

- identify the various parts of stairs.
- describe the construction of stairs.
- determine the quantity of material required to build a stairway.

All houses, except those with only one story and no basement, have a stairway. Normally when the cellar is unfinished the stairs from the first floor to the cellar are of rough construction with open risers. This means that there is no piece covering the opening from one step (*tread*) to the next. The *stringers* (pieces at the sides supporting the stairs) are commonly made of 2 x 10's on this type of stairs; sometimes 2 x 12's are required so that the notch does not reduce the material to less than 3 1/2" per code.

Each part of the stairway has a specific name and purpose. In order for an estimator to discuss stairs and list the materials to construct them, it is important to know the name and purpose of each part.

TREAD

The tread, figure 33-1, is the part one steps on when using the stairs. Treads are usually made of 1 1/16 inch thick oak or pine and may be purchased in several widths and lengths. The front edge of the tread, called the nosing is rounded. The width of the tread is measured from the face of one riser to the face of the next and does not include the nosing. The nosing projects 1 1/8 inches beyond the face of the riser. This means that if a stairway has 10-inch treads, the actual width of the treads is 11 1/8 inches.

RISERS

The space at the back of the treads is enclosed by the risers, figure 33-1. The risers are usually constructed of clear pine. Do not confuse the riser with the *rise of the stairs*. The rise is the vertical distance from the top of one tread to the top of the next.

A 5/8-inch by 3/4-inch cove molding, called the stair cove, is nailed to the bottom of the tread nosing and the face of the riser. It is for appearance only and has no structural purpose.

The stringers are the inclined pieces that support the treads. They are frequently made of 1 x 10 pine. A housed stringer, figure 33-2, has *dados*

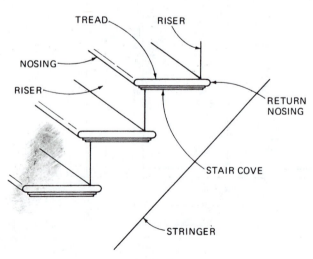

Fig. 33-1

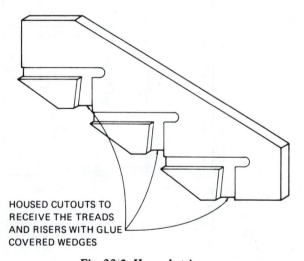

HOUSED CUTOUTS TO
RECEIVE THE TREADS
AND RISERS WITH GLUE
COVERED WEDGES

Fig. 33-2 Housed stringer

(grooves) cut 1/2 inch deep on the inside to receive the treads and risers. These grooves are tapered, so wedges can be glued into them after the treads and risers are in place. This assures a tight fit.

One type of construction uses a *carriage*, figure 33-3, to support the treads and risers. The carriage is cut from a 2 x 10 and is covered by a skirtboard of 1-inch pine. Housed stringers are generally recommended however, as they are self-supporting and are stronger.

TYPES OF STAIRS

Stairs may be either closed, half-open, or open. They may be *straight*, have *landings*, or turn with *winders*. Straight stairs are those which do not turn from top to bottom. A landing is a platform partway up the stairs and is generally used where the stairs make a turn. Winders, figure 33-4, are tapered treads, used to make a turn without a landing. Closed stairs are those with a partition or wall at both sides, concealing the ends of the treads. Open stairs, figure 33-5, have one or both ends of the treads exposed to view and half-open stairs are those which are open for part of the run and closed for part of the run. On half-open stairs two to five treads, not counting the starting step, are left open.

Half-open and open stairways require more parts than do closed stairways. The stringer is mitered where the riser is joined to it, figure 33-6,

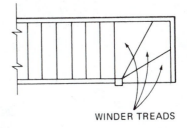

WINDER TREADS

Fig 33-4

and the tread nosing is continued around the open end of the tread. On stairs with a housed stringer, such as closed stairs, the cove molding extends from one stringer to the other. On open stairs the cove molding follows the return nosing around the end of the tread.

Open stairs frequently use a special *starting step*, figure 33-5, to which the newel post is attached. Starting steps are available in several styles, including circle end, quarter circle, and scroll. They are reversible, so they can be used for either right open or left open stringers. The vertical pieces which support the handrail at each tread are called *balusters*, figure 33-5. There are usually two or three balusters on each tread and they are doweled into the tread and the handrail. Balusters are available in 30, 33, 36, and 42-inch lengths. If two balusters are used on each tread they should be 30 inches and 33 inches long. If three are used they should be 30 inches, 33 inches, and 36 inches long. Balusters under volutes should be 36 inches long, under goosenecks they are 42 inches long, and under the railing at the second floor they are 33 inches long, figure 33-5. These dimensions will allow the handrail to be 30 inches high above the treads and 34 inches high on the level.

If the stairway includes a landing, a landing nose must be installed at its edge. Landing nosing is 1 1/16 inches thick by 3 1/2 inches wide with a rabbet to allow it to match 13/16-inch flooring.

Estimating Stairs

Information pertaining to the stairs is included in Division 8, Section H of the specifications for the sample house and on detail drawings, figure 33-7, and floor plans with the working drawings.

Most builders purchase factory-built stairs from a millwork manufacturer. In this case it is

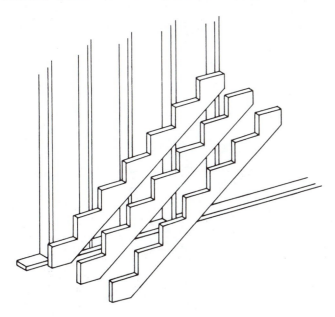

Fig. 33-3 Stair carriage

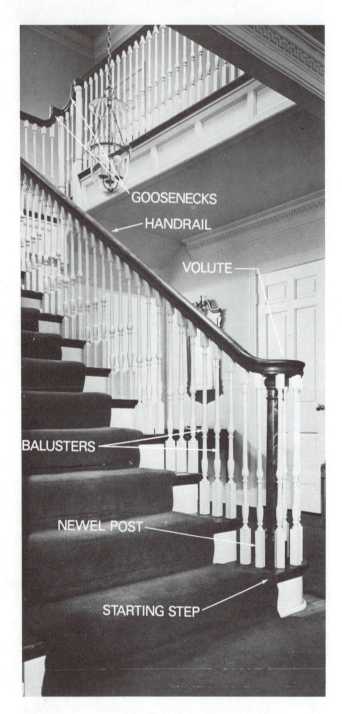

Fig. 33-5 Open stairs

addition to the starting steps there are twelve risers. The stairs are of housed-stringer construction, so two sets of housed stringers are required. Treads do not have to be ordered for the landings, starting steps, or second floor, so nine oak treads are required. Three pieces of landing nosing are needed for the landings and edge of the second floor. The handrail includes twenty feet of birch railing, two volutes for the bends, and six handrail wall brackets. Remember to include the wedges and glue used to assemble the stairs. This material list is shown at the top of the next page.

Normally basement stairs are in a straight run and are built from stock material. The following is a list of materials needed to construct a plain stairway to a seven-foot basement:

2 pcs. 2 x 10 x 14′ stringers
4 pcs. 2 x 10 x 12′ treads
1 pc. 2 x 10 x 12′ tread
4 pcs. 1 x 8 x 12′ risers (if required)

The cellar stairs in the sample house are not in a straight run. The sample plan shows two landings with the stairs resting on these platforms. To provide strong support for the stairs, each platform is framed using 2 x 4's for the shoes, plates, studs, and joists. Plywood is used for the subflooring and

only necessary to list the size and style of the stairs. The sample house being estimated here requires the carpenter to build the stairs. In estimating these stairs the student will become more familiar with the construction and terminology.

Looking at the first floor plan and the detail drawing, it can be seen that there are two starting steps. These are supplied complete with risers. In

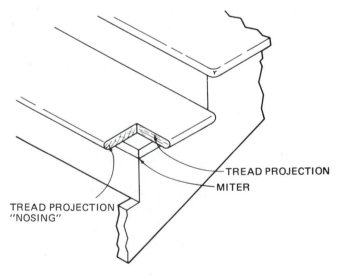

Fig. 33-6 Mitered stringer and riser

QUARTER - CIRCLE				
STARTING TREADS			2	
PINE STAIR RISERS	¾"x7½"x3'		12	
HOUSED STRINGERS - 9 RISERS			2	
OAK STAIR TREADS	1¹⁄₁₆"x10"x3'		9	
OAK NOSING		LIN.FT.	3	
COVE MOLDING	¾"	LIN.FT.	36	
BIRCH HANDRAIL		LIN.FT.	20	
BIRCH VOLUTES			2	

pine is used for the risers and treads of the stairs. The number of risers and treads can be found on the Basement Plan. A handrail and handrail hardware must also be included. Add the following items to the material list for the basement stairs.

SUMMARY

The basic parts of stairs are treads, risers, and stringers. The treads are the tops of the steps. Treads are normally 1 1/16 inch thick oak with the exposed edges rounded to form the nosing. The

FIR-CELLAR STAIRS	2x4x12'		6	
FIR-CELLAR STAIRS	2x4x8'		12	
FIR-CELLAR STAIR PLATES	2x4x12'		3	
FIR-CELLAR STAIR STUDS	2x4x8'		12	
FIR-CELLAR LANDING JOIST	2x4x12'		3	
CD PLYWOOD-LANDING FLOOR	⅝"x4'x8'		1	
PINE-CELLAR STAIR RISERS	1x8x12'		3	
PINE STAIR TREADS	3'-0"		10	
PINE HANDRAIL		LIN.FT.	18	
HANDRAIL BRACKETS			8	
WEDGES FOR STAIRS			42	
WHITE GLUE	PINT		1	

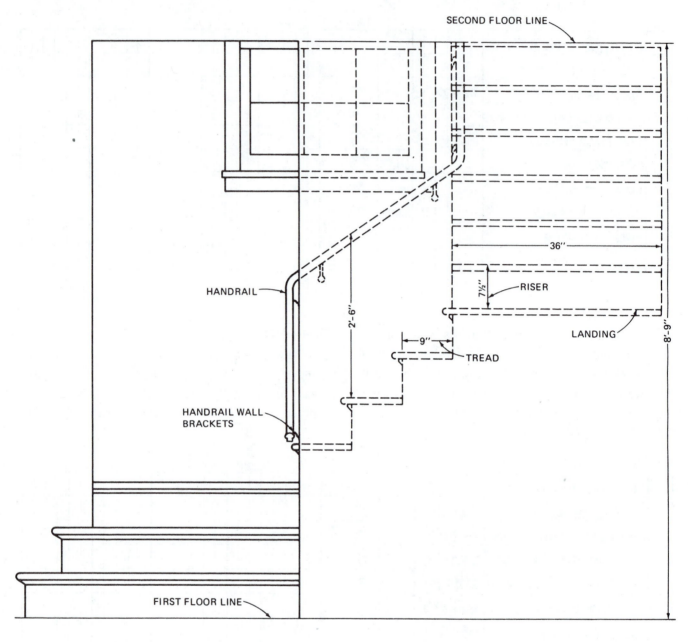

SECOND FLOOR LINE

HANDRAIL

36"

7½" RISER

2'-6"

LANDING

8'-9"

9"

TREAD

HANDRAIL WALL
BRACKETS

FIRST FLOOR LINE

SCALE 3/4" = 1'-0"

Fig. 33-7 Main stair details

risers are the pine boards which enclose the vertical space between the treads. The treads and risers are held in place by the stringers, which are usually made of pine. Housed stringers are generally considered strongest, but for open stairs, a carriage may be used underneath the treads and risers. On open stairs, newel posts and balusters are used to support the handrail. In addition, a special starting step is used with open stairs.

Information pertaining to stair construction is found in the specifications under millwork and on the working drawings. The drawings usually include a stair detail drawing. The number of risers is normally given on a floor plan or it can be counted on the special detail. Most stair parts are available as standard items from millwork manufacturers.

REVIEW QUESTIONS

Select the letter preceding the best answer.

1. What is the name of a stringer which covers the ends of the treads?

   a. Open   c. Baluster
   b. Housed   d. Riser

2. What kind of wood are stair treads usually made of?

   a. Pine   c. Redwood
   b. Birch   d. Oak

3. What kind of wood are risers and stringers usually made of?

   a. Pine   c. Redwood
   b. Birch   d. Oak

4. What is the proper term for the vertical distance from the top of one tread to the top of the next?

   a. Pitch   c. Climb
   b. Tread   d. Rise

5. What is the name of a tapered tread which allows the stairs to go around a corner?

   a. Winder   c. Newel
   b. Turn tread   d. Corner tread

6. How many treads does a stairway with 15 risers have?

   a. 13   c. 15
   b. 14   d. 16

7. How long should the balusters be if there are two on a tread?

   a. 33 inches and 36 inches   c. 36 inches and 42 inches
   b. 30 inches and 36 inches   d. 30 inches and 33 inches

8. Which of the following might be the size of the lumber used to construct a stair carriage?

   a. 2 x 10   c. 1 x 6
   b. 1 x 10   d. 2 x 8

9. Which of the following uses wedges?

   a. Open stringer   c. Carriage
   b. Half-open stringer   d. Housed stringer

10. Which of the following parts is a newel post attached to?

    a. Starting step   c. Riser
    b. Baluster   d. Housed stringer

unit 34 ceramic tile

OBJECTIVES

After studying this unit the student will be able to

- explain the specifications for ceramic tile.
- describe the application of ceramic tile.
- determine the quantity of ceramic tile required for a house.

There are many materials which may be used on the bathroom walls and floors. The most common is ceramic tile, figure 34-1, because it is durable and easy to clean. Other materials that might be used on the bathroom walls are plastic laminates, vinyl-coated wallpaper, and paint. Carpeting or resilient flooring might be used on the floors. See Unit 35 for slate and resilient flooring. The material to be used is indicated in the specifications.

DIVISION 9: FINISHES

A. **WORK INCLUDED**

This work shall include but shall not be limited by the following:

1. Ceramic and quarry tile
2. Slate flooring
3. Resiliant flooring
4. Painting

B. **CERAMIC TILE**

All ceramic tile shall be standard grade of approved quality. The contractor shall submit samples of all ceramic and quarry tile to the owner for approval as to color and texture.

1. *Wall tile* shall be matte faced 4 1/4" x 4 1/4" in size. Setting shall be by conventional tile adhesive over waterproof wallboard. All cove, bullnose, and grim pieces shall be furnished and neatly finished to drywall construction according to the drawings.

2. *Floor tile* shall be unglazed, 1 inch x 1 inch, smooth mosaic tile, of a similar color to wall tile.

3. *Accessories* shall be vitreous china to match wall tile. The second-floor bath is to have a soap dish and grab bar, and one grab bar in the tub area.

4. *Marble thresholds* shall be installed the same width as the bathroom door opening. Thresholds shall be of approved quality, 7/8" thick and 3 1/2" wide.

5. *Installation*; Lay out ceramic tiles on floors and walls so that no tiles less than one-half size occur. Maintain full courses to produce nearest obtainable heights as shown on drawings without cutting the tile. Align joints in the wall and trim vertically and horizontally without staggered joints.

C. QUARRY TILE

1. Quarry tile for the fireplace hearth shall be 4 inch x 4 inch x 12 inch standard red.

2. Set quarry tile on approved mortar bed with 1/4 inch aligned joints in both directions. Mortar bed on cement base shall bring the tile surface to the level of the finished floors. Joints shall be tooled until smooth and flush with the face of the tile surface.

ESTIMATING CERAMIC TILE

According to the Second Floor Plan, the tub area is the only area to be covered with wall tile. The standard size for a bathtub is 2'-6'' wide by 5'-0'' long. This means that 10 linear feet of wall must be tiled. According to the specifications, the tile is to be set on waterproof wallboard. Each sheet of wallboard is 4 feet wide, so three sheets are required for the wall area to be tiled. Add this to the material list.

tiply this height by the linear feet to be covered to find that 60 square feet of wall tile is required.

Tile caps are used to finish the exposed edge of the ceramic tile. These are 2-inch by 6-inch tiles with a rounded edge, figure 34-2. The cap runs along both ends and across the top of the tiled area. This means that 22 feet of cap is needed (6 feet for each end and 10 feet for the top edge). Two inside corners are needed where the vertical edge meets horizontal edge. The tiles are set with tile adhesive.

WATERPROOF GYPSUM					
WALLBOARD	½"x4'x8'	SHTS.	3		

Ceramic tile is sold by the square foot, so find the area of the walls to be covered. The tile is to be applied to a height six feet above the tub. Mul-

The specifications also call for a vitreous china grab bar and a soap dish with a grab bar. These are to be the same color as the tile.

When the tiles are in place, the joints between them are to be filled with white *grout* (special cement for tile joints). The amount of needed

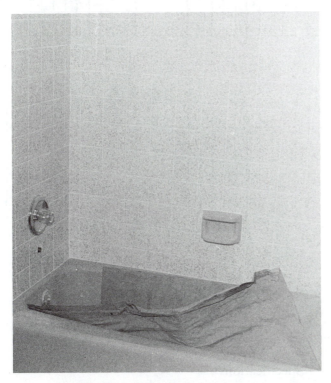

Fig. 34-1 Ceramic tile in tub area

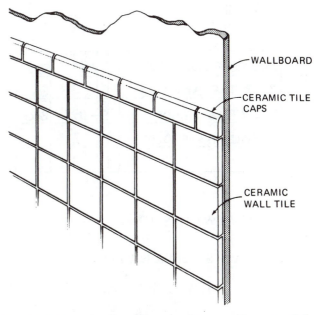

Fig. 34-2 Ceramic tile caps are used to provide a rounded edge at the top or end of a tiled area.

grout varies depending on the size of the joints and the way the grout is used, but 5 pounds should be sufficient for 60 square feet of tile. Add these to material list (A).

The bathroom floor is also to be covered with ceramic tile. These tiles are different from those used on the walls, but the method of application is the same. The bathroom floor is 7 feet by 10 feet, so 70 square feet of floor tile is needed. In addition, the doorway is to have a marble threshold.

Add the materials for the ceramic floor to material list (B).

SUMMARY

Ceramic tile is available in an assortment of colors and patterns and may be either glazed — for use on walls, or unglazed — for floors. Accessories, such as toothbrush holders, soap dishes, and grab bars, are available in matching colors. When ceramic tile is used for the floor, a marble threshold is normally placed in the doorway. Tile adhesive is used to secure the tile and accessories to the wall or subfloor. Once the tile is set in place, the joints between the individual tiles are filled with grout.

A

CERAMIC WALL TILE	4¼ x 4¼	SQ. FT.	60	
TILE ADHESIVE		GAL.	1	
CERAMIC TILE CAPS		LIN. FT.	22	
CERAMIC TILE CAP–				
INSIDE CORNERS			2	
CERAMIC TILE CAPS–			2	
OUTSIDE CORNERS				
CHINA SOAP DISH				
W/ GRAB BAR			1	
CHINA GRAB BAR			1	
CERAMIC TILE GROUT	5-lb. BAG		1	

B

MOSAIC FLOOR TILE	1" x 1"	SQ. FT.	70	
FLOOR TILE ADHESIVE		GAL.	1	
FLOOR TILE GROUT	5-lb.	BAG	1	
MARBLE THRESHOLD	3½" x 2'-4"		1	

REVIEW QUESTIONS

A.   Select the letter preceding the best answer.

1. Where is the material to be used on the bathroom walls indicated?

a. Floor Plan   c. Specifications
b. Elevation   d. Special detail drawings

2. What is the smallest part of a tile that should be allowed when laying out ceramic tiles?

   a. 1/2 of a tile

   b. 1/4 of a tile

   c. 3/4 of a tile

   d. All tiles should be full-size.

3. Which of the following measures is used for estimating ceramic tile?

   a. Cubic feet

   b. Square feet

   c. Linear feet

   d. Piece

4. Which of the following is not characteristic of ceramic tile?

   a. It is attractive.

   b. It is durable.

   c. It is inexpensive.

   d. It is easy to clean.

5. If only the bottom portion of a wall is tiled, what is used at the top edge of the tilework?

   a. Tile edging

   b. Cap tiles

   c. Marble trim

   d. The tiles form a decorative edge.

6. When ceramic tile is used on a floor what material is usually used for the thresholds?

   a. Oak

   b. Stone

   c. Marble

   d. Thresholds are not used with ceramic tile.

7. How are ceramic tiles attached to a wall?

   a. Special adhesive

   b. Grout

   c. Special mortar

   d. None of these

8. What is used to fill the joints between ceramic tiles?

   a. Special adhesive

   b. Grout

   c. Common mortar

   d. Wax

B. Using the following given information, construct a material list for tiling the tub area in figure 34-3 The tile is to extend six feet above the tub.

Quantity	Material
a._____sq. ft.	Ceramic wall tile
b._____lin. ft.	Ceramic tile cap
c._____	Inside corner cap tiles
d._____	Outside corner cap tiles
e._____gallons	Ceramic tile adhesive.
f._____lbs.	Tile grout

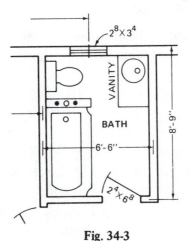

Fig. 34-3

unit 35 slate and resilient flooring

OBJECTIVES

After studying this unit the student will be able to

- describe the application of slate and resilient flooring.
- explain the specifications for flooring materials.
- determine the quantity of material required to cover floors with slate or resilient flooring.

There are a great number of materials which are commonly used for floors. The most common materials are hardwood, carpet, ceramic tile, vinyl tile, vinyl asbestos tile, and slate. The use for which the room is intended is an important factor in deciding which material to use.

Although hardwood flooring is a popular material for many rooms, it is not normally used in entrances. Hardwood may be stained from the mud and water which are tracked into the entrance and the heavy traffic can cause hardwood to show wear. The most common materials for an entrance floor are brick, quarry tile, slate, stone, and vinyl tile. These are durable and are not affected by water. See Unit 34 for ceramic tile.

Division 9, Section D of the specifications for the sample house indicates that the entrance floor is to be slate. Slate, figure 35-1, is set in much the same manner as ceramic tile; individual pieces of slate are set in adhesive, then the joints are grouted.

Fig. 35-1 Slate floor

D. SLATE FLOORING

1. The front entrance floor shall be 3/8" thick slate flooring and shall be laid in a random pattern.

2. Slate shall be set in conventional slate adhesive with 3/8-inch joints. Slate flooring shall be placed in both the entrance foyer and the closet adjacent to the entrance.

Estimating Slate Flooring

Slate flooring is sold by the square foot. To determine how much is needed, find the dimensions on the floor plan. The front entrance foyer, including the closet, is 5 feet by 7 feet. One gallon of adhesive will apply 100 square feet of slate flooring. Five pounds of white grout, similar to that used for ceramic tile will also cover 100 square feet. Add the materials for the slate floor to the material list as shown at the top of the next page.

RESILIENT FLOORING

Kitchen floors are usually covered with either carpeting or some type of resilient flooring. Indoor/outdoor carpeting is popular because it resists staining and may be easily cleaned.

SLATE FLOORING	3/8"	SQ.FT.	35												
SLATE ADHESIVE		GAL.	1												
GROUT - SLATE FLOORING	5 LB.	BOX	1												

Resilient flooring (vinyl tile) is also popular because of the ease with which it may be cleaned and the many attractive patterns that are available, figure 35-2.

Division 9, Section E of the specifications for the sample house indicates that the kitchen floor is to be vinyl tile. In addition, a vinyl base is to be installed in the kitchen.

for doorways. The perimeter of the kitchen is approximately 42 feet and it has three doorways, so 36 feet of vinyl base is required.

Vinyl tile adhesive covers approximately 175 square feet per gallon, so one gallon of adhesive is sufficient. Add the materials for the resilient flooring to the material list at the top of the next page.

E.   RESILIENT FLOORING

1. The floors in the kitchen, half bath, and closet shall be 1/8-inch thick, 12" x12" vinyl tile. Color and style are to be selected by the owner.

   The base in the kitchen shall be vinyl wall base.

2. Finished resilient flooring materials shall be installed with adhesives as recommended by the manufacturer.

Estimating Resilient Flooring

Most resilient floor covering materials are sold by the square foot. The specifications indicate which rooms are to have resilient flooring and the floor plan gives the size of these rooms. For the sample house these sizes are as follows:

Kitchen and half bath - 12 feet x 12 feet = 144 square feet

Cellarway and closet - 3 feet x 7 feet = 21 square feet

Cellarway closet - 1 1/2 feet x 6 feet = 9 square feet

This makes a total of 174 square feet, but it is not necessary to tile the area under the kitchen cabinets. The built-in cabinets cover an area of approximately 28 square feet, so deduct this from the overall area to find that 146 square feet of tile is needed. If sheet vinyl or carpeting is used, the estimator must be sure to allow for waste because these materials are generally sold in 6-foot and 12-foot widths only.

To find the amount of *vinyl base* (a 4-inch vinyl molding used in place of baseboard) needed, measure the perimeter of each room and deduct

In many cases the estimator is called upon to estimate the cost of materials which are sold by the

Fig. 35-2 Resilient flooring is frequently used on kitchen, bathroom, and utility room floors.

VINYL TILE - KITCHEN	12"x12"		SQ.FT.	146												
VINYL TILE ADHESIVE			GAL.	1												
VINYL BASE - KITCHEN	4"		LIN.FT.	36												

square yard, such as carpeting and vinyl-sheet floor covering. To determine the amount of these materials required to cover a floor, find the number of square feet to be covered and divide by 9 (the number of square feet in a square yard). If carpeting is used, a pad is normally installed under it. If sheet vinyl is used, it is cemented down in much the same way as vinyl tile.

SUMMARY

The materials used most often for floors are hardwood flooring, carpeting, sheet vinyl, vinyl tile, vinyl asbestos tile, ceramic tile, and slate. The intended use of the room should be considered in selecting the flooring material. Areas which are subjected to heavy traffic or water should be covered with something other than hardwood. The most common floor covering for entrances is slate or tile. Slate is glued to the subfloor with a special adhesive and then grouted. In kitchens, bathrooms, and utility rooms resilient flooring is common. Resilient flooring is also cemented down with a special adhesive. Most of these materials are sold by the square foot. If the floor covering is a type which is sold by the square yard, the quantity is found by dividing the number of square feet by nine.

The type of flooring to be used is generally indicated in the specifications. The size of the rooms or areas to be covered can be determined from the floor plans.

REVIEW QUESTIONS

A. Select the letter preceding the best answer.

1. Which of the following is not used frequently for entrance floors?

    a. Slate   c. Quarry tile

    b. Brick   d. Hardwood

2. How is slate attached to the subfloor?

    a. Grout   c. Special adhesive

    b. Special mortar   d. Common mortar

3. What material is used to fill the joints between the individual pieces of slate?

    a. Grout   c. Special adhesive

    b. Special mortar   d. Common mortar

4. Which of the following materials is commonly used on kitchen floors?

    a. Indoor/outdoor carpeting   c. Sheet vinyl

    b. Vinyl tile   d. All of these

5. Where is the type of material to be used on a kitchen floor indicated?

    a. Floor plan   c. Special detail drawings

    b. Specifications   d. Elevations

6. How many square yards of vinyl are required to cover a kitchen floor measuring 12 feet by 15 feet?

    a. 180   c. 20

    b. 18   d. 6 2/3

7. How much adhesive is required to set 350 square feet of vinyl tile?

a. 1 gallon

b. 2 gallons

c. 3 gallons

d. 3 1/2 gallons

B. Using the following given information, construct a material list for the materials used to cover the kitchen floor in figure 35-3 with vinyl tile and 4-inch vinyl base. Vinyl base is to be installed only in the area which has vinyl tile.

Quantity	Size	Material
a._____sq. ft.	12″ x 12″	Vinyl floor tile
b._____gallons		Tile adhesive
c._____lin. ft.	4″	Vinyl base

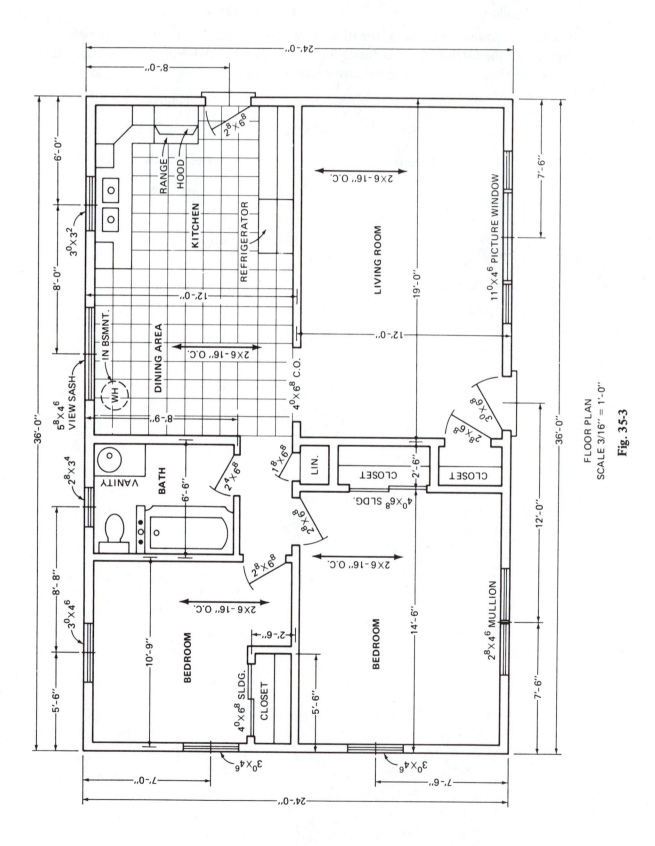

FLOOR PLAN
SCALE 3/16" = 1'-0"

Fig. 35-3

unit 36 painting and decorating

OBJECTIVES

After studying this unit the student will be able to

- explain the specifications for painting and decorating.
- describe the various types of painting and finishing materials used for houses.
- list the quantities of painting and finishing materials required for the construction of a house.

Before subcontractors can bid on a job they must estimate the cost of labor and materials for their part of the work. In many cases the general contractor relies on the subcontractor to estimate that portion of the work. However, this is not always true and the estimator should have knowledge of all phases of the construction process.

Division 9, Section 6 discusses the painting of the sample house.

Specifications

F. PAINTING

1. All exterior paint shall be nonchalking and shall be guaranteed not to stain or otherwise discolor any adjacent work. All materials shall be used only as specified by the manufacturer. All knots and sap spots shall be treated with two coats of pure white shellac where paint or enamel is to be applied.

2. Schedule of Painting

   a. Priming: Prime all exterior work which has not been primed by the manufacturer. Prime exterior trim with an approved trim primer before installation. All primer is to be of the same manufacturer as that of the finish material which is to be applied.

   b. Exterior woodwork: Paint all exterior doors, windows, and trim with two coats of approved trim and shutter paint. Paint all other exterior woodwork with two coats of an approved nonchalking, white, house paint.

   c. Interior woodwork: Paint all interior woodwork with one coat of enamel undercoating and one coat of semigloss enamel.

   d. Walls and ceilings: All walls and ceilings, except those in the bathrooms and kitchen, shall be painted with two coats of flat wall paint. The walls and ceiling in the bathrooms and kitchen shall be painted with two coats of semigloss wall paint.

It is quite common for specifications to indicate the name of a paint manufacturer and specific grades of materials to be used. This is done to indicate the quality expected; not to restrict the contractor from using other brands. Most manufacturers supply paint, varnish, and other finishing materials with approximately the same characteristics.

Paint used in building construction is an opaque (not transparent) material which is applied in a thin film for protection or decoration. Paint consists of two main parts: the *pigment*, which supplies the color; and the *vehicle*, which carries the pigment to the surface being painted. The vehicle may also contain a drier to quicken the

drying process. There are two general classifications of paint, depending on the vehicle used.

Oil-Based Paint

Oil-based paints use either linseed oil, soybean oil, or tung oil for the vehicle. Soybean oil is the most widely used, because of its nonyellowing characteristics. Mineral spirits or turpentine is normally used to thin oil-based paints and to clean equipment which is used with these paints. The following is an example of the contents of oil-based paint:

Pigment - 53.0%

Titanium Dioxide 13.5%
Calcium Carbonate 35.0%
Magnesium Silicate 4.5%

Vehicle - 47.0%

Alkyd Resin Solution 32.0%
Mineral spirits & driers 14.2%
Phenyl Mercury Oleate 0.8%

Water-Based Paint

These paints use a combination of water and synthetic resin for the vehicle. Latex was the resin used in the first water-based paints, therefore, they are sometimes referred to as "latex-based paint." even though other resins, such as acrylic, are more often used. These paints use pigments which are similar to those used in oil-based paints. One of the chief advantages of water-based paint is that is can be cleaned up with water, yet still provides a tough, water-resistant surface when dry. The following is an example of the contents of water-based paint:

Pigment - 36%

Titanium Dioxide 22%
Silicates. 12%
Tinting colors 2%

Vehicle - 64%

Acrylic Resin. 10%
Additives. 2%
Water. 44%
Ethylene Glycol 8%

Primer

Some surfaces hold paint better than others, so special *primer* may be used for the first coat. For example, wood grows more rapidly in the spring than in the summer, so the summer wood is less porous and does not hold paint as well. By painting the entire surface with a coat of primer, the paint has a uniform surface on which to adhere. Most paint manufacturers make primer especially for use with their paint. It is generally best to use primer and paint made by the same manufacturer.

Frequently, wood products contain knots and sap pockets which release sap or pitch and bleed through the finished paint, figure 36-1. This can be prevented by painting knots and sap pockets with shellac before the wood is painted. This seals the surface and prevents the pitch from bleeding through into the paint. Shellac can be thinned and cleaned up with denatured alcohol.

Stain

Frequently wood siding, shingles, and trim are stained a darker color. This is done by applying wood stain, which is similar to paint in that it con-

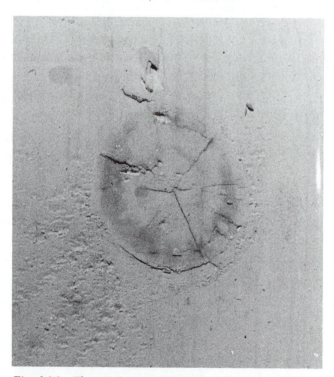

Fig. 36-1 The pitch or sap contained in knots can bleed through paint, unless it is sealed with shellac or some other sealer.

tains a vehicle and pigment, but it is not opaque. Stain darkens the wood while allowing the grain to remain visible. When woodwork is stained, it is normally varnished afterwards.

Varnish

Varnish is a combination of resin, oil, thinner, and drier which makes a durable transparent coating. Most varnish does not contain pigment, although additives may be included to make satin varnish. Satin varnish is transparent, but does not provide the high gloss that would result if no additives were included.

Some specifications call for the use of polyurethane, which is sometimes referred to as varnish. *Polyurethane* is a clear, plastic coating material which provides the appearance of varnish, but is even more durable than varnish. Most polyurethane and varnish can be cleaned up with turpentine or mineral spirits.

Coverage

Most paints and finishing materials cover about 450 square feet per gallon, if applied to unpainted wood; and about 600 square feet per gallon, if applied to a previously painted surface. This is only an approximation; more accurate coverage rates can be found by reading the manufacturer's label on the container. Before using any finishing material, it is important to read the manufacturer's directions carefully.

ESTIMATING PAINT

Siding

The area of the exterior walls of the house can be found from the number of sheets of plywood estimated for the sheathing. Each sheet of plywood covers an area of 32 square feet, so if the sample house used 64 sheets of plywood for the sheathing, the area of the walls is 2048 square feet (64 x 32 = 2048). Allowing 450 square feet per gallon for the primer, it takes 4 1/2 gallons to cover 2048 square feet. The specifications indicate that the siding and trim are to be primed on the back, so this figure must be doubled. Nine gallons of exterior primer are required for the siding.

The specifications indicate that the house is to be painted with two coats of house paint. The siding has already been covered once with primer, so the house paint should cover approximately 600 square feet per gallon. If the wall area of the house is 2048 square feet, it will require approximately 3 1/2 gallons for one coat (2048 ÷ 600 = 3.4 or 3 1/2) and 7 gallons for two coats.

Cornice

The box cornice is made up of 3 5/8" crown molding, 1 x 4 fascia, 1 x 8 soffit, 1 3/4" bed molding, and 1 x 8 frieze. Adding all of these together, the total width of the cornice is approximately 2 feet, so for every linear foot of cornice there are two square feet to be painted. The length of the cornice was found to be 72 feet when the bed molding was estimated, so the total area of the cornice at the eaves is 144 square feet (72 x 2 = 144).

The rake is made up of a piece of 5/4 x 6 stock and a piece of 3 5/8" crown mold, which amounts to approximately 6 inches when it is installed. This means that for every linear foot of rake, one-half of a square foot must be painted. The rafters on the sample house are 14 feet long and there are 3 gables, so the total length of the rakes is 84 feet (two 14-foot rafters on each gable). The area of the rakes is 42 square feet.

Adding the area of the box cornice and the area of the rakes, there is a total area of 186 square feet to be painted. This is to be primed on the back as well, so 1 gallon of primer is required (186 x 2 = 372; 1 gallon of primer covers 450 square feet). One gallon of house paint is sufficient to apply two coats to the outside of the cornice and rake.

Porch

The cornice at the eaves of the side porch is 18 feet long and the front porch has two 5-foot eaves, for a total of 28 linear feet of box cornice. Allowing 2 square feet for every linear foot of box cornice, the area to be painted is 56 square feet. The total of the rakes on the two porches is 30 linear feet or 15 square feet (1/2 square foot per linear foot). The porch posts are 5 inches on a side, so allow 1 1/2 square feet per linear foot of post (5 x 4 = 20 inches). There are five 8-foot posts, so the

area of the posts is 60 square feet (5 x 8 = 40; 40 x 1 1/2 = 60). The cornices are to be back primed, so the total area to be primed is 202 square feet (56 + 15 = 71; 71 x 2 = 142; 142 + 60 = 202). This requires approximately 2/3 of a gallon, so allow 1 gallon of primer on the estimate. The area of the porch trim to be covered with two coats of exterior paint is 131 square feet (56 + 15 + 60 = 131). At 600 square feet per gallon, two coats require 1/2 gallon of paint.

Windows and Doors

As a "rule of thumb" the sash and frame of one side of an average-size window is equal to about 25 square feet. Most doors are also about 25 square feet. The sample house has 17 windows and 3 exterior doors, for a total of 500 square feet (17 + 3 = 20; 20 x 25 = 500). Two gallons of paint are required for two coats on the windows and doors.

Add all of the paint for the exterior of the house to the material list.

interior doors which must be painted on both sides, so their area is 750 square feet (15 x 25 = 375; 375 x 2 sides = 750). The total area of doors and windows inside the house is 1250 square feet. In addition to this there are three sets of jambs and casings with no doors. Allow about 20 square feet for each opening with no door.

The baseboard and carpet strip are about 4 inches wide, so allow 1 square foot for every three linear feet. The baseboard has already been estimated as 566 linear feet, so the area is approximately 188 square feet (566 ÷ 3 = 188).

Add all of the interior trim areas to get a total of 1,498 square feet (1,250 + 60 + 188 = 1,498). The specifications indicate that interior trim is to have one coat of enamel undercoating and one coat of enamel finish. This requires 3 gallons of each type of enamel.

Hardwood Flooring

Oak flooring is normally sanded smooth, filled with wood filler, then varnished. The area of

EXT. PRIMER				GAL	11															
HOUSE PAINT				GAL.	9															
TRIM & SHUTTER PAINT				GAL.	2															

Interior Walls and Ceilings

All rooms, except the kitchen and bathrooms, are to have two coats of flat wall paint. The total wallboard area of the house was determined to be 6500 square feet when the wallboard was estimated. Subtract the area of the walls and ceilings in the bathroom and kitchen from the total wallboard area to find the area to be covered with flat wall paint. (6500 – 944 = 5556). Approximately 10 gallons of flat wall paint are required for each coat, so 20 gallons of this paint is added to the material list.

Interior Trim

The paint for the interior trim is estimated the same way as the paint for the exterior trim. The cove molding at the ceiling need not be included because this is normally painted with the ceiling. The area of exterior doors and windows has already been estimated as 500 square feet. There are 15

the hardwood floors can be estimated by subtracting the allowance for matching (1/4 of the total) from the total amount of flooring included on the material list. For the sample house, 1,460 square feet of flooring is listed, so the area of the floor to be varnished is approximately 1,095 square feet (1/4 of 1,460 = 365; 1,460 – 365 = 1,095). Wood filler covers about 600 square feet per gallon, so approximately 2 gallons are required. Varnish covers at about the same rate as paint, therefore 4 gallons of varnish are sufficient for two coats on the hardwood floors.

Add the paint and varnish for the interior of the house to the material list at the top of the next page.

SUMMARY

The amount of paint needed for a house is usually estimated by dividing the area to be covered by the coverage rate for the paint being used. Most of the areas to be covered have been measured before

				GAL.	20													
FLAT WALL PAINT				GAL.	20													
SATIN WALL PAINT				GAL.	3													
ENAMEL UNDERCOAT				GAL.	3													
ENAMEL PAINT				GAL.	3													
PASTE WOOD FILLER				GAL.	1½													
VARNISH				GAL.	4													

the paint is estimated. Areas which have not been previously determined can be scaled from the drawings. Although different materials cover at different rates, they are generally similar enough to be estimated at the same rate. Unpainted surfaces normally require approximately one gallon for every 450 square feet. Previously painted or primed surfaces usually require one gallon for every 600 square feet.

Different finishes and different grades of paint are normally specified for different parts of the house. The specifications include a schedule of paints and finishing materials. The estimator must consult this schedule to determine what is needed.

REVIEW QUESTIONS

A. Select the letter preceding the best answer.

1. What is the reason for specifying the brand name of paint in the specifications?

   a. To ensure that the proper brand is used
   b. To indicate the quality that is to be used
   c. To ensure that the proper colors are used
   d. None of the above

2. Why is it necessary to shellac some areas before painting?

   a. So the paint will adhere better
   b. So the wood does not absorb too much paint
   c. To stop the paint from blistering
   d. To prevent knots and sap spots from bleeding through the paint

3. Which of the following is primed?

   a. Hardwood floors
   b. The back of the siding
   c. Window sash
   d. The back of the cornice

4. Which of the following is a part of all paints?

   a. Mineral spirits
   b. Water
   c. Denatured alcohol
   c. Pigment

5. What is used to clean up oil-based paint?

   a. Mineral spirits
   b. Denatured alcohol
   c. Water
   d. Soybean oil

6. What is used to clean up water-based paint?

   a. Mineral spirits
   b. Denatured alcohol
   c. Water
   d. Soybean oil

7. What is the purpose of primer?

    a. To prevent the wood from absorbing too much paint

    b. To ensure that the color is uniform

    c. To prevent knots and sap spots from bleeding through

    d. So the paint will adhere better

8. What is the area to be painted on one side of 5 windows?

    a. 25 square feet

    b. 125 square feet

    c. 250 square feet

    d. 500 square feet

9. How much varnish is required to cover a hardwood floor of 900 square feet with two coats?

    a. 3 gallons

    b. 3 1/2 gallons

    c. 4 gallons

    d. 4 1/2 gallons

10. What is the area to be painted on a set of jambs and casing with no door?

    a. 10 square feet

    b. 20 square feet

    c. 30 square feet

    d. None of the above

B. Using the following given information, construct a material list to paint the house in figure 36-2. The exterior siding and trim are to be primed on all surfaces. The siding is to have 2 coats of house paint and the exterior trim is to have two coats of exterior trim paint. All interior walls and ceilings except the kitchen ceiling and walls are to have two coats of flat wall paint. The kitchen ceiling and walls are to have two coats of satin wall paint. All interior trim is to have 2 coats of satin enamel. It is not necessary to estimate paint and varnish for the doors or floors in this house.

The following materials are listed on the material list for this house:

a) 39 sheets of plywood for sheathing

b) 80 feet of bed molding for cornice at eaves

c) 64 feet of molding for rake cornice

d) 3376 square feet of gypsum wallboard

e) 288 feet of baseboard

1. _____ gallons of exterior primer

2. _____ gallons of house paint

3. _____ gallons of exterior trim paint

4. _____ gallons of flat wall paint for ceilings and interior walls

5. _____ gallons of satin wall paint

6. _____ gallons of satin enamel

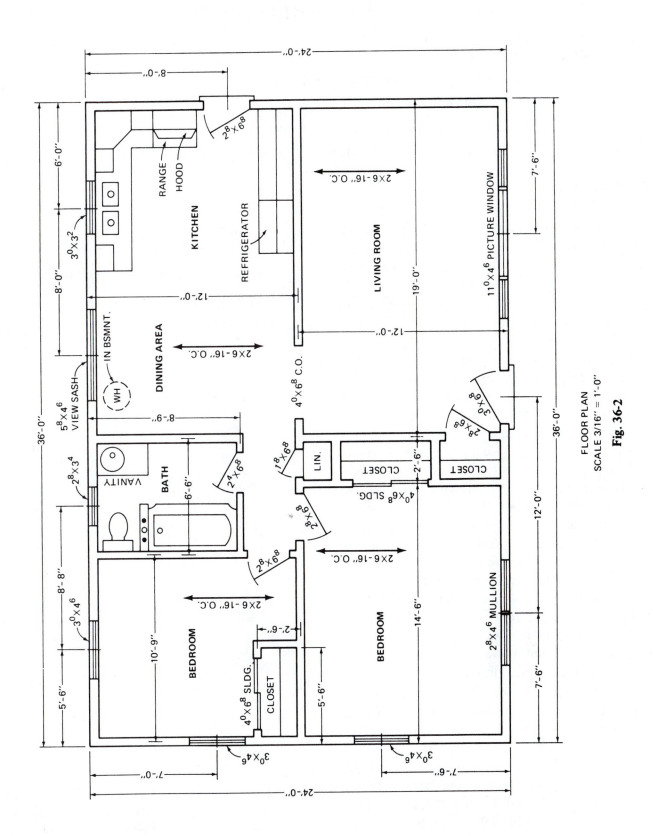

FLOOR PLAN
SCALE 3/16" = 1-0"

Fig. 36-2

unit 37 hardware

OBJECTIVES

After studying this unit the student will be able to

- describe the common items of finish hardware.
- describe the most commonly used types of nails and their sizes.
- list the items which should be included on a finish-hardware allowance.

FINISH HARDWARE

Finish hardware is generally considered to be all of the hardware which is exposed to view in the completed house. This includes such things as door lock sets, hinges for swinging doors, push plates, kick plates, and door stops.

A *lock set* is an assembly consisting of the latching mechanism, lock, and knob for a door. Lock sets are available in a variety of styles and with a variety of locking or latching mechanisms, for use in different areas of the house, figure 37-1. *Exterior lock sets* have a cylinder which can be locked from the outside with a key. Interior lock sets include *bathroom lock sets, bedroom lock sets,* and *passage latch sets.* Bathroom and bedroom lock sets have a turnbutton or some other means for locking the door from the inside, but do not have cylinders for keys. The only difference between the two is that bedroom lock sets usually have a brass finish and bathroom lock sets have a chrome finish. A passage latch set has no provision for locking the door from either side.

Butt hinges (sometimes called butts) are the hinges on which the doors swing. They may be either *loose pin,* figure 37-2, meaning that the pin may be slipped out to separate the two leaves, or *fast pin,* meaning that the pin cannot be removed. Loose-pin hinges are normally used in residential construction. Butts are available in sizes ranging from the very small ones used on miniature furniture to large ones for exterior doors. Usually 1 3/8-inch doors are hung on one pair of 3 1/2-inch by 3 1/2-inch butts and 1 3/4-inch doors are hung on 1 1/2 pairs (3 hinges) of 4 1/2-inch by 4 1/2-inch butts. Butt hinges are available with a variety of finishes, but most are either dull brass or polished brass.

Fig. 37-1 Several styles of lock sets. A & B Exterior lock sets; C. Paggage latch set; D. Bedroom or bathroom lock set; E. Auxiliary lock set (dead bolt); F. Screen door latch set.

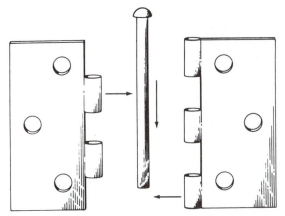

Figure 37-2 A loose-pin hinge can be taken apart by slipping the pin out.

Finish-Hardware Allowance

Because of the great variety of styles and grades of hardware available for residential construction, it is impractical to estimate the cost of finish hardware for a house. Instead, it is customary to specify an allowance for these items. This allowance should be sufficient to purchase all of the finish hardware required. If the owner selects more or less expensive hardware, the difference is either deducted from or added to the contract price. If pre-hung doors are specified the hardware is installed at the factory, but the style must be indicated when the doors are ordered.

SPECIFICATIONS

Division 10 of the specifications covers specialties, including finish hardware.

NAILS

Due to the many sizes and types of nails used in the construction of a house, it is impractical to include nails on an estimate. Normally an amount of money is allowed for the purchase of all of the nails for the building. This is unlike a finish-hardware allowance in that the owner does not make up any difference in cost. Although the estimator is not required to list each type of nail some knowledge of the types and sizes used is desirable.

Nail sizes are listed by the old English term *penny* (abbreviated d). The penny-size of a nail refers to its length. For example, a 2d nail is one inch long and a 16 d nail is 3 1/2 inches long. The chart in figure 37-3 shows the most common sizes of nails. Most nails are made of steel, but other materials may be used for special-purpose nails. Nails for aluminum siding and trim are made of aluminum and nails for copper flashing are made of copper. Nails which are exposed to the weather, where they might rust and stain the surrounding paint, are usually galvanized.

Kinds of Nails

There are a great number of types of nails for general construction and special purposes, figure 37-4. Some have smooth shanks and some have screw shanks, ring shanks, or barbed shanks for greater holding power. Only the most commonly used nails are discussed here.

Specifications

DIVISION 10: SPECIALTIES

A. WORK INCLUDED

This work shall include but shall not be limited by the following:

1. Finish hardware
2. Kitchen cabinets
3. Bathroom vanities
4. Kitchen appliances

B. FINISH HARDWARE

The contractor shall allow in the proposal the sum of Five Hundred Dollars ($500.00) for the purchase of all hardware. This allowance covers the net cost to the contractor and does not include any labor, overhead or profit. Hardware shall be selected by the owner but will be purchased by the contractor where directed. The net difference in cost, if any, shall be added to or deducted from the contract as the case may be. Hardware shall be left in perfect working order, upon completion. All keys shall be delivered to an authorized representative of the owner.

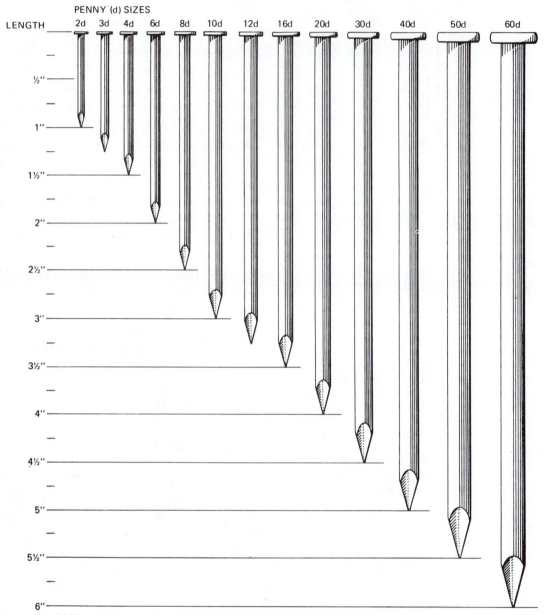

PENNY (d) SIZES

NOTE: All "d" sizes are same length. Only diameter changes between Common and Box nails.

Fig. 37-3 Nail sizes

Common or coated nails and box nails have smooth shanks and flat heads. The difference between them is the diameter of their shanks. All are used for general building construction, such as framing, however, box nails bend easily, so they are usually used only where splitting is a problem. Most framing members, such as studs, joists, beams, and sills are nailed with 16d common nails or coated nails. For toenailing studs to plates, rafters to top plates, or box sills to sills either 8d or 10d common nails or coated nails are used. Subflooring, wall sheathing, and roof sheathing are usually nailed with either 6d or 8d common coated or box nails.

Casing and finishing nails have very small heads. Casing nails are slightly larger in diameter than finishing nails. These nails are used for nailing trim, such as casing, cornices, and siding, where the nail-head is to be concealed. They may be driven slightly below the surface of the wood with a nail set, so that the head can be covered with putty or plastic wood. The sizes used most often are 4d, 6d, and 8d.

Roofing nails are special nails used for applying roofing materials. They are available in varying

Underlayment and flooring nails are special nails for use on those flooring materials. Underlayment nails have ring shanks to prevent them from loosening. For 3/8-inch underlayment, 3d underlayment nails should be used and for 5/8-inch underlayment, 4d nails should be used. For hardwood strip flooring, 8d hardened nails are used.

Duplex-headed nails have two heads so that they may be removed easily. These are useful for building scaffolding which is to be taken apart after it is used.

SUMMARY

Finish hardware is exposed metal which is attached to windows, doors, and cabinets. Because of the great variety of styles, finishes, and grades of finishes, and grades of finish hardware which is available, an allowance is usually specified to cover this. The finish hardware is then selected by the owner, who either pays the excess cost or is reimbursed if the hardware costs less than the allowance.

Door locks may be exterior lock sets, bedroom or bathroom lock sets, or passage latch sets. Exterior lock sets may be locked and unlocked from the outside with a key. Bedroom and bathroom lock sets may be locked from the inside. Passage latch sets cannot be locked.

Doors are hung on butt hinges. Usually 1 1/2 pairs are used for exterior doors and 1 pair for interior doors. Butt hinges are available with a polished-brass or dull-brass finish.

Nails are specified according to penny size. While the diameter of the shank may vary, the length is the same for all types of nails of one penny size. Common and box nails have a smooth shank and a flat head. Common nails are usually used for framing. Casing and finishing nails have a very small head. They are used where it is desirable to conceal the head of the nail. Special nails are available for such work as underlayment, flooring, roofing, wallboard, and nailing into masonry.

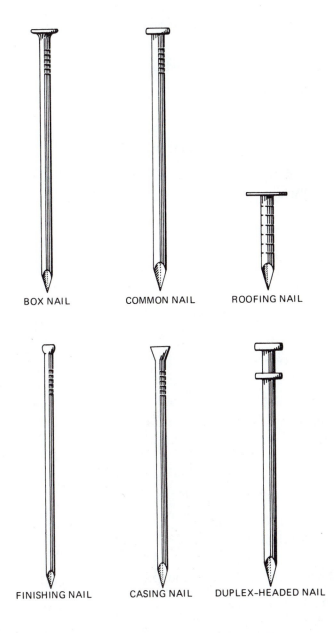

BOX NAIL   COMMON NAIL   ROOFING NAIL

FINISHING NAIL   CASING NAIL   DUPLEX-HEADED NAIL

Fig. 37-4 There are a great number of types of nails available for various applications. This illustration shows some of the most common types.

lengths to suit the thickness of the roofing material. Roofing nails are normally galvanized to prevent them from rusting.

Drywall nails are special nails used for applying gypsum wallboard. Most drywall nails have ring shanks to prevent *nail popping*. Nail popping is a term used to describe the loosening of nails in gypsum wallboard, causing the nailhead to show through the joint compound.

REVIEW QUESTIONS

A. Select the letter preceding the best answer.

1. Which of the following types of door hardware cannot be locked?

   a. Bathroom   c. Bedroom
   b. Exterior   d. Passage

2. Which of the following types of door hardware might be chrome plated?

   a. Bathroom   c. Bedroom
   b. Exterior   d. Passage

3. What size hinges should be used on an exterior door?

   a. 3 inch   c. 4 inch
   b. 3 1/2 inch   d. 4 1/2 inch

4. What size hinges should be used on a 1 3/8-inch interior door?

   a. 3 inch   c. 4 inch
   b. 3 1/2 inch   d. 4 1/2 inch

5. What is the difference between a box nail and a common nail?

   a. Box nails are longer.   b. Box nails are shorter.
   c. Box nails have a smaller diameter shank.
   d. Box nails have a larger diameter shank.

6. What does the lower case d refer to in the description of a nail?

   a. Diameter of the head   c. Length of the nail
   b. Diameter of the shank   d. None of these

7. Which of the following nails should be used for nailing the stool on a window?

   a. 8d common nail   c. 12d casing nail
   b. 6d box nail   d. 8d finishing nail

8. Which of the following nails normally has a ring shank?

   a. Roofing   c. Casing
   b. Underlayment   d. Duplex headed

9. Which type of hinge is normally used to hang doors in residential construction?

   a. Loose pin   c. Strap
   b. Fast pin   d. Offset

10. Which of the following types of nails is best to use where splitting of the wood is a problem?

    a. Ring shank   c. Casing
    b. Box   d. Duplex headed

B. Construct a material list that includes the finished hardware for the house in figure 37-5. It is not necessary to indicate finish or style for this problem. Be sure to specify sizes where necessary. Include all lock sets and hinges. Sliding door closets are to be pre-hung, so it is not necessary to include finish hardware for them.

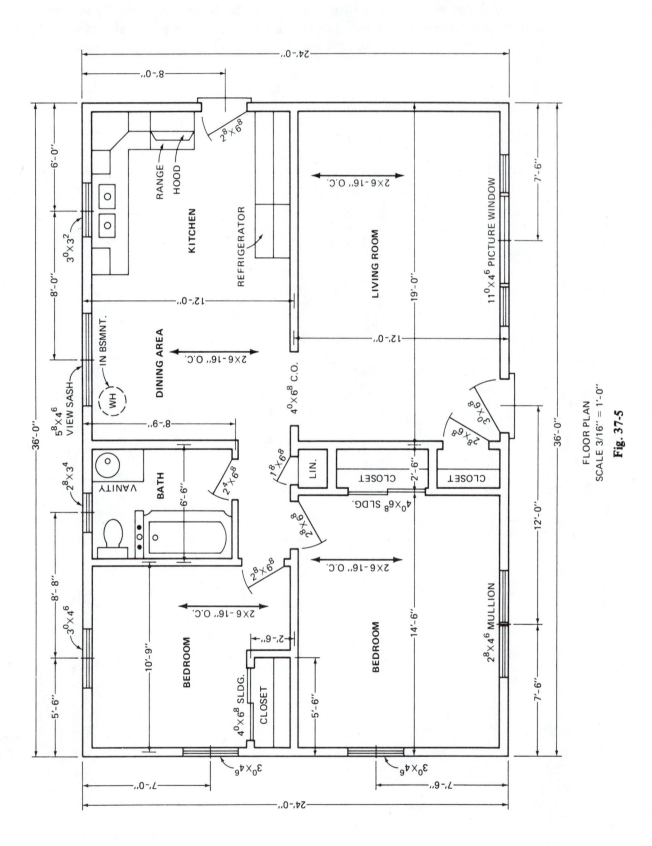

FLOOR PLAN
SCALE 3/16" = 1'-0"

Fig. 37-5

unit 38 cabinets

OBJECTIVES

After studying this unit the student will be able to

- locate kitchen cabinets and bathroom vanities on working drawings.
- explain the specifications for cabinets.
- determine the quantity and style of cabinets and countertop for a house, by reading the working drawings and specifications.

The kitchen cabinets and bathroom vanities for a job may be *site-constructed* (built on the job by the carpenter) figure 38-1, or factory-built and finished. The most common practice is to use factory-built cabinets, except when special requirements are needed. Information about the cabinets may be found in several places. The floor plan of the kitchen and bathrooms usually indicates where cabinets are to be placed in the room. Most sets of working drawings also include special detail drawings or cabinet elevations to show more precise information. These special details or cabinet elevations usually indicate the size of each cabinet. Most often, the manufacturer, style, and finish is included in the specifications for the house.

WALL CABINETS

All of the cabinets that are mounted on the wall, above the countertop, in the kitchen are considered *wall cabinets*. Although wall cabinets are manufactured in a variety of sizes to suit the needs of the kitchen, they are standardized in some ways. Usually the tops of all of the wall cabinets are positioned 84 inches from the floor. The space above the cabinets is usually enclosed and finished with the same material as the kitchen walls, figure 38-2. The height of the cabinet is varied according to the kitchen layout. Where the cabinet is mounted directly above the counter, the standard cabinet height is 30 inches. Usually cabinets which are mounted above a range or cooking surface are 18 to 21 inches high. Frequently a metal hood with an exhaust fan is mounted below these cabinets, figure 38-3. The hood is fastened to the bottom of the wall cabinet and the exhaust fan is ducted through the wall or up through the cabinet. Above a refrigerator, the wall cabinets are 12 or 14 inches high.

Fig. 38-1 Site-constructed cabinets

Fig. 38-2 The space above the wall cabinets is usually enclosed with a soffit.

Fig. 38-3 A metal exhaust hood may be mounted above the cooking surface.

All wall cabinets are usually 12 inches deep, but their width varies in 3-inch increments. They most frequently range from 9 inches wide to 48 inches wide. In addition, blank filler pieces are available to allow for minor adjustments.

BASE CABINETS

All of the kitchen cabinets that rest on the floor are considered *base cabinets*. Base cabinets vary in width only. The standard height for base cabinets is 34 1/2 inches from the floor to the top of the cabinet. The standard depth is 24 inches. Like wall cabinets, base cabinets are also available in widths from 9 inches to 48 inches, in 3-inch increments.

It is common for most base cabinets to have one drawer in the top and doors below. Most kitchens also include one base cabinet with four or five drawers and no doors. Special units are available for storage of brooms and cleaning supplies. These broom cabinets extend from the floor to the top of the wall cabinets.

COUNTERTOPS

The countertop may be either site-constructed or factory-built, regardless of how the cabinets are

constructed. The surface is usually made of plastic laminate. At the rear of the countertop is a 4-inch *backsplash* (a raised lip to prevent water from running down the back of the cabinets). Most countertops are 25 inches wide, including the backsplash.

Manufactured countertops are usually 1 1/2 inches thick and have a rounded edge and backsplash, figure 38-4. Site-constructed countertops are usually made of 3/4-inch plywood, with plastic laminate glued to the surface. The front edge of a site-constructed countertop is usually built up by adding a 3/4 inch thick piece of wood to the bottom edge.

In some kitchens the space between the base cabinets and the wall cabinets is covered with the same plastic laminate as the countertop, figure 38-5. In this case the backsplash may be eliminated.

Fig. 38-4 Manufactured countertops usually have rounded edges and backsplash.

Fig. 38-5 The wall space between the upper and lower cabinets is sometimes covered with plastic laminate.

VANITIES

Lavatories are frequently installed in a base cabinet called a vanity, figure 38-6. Vanities are normally 29 1/2 inches high by 21 inches deep and vary in length from 12 inches to 60 inches. The top may be a laminated plastic countertop, similar to that used in a kitchen, or a single cast piece, including the lavatory.

SPECIFICATIONS

Division X, Specialties, Section C are the specifications for the kitchen cabinets and bathroom vanities in the sample house.

Specifications

DIVISION 10

C. CABINETS

1. Kitchen cabinets shall be Kingswood Oakmont, as manufactured by the B. J. Sutherland Company of Louisville, Kentucky, or equal. Sizes and styles are to be as shown on the special detail drawings.

2. Bathroom vanity shall be RV-48 Moonlight, as manufactured by the B. J. Sutherland Company of Louisville, Kentucky, or equal.

3. Kitchen cabinet and vanity countertops shall by 1/16" laminated plastic bonded to 3/4" plywood. Countertops shall be of one-piece molded construction, with 4-inch backsplash and no back seams. The color and pattern are to be selected by the owner.

Fig. 38-6 Bathroom vanity with two lavatories

ESTIMATING CABINETS

The manufacturer and style of the cabinets is indicated in the specifications. The sizes and quantities of kitchen cabinets can be determined from the kitchen cabinet plan and the cabinet elevations for the sample house. The numbers on these special detail drawings indicate the size and type of cabinets. On wall cabinets, the first two digits indicate the width in inches, and the last two digits indicate the height in inches. For example, a 2418 wall cabinet is 24 inches wide and 18 inches high. All base cabinets are the same height, so only two digits are shown, to indicate the width. The letters

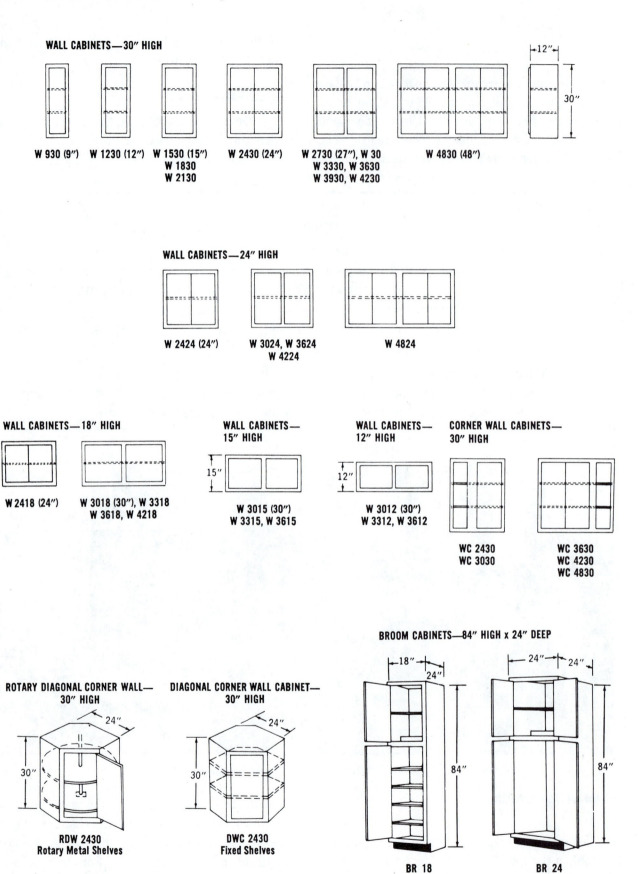

WALL CABINETS—30" HIGH

W 930 (9")   W 1230 (12")   W 1530 (15")   W 2430 (24")   W 2730 (27"), W 30   W 4830 (48")
                      W 1830   W 3330, W 3630
                      W 2130   W 3930, W 4230

WALL CABINETS—24" HIGH

W 2424 (24")   W 3024, W 3624   W 4824
            W 4224

WALL CABINETS—18" HIGH

W 2418 (24")   W 3018 (30"), W 3318
         W 3618, W 4218

WALL CABINETS— 15" HIGH

W 3015 (30")
W 3315, W 3615

WALL CABINETS— 12" HIGH

W 3012 (30")
W 3312, W 3612

CORNER WALL CABINETS— 30" HIGH

WC 2430   WC 3630
WC 3030   WC 4230
         WC 4830

ROTARY DIAGONAL CORNER WALL— 30" HIGH

RDW 2430
Rotary Metal Shelves

DIAGONAL CORNER WALL CABINET— 30" HIGH

DWC 2430
Fixed Shelves

BROOM CABINETS—84" HIGH x 24" DEEP

BR 18   BR 24

Fig. 38-7   Letters are used to indicate types of cabinets. These may vary from one manufacturer to another.

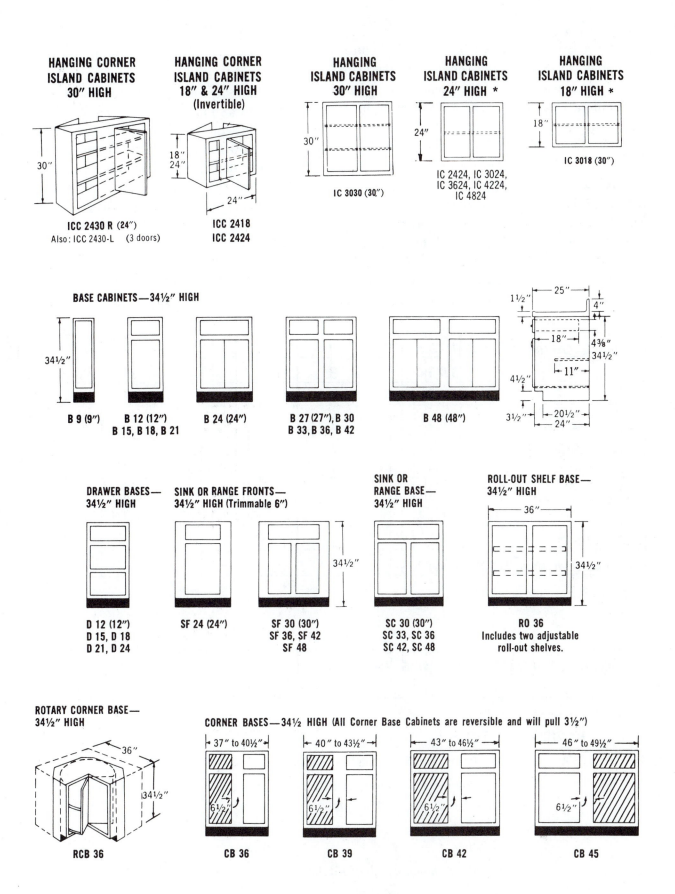

HANGING CORNER ISLAND CABINETS 30" HIGH
30"
ICC 2430 R (24")
Also: ICC 2430-L (3 doors)

HANGING CORNER ISLAND CABINETS 18" & 24" HIGH (Invertible)
18" 24"
24"
ICC 2418
ICC 2424

HANGING ISLAND CABINETS 30" HIGH
30"
IC 3030 (30")

HANGING ISLAND CABINETS 24" HIGH *
24"
IC 2424, IC 3024, IC 3624, IC 4224, IC 4824

HANGING ISLAND CABINETS 18" HIGH *
18"
IC 3018 (30")

BASE CABINETS—34½" HIGH
34½"
B 9 (9")
B 12 (12")
B 15, B 18, B 21
B 24 (24")
B 27 (27"), B 30
B 33, B 36, B 42
B 48 (48")

1½" — 25" — 4"
18"
4⅜"
34½"
11"
4½"
3½" — 20½" — 24"

DRAWER BASES— 34½" HIGH
D 12 (12")
D 15, D 18
D 21, D 24

SINK OR RANGE FRONTS— 34½" HIGH (Trimmable 6")
SF 24 (24")
SF 30 (30")
SF 36, SF 42
SF 48
34½"

SINK OR RANGE BASE— 34½" HIGH
SC 30 (30")
SC 33, SC 36
SC 42, SC 48

ROLL-OUT SHELF BASE— 34½" HIGH
36"
34½"
RO 36
Includes two adjustable roll-out shelves.

ROTARY CORNER BASE— 34½" HIGH
36"
34½"
RCB 36

CORNER BASES—34½ HIGH (All Corner Base Cabinets are reversible and will pull 3½")
37" to 40½"
6½"
CB 36

40" to 43½"
6½"
CB 39

43" to 46½"
6½"
CB 42

46" to 49½"
6½"
CB 45

Fig. 38-7 (Continued)

preceding the numbers indicate the type of cabinet. These letters may vary from one manufacturer to another, so the estimator should have the manufacturers' catalogs available. Some of the most common cabinet designations are shown in figure 38-7.

The following cabinets on material list A are shown on the detail drawings for the sample house.

According to the specifications, the countertop is to have a molded edge and backsplash. This type of countertop is usually ordered cut to size, so the exact length must be listed. To allow for mitered corners (cut at a 45 degree angle) measure the length at the back edge of the counter. Add the length of the countertop to material list B.

SUMMARY

Most cabinets are factory-built, but they may be built by the carpenter in special cases. Factory-built cabinets are available in widths ranging from 9 inches to 48 inches. Base cabinets are usually a standard height, but wall cabinets vary from 14 inches to 30 inches in height, depending on their location. Bathroom vanity cabinets, which are constructed like kitchen cabinets, are usually 29 1/2 inches high by 21 inches deep.

The specifications list the manufacturer and style of the cabinets. Most of the time the sizes are shown on special detail drawings. The material to be used for the countertop is indicated in the specifications.

A

WALL CABINETS	1233		1	
	3030		3	
	3036		1	
	1830		1	
BASE CABINETS	B18		1	
	D18		1	
	C42		2	
	SF24		1	
FILLER STRIP	3"		2	
VANITY w/ COUNTERTOP	RV48		1	

B

MOLDED COUNTERTOP	25" WIDE	LIN.FT.	19	

REVIEW QUESTIONS

A. Select the letter preceding the best answer.

1. What is the most common height for wall cabinets that are mounted above a clear counter space, with no appliances beneath?

   a. 21 inches   c. 30 inches
   b. 24 inches   d. 36 inches

2. What is the standard depth of kitchen wall cabinets?

   a. 12 inches   c. 18 inches
   b. 15 inches   d. 24 inches

3. What is the most common height of base cabinets?

    a. 36 inches   c. 34 1/2 inches

    b. 30 inches   d. 32 1/2 inches

4. What is the standard depth of base cabinets?

    a. 24 inches   c. 26 inches

    b. 25 inches   d. 30 inches

5. Which of the following is a common width for a base cabinet?

    a. 14 inches   c. 19 inches

    b. 16 inches   d. 21 inches

6. Which of the following descriptions best fits a cabinet number D24?

    a. A 24-inch wide wall cabinet with a door

    b. A dishwasher cabinet

    c. A 24-inch wide double wall cabinet

    d. A 24-inch wide base cabinet with drawers

7. What is the standard width for a kitchen countertop?

    a. 24 inches   c. 26 inches

    b. 25 inches   d. 30 inches

8. What is the standard height of a vanity cabinet?

    a. 29 1/2 inches   c. 34 1/2 inches

    b. 32 1/2 inches   d. 36 inches

9. Where are the sizes of the kitchen cabinets indicated?

    a. Specifications   c. Special detail drawings

    b. Floor plan   d. Section view

10. How is the color of the countertop in the sample house determined?

    a. By the contractor   c. By the supplier

    b. By the architect   d. By the owner

B.   Construct a material list for the cabinets for the kitchen plan in figure 38-8. Only the width of the cabinets is shown on the plan, so it will be necessary to determine the proper height for each wall cabinet and the type of each base cabinet. Use the abbreviations in figure 38-7 for the various types of cabinets.

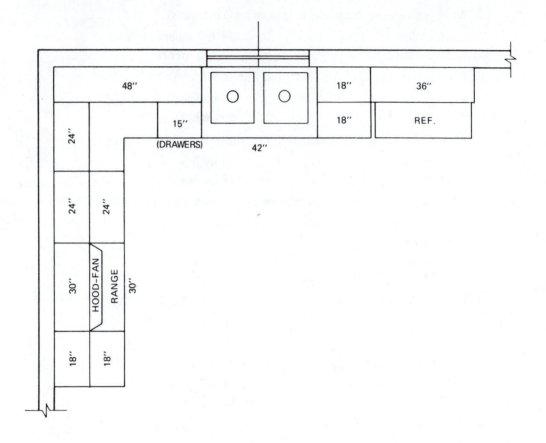

Fig. 38-8

unit 39 material estimate check list

OBJECTIVES

After studying this unit the student will be able to

- understand plans and specifications
- know how to use the plans and specifications to make a material list
- follow a procedure to assure an accurate estimate of materials

Specifications are written in the sequence in which the building would be constructed, with a few exceptions. (For example, the section that deals with moisture protection covers foundation dampproofing or waterproofing and also roofing. These operations are completed at different times during the construction of a house.) Specifications are written to spell out in detail the information that cannot be clearly expressed on the working drawings (plans). For example, if the working drawings show wood floors, they might be oak, maple, or vertical grain fir. Tile flooring on the working drawings might be ceramic, asphalt tile, or vinyl tile. Where the working drawings show flashing, it could be galvanized iron, aluminum, or copper. Roof shingles could be asphalt or wood. The specifications describe the quality and type of materials, colors, finishes, and workmanship required. A typical set of specifications would include the following divisions for house construction:

Instruction to Bidders
General Conditions

Division I:	General Requirements
Division II:	Site Work
Division III:	Concrete
Division IV:	Masonry
Division V:	Metals
Division VI:	Carpentry and Millwork
Division VII:	Moisture Protection
Division VIII:	Doors, Windows, and Glass
Division IX:	Finishes
Division X:	Specialities
Division XI:	Mechanical
Division XII:	Electrical

You must constantly refer to the specifications as you study and make your material esti-mate from the working drawings. The following schedule will help you in making your material list and also in figuring the necessary labor to put that material in place according to the specifications and the working drawings.

We will start our estimate for the materials, starting at the site with the batter boards and stakes.

SITE WORK

Figure 39-1 shows a typical right angle batter board. Depending upon the number of corners and the layout of the house, you should figure three 2 x 4's approximately 6 feet long, two pieces of shiplap 1 x 8 about 5 feet long, and one stake for each corner of the house. You will also need a few nails and a roll of twine for the lines.

Figure the number of corners of the building. Multiply this by 3 for the number of 2 x 4's for the batter boards. Take the number of corners and multiply by 2 and this gives the number of pieces of 1 x 8 shiplap for the corner batter boards. Next count the number of corners and this will give you the number of stakes to lay out the house. We now start our material list as follows:

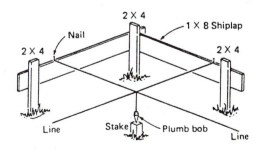

Fig. 39-1 Batter boards

ESTIMATE FOR _____ Page _____
_____ Date _____
Job _____

Pieces	Feet	Description	Cost	Total
		2 X 4 X 6' BATTER BOARDS		
		1 X 8 X 5' SHIPLAP - BATTER BOARDS		
		STAKES FOR BATTER BOARDS		
		TWINE FOR LINES ON BATTER BOARDS		

MASONRY ESTIMATE

Concrete Work — Footing Forms

Forms for the sides of the footings in residential construction are usually constructed using 2 x 8 low-grade lumber. To determine the amount of lumber needed, double the perimeter (distance around the building). Make sure that you have included any areas for support posts, fireplaces, and porches. Wood stakes are required to hold the forms in place until the concrete is poured. Figure two stakes for every 4 linear feet of perimeter and add two for each corner.

Pieces	Feet	Description	Cost	Total
		LINEAR FEET 2 X 8 - FORMS FOR FOOTINGS		
		BUNDLES WOOD STAKES - FOR FORMS (12 TO BUNDLE)		

Concrete Work — Footings

To estimate the amount of concrete necessary for the footings, determine the linear feet around the perimeter of the building including the length of any other walls such as porches. Multiply the width of the footings in feet times the thickness in feet. This product is then multiplied by the total length of the footings in feet. This will give you the number of cubic feet in the footings. Divide this number of cubic feet by 27 to find the number of cubic yards. Transit mix is sold by the cubic yard. The amount of concrete to be used for the supporting posts and fireplace footings must be added to this amount. To figure this area, multiply the length in feet times the width in feet times the thickness in feet and divide by 27.

Pieces	Feet	Description	Cost	Total
		CUBIC YARDS CONCRETE - FOOTINGS		

Concrete Work — Floors and Porches

To estimate the amount of concrete needed for the floors and porches, determine the total square-foot area of all the places to be covered with concrete. Multiply this area by the thickness in feet and divide by 27 to get the number of cubic yards.

Example: You have 908 square feet to be covered, including all floors and porches, and the thickness is 4 inches. You multiply 908 by 1/3 (4 inches is 1/3 foot) and you get 303. Divide 303 by 27 and you get 11 6/27 or rounded off 11 1/3 cubic yards of concrete is required.

Pieces	Feet	Description	Cost	Total
		CUBIC YARDS CONCRETE - FLOORS AND PORCHES		

Concrete Work — Sidewalks

Sidewalks are usually shown on the plot plan and are in the specifications. You estimate the concrete for the sidewalk the same as you did for the floors and porches. Length in feet times width in feet times thickness in feet divided by 27 gives you the number of cubic yards.

Example: Your plot plan and specifications call for a sidewalk 4 feet wide and 3 inches thick. The plot plan shows the house sets back 30 feet so the walk will be 30-feet long. You multiply 30 (length) times 4 (width) = 120 square feet. Three inches thick means 1/4 foot; 1/4 X 120 = 30, and 30 ÷ 27 = 1 3/27 or 1 1/9 yards concrete required for the walk.

Pieces	Feet	Description	Cost	Total
		CUBIC YARDS CONCRETE - SIDEWALK		

Concrete Work — Gravel Fill

The only place you will find this information on the working drawings will be on the section view, and not enough information will be given. You must refer to the specifications under Concrete Work to find this. To determine the amount required, take the square-foot area where the concrete is to be placed and divide by 2, or the proportional part of a foot that the specifications call for in thickness. (In house construction, a 6-inch thick layer of gravel fill is usually specified.) This gives the amount in cubic feet. For cubic yards, divide by 27. Alternatively, length in feet times the width in feet times the thickness in feet divided by 27 gives the number of cubic yards needed.

Pieces	Feet	Description	Cost	Total
		CUBIC YARDS GRAVEL FILL ALL FLOORS		

Masonry Work — Cement Block

There are two methods of figuring cement block. In the first method, find the area of the wall. Each 8″ x 16″ block covers 0.888 square feet, so figure 112 1/2 blocks for every 100 square feet of wall. Add 5 to 10 percent for waste.

The second method is to figure 3 blocks for every 4 linear feet per course. Add the perimeter of the building in feet. Add the porches and fireplace base, if any. You multiply this by three and divide by four. This gives you the number of blocks for one course. You get 3 courses for every 2 feet in height. The standard height of a foundation in residential construction is 7 feet or 84 inches. This would be 10 1/2 courses high. You multiply the number of blocks required for one course by ten for the number of full-size blocks needed. The wall usually is capped with 4-inch solid cap blocks, so list the number of blocks required for one course for the cap. Corner blocks are listed separately, so count the number of corners in the foundation and multiply this by 10. You subtract this total from the total number of regular blocks you estimated.

You should figure two half sash blocks and two whole sash blocks for every 15 x 12 two-light cellar sash.

Example: Our total perimeter including porches and fireplace base is 139 linear feet. 139 X 3 = 417; 417 ÷ 4 = 106 blocks for one course. You multiply this by 10 for a total of 1,060 regular blocks and 106 4-inch cap blocks. If we have 10 corners in the foundation, we need 10 X 10 = 100 corner blocks. We would subtract this 100 from our 1,060. You count the number of cellar sashes and then figure two whole and two half sash blocks for each window.

Pieces	Feet	Description	Cost	Total
		8 X 8 X 16 REGULAR CEMENT BLOCK		
		8 X 12 X 16 REGULAR CEMENT BLOCK		
		8 X 8 X 16 REGULAR CORNER BLOCK		
		4 X 8 X 16 REGULAR CAP BLOCK		
		WHOLE SASH BLOCK		
		HALF SASH BLOCK		

Masonry Work — Poured Foundation Walls

If the foundation is to be poured concrete instead of cement block, the number of cubic yards of concrete in the wall must be computed. To do this, multiply the length of the foundation times its thickness in feet. Multiply this times its height in feet. This is the number of cubic feet in the wall. Divide this by 27 to find the number of cubic yards needed.

Masonry Work — Basement Sash

To determine the number of cellar sashes needed, you count the number of windows shown on the foundation and basement plan. Check your specifications to make sure you are estimating what is called for. Also check to see if storm sash and screens are to be included.

Pieces	Feet	Description	Cost	Total
		15 X 12 STEEL CELLAR SASH COMPLETE WITH STORM AND SCREEN		

Masonry Work — Metal Areaways

Where the building sets close to the finished grade, an areaway has to be placed around the basement windows to allow for ventilation and to keep the earth away from the window. The areaways are shown on the foundation and basement plan and are also found in the specifications as to the type and material. List them as specified and the number should be the same as the basement cellar sash.

Pieces	Feet	Description	Cost	Total
		GALVANIZED IRON AREAWAYS		

Masonry Work — Mortar for Block

In estimating mortar for cement block, allow one bag of cement for every 28 blocks. Check the specifications to find out if mortar is to be made from masonry cement or with portland cement and mason's lime.

Pieces	Feet	Description	Cost	Total
		BAGS MASONRY CEMENT-MORTAR FOR BLOCK		
		BAGS REGULAR CEMENT-MORTAR FOR BLOCK		
		BAGS MASON'S LIME -FOR MORTAR BLOCK WORK		

Masonry Work — Washed Sand

A good grade washed sand is required to make a good mortar. You should figure 3 cubic feet of sand for every bag of cement. The easiest way is to figure one cubic yard of sand for every 10 bags of cement. This will allow for waste.

Pieces	Feet	Description	Cost	Total
		YARDS WASHED SAND - MORTAR FOR BLOCKWORK		

Masonry Work — Wall Reinforcing

This information will be given on the section view and also in the specifications.

Determine the number of courses which require reinforcing. Take the linear feet where the reinforcing is required and multiply by the number of courses where it will be placed. This will give the linear feet required.

Pieces	Feet	Description	Cost	Total
		LINEAR FEET DUR-O-WALL REINFORCING 8"		
		LINEAR FEET DUR-O-WALL REINFORCING 12"		

Masonry Work — Steel Support Posts

These are found on the basement plan and are listed according to size and the number required. Check specifications to make sure both working drawings and specifications agree.

Pieces	Feet	Description	Cost	Total
		STEEL POST W/BASE AND CAP		

Masonry Work — Chimney

For a concrete block chimney, find the total height of the chimney in feet from the basement to the top of the chimney. (See the elevations.) To determine the number of flue liners needed, divide this height by 2. (Flue liners are 2 feet long.) A chimney is usually built of whole chimney block and flue liners up to the roof line. Brick is used where the chimney comes through the roof. From the top of the roof to the top of the chimney should be at least two feet above the ridge. To build an 8" x 8" furnace chimney, 3 whole chimney blocks are required for each 2 feet of height.

Example: Consider a 24-foot high chimney which extends 6 feet through the roof with a single 8 x 8 flue liner. The total height of 24 feet requires 12 flue liners. Subtract the 6 feet from the total height and we have 18 linear feet of whole chimney block. Two divided into 18 gives us 9, which, multiplied by 3, gives 27 whole chimney

blocks required. From the roof line to the top of the chimney is 6 feet. We multiply 6 by 27 (number of bricks needed for each foot of height for a single 8 x 8 flue chimney), for a total of 162 bricks. We would also list the cement and flashing for the chimney.

Pieces	Feet	Description	Cost	Total
		8 X 8 FLUE LINERS - CHIMNEY		
		WHOLE CHIMNEY BLOCK - CHIMNEY		
		BRICK - CHIMNEY		
		BAGS MASONRY CEMENT - CHIMNEY		
		BAGS REGULAR CEMENT - CAP		
		LINEAR FEET FLASHING - CHIMNEY		

Masonry Work — Fireplace

In estimating a fireplace you must determine the location of the chimney to know what to figure. If the chimney goes up through the house, the only finished brick work would be the fireplace face and the chimney above the roof. If the chimney is built on the outside of the house, it requires face brick on the exposed face. The flue liners are figured the same, one for every two feet in height. You will need about four fewer of the fireplace flues if the fireplace is located on the first floor. Check your plans to make sure how many fireplaces are on the chimney. In many plans, a fireplace is located in the recreation area in a basement and also in the living room or library on another floor.

Estimating a fireplace, with a prefabricated fireplace unit (which most fireplaces now use) you figure 30 firebricks for the firebox base. Your fireplace base is usually figured in with the foundation. You will need an ash dump, a 12" x 15" cleanout door for the ash pit, and an 8" x 8" cleanout for the flue liners of the furnace flue.

List the prefabricated fireplace unit by size and manufacturer. Each unit will need an angle iron to hold the brick across the fireplace face opening. This length should be at least 8 inches wider than the opening. Estimate brick by the square foot, adding any necessary fills. You should also figure the hearth material in your list. This information is in the specifications and the floor plans.

Pieces	Feet	Description	Cost	Total
		REGULAR FIREBRICK - FIREPLACE BASE		
		ASH DUMP		
		CLEANOUT DOORS - 12"x15" - 8" X 8"		
		NO. PREFABRICATED FIREPLACE UNIT		
		8 X 8 FLUE LINERS		
		8 X 12 FLUE LINERS		
		$3\frac{1}{2}$ X $3\frac{1}{2}$ ANGLE IRON X FEET		
		BRICK FOR FIREPLACE FACE AND EXTERIOR FACE		
		QUARRY TILE FIREPLACE HEARTH		
		BAGS REGULAR CEMENT - HEARTH AND CAP		

Masonry Work — Fireplace Mortar

The amount of mortar cement that is required to build a fireplace depends partly on the mix that is used. A typical mix would be 1 part cement, 1 part mason's lime, and 6 parts sand. If a 3/8-inch mortar joint is used, then 21 cubic feet of mortar is required to lay approximately 1,000 bricks. One bag of portland cement, one bag of mason's lime, and 48 shovels of sand will lay approximately 400 bricks. Allow one-third for filling between the brick and flue liners and for fills.

Pieces	Feet	Description	Cost	Total
		BAGS REGULAR CEMENT - FIREPLACE		
		BAGS MASON'S LIME - FIREPLACE		
		BAGS MASONRY CEMENT - FIREPLACE		
		CUBIC YARDS WASHED SAND - MORTAR FOR FIREPLACE		

Masonry Work — Brick Veneer

To estimate the quantity of common or standard face brick in veneer work, 675 bricks are needed for every 100 square feet of wall. Allow 5-10 percent for waste. You should also figure 70 wall ties for every 100 square feet of brick.

Pieces	Feet	Description	Cost	Total
		BRICK - VENEER WORK		
		GALVANIZED IRON WALL TIES - BRICK VENEER WORK		

Masonry Work — Brick Veneer Mortar

Figure one bag of regular cement and one bag of mason's lime for every 400 bricks. Add any angle irons for lintels, and flashing material.

Pieces	Feet	Description	Cost	Total
		BAGS REGULAR CEMENT - VENEER		
		BAGS MASON'S LIME - VENEER		
		3½" x 3½" x "ANGLE IRON OVER DOORS AND WINDOWS - VENEER		
		GALVANIZED IRON 30" x 96" - FLASHING, VENEER AND CHIMNEY		

Masonry Work — Foundation Plastering

One bag of portland cement and one bag of mason's lime with 48 shovels of sand will cover approximately 40 square feet. Determine the square-foot area of the foundation to be covered. Divide this by 40 to get the number of bags of cement and lime needed.

Pieces	Feet	Description	Cost	Total
		BAGS REGULAR CEMENT - PARGING		
		BAGS MASON'S LIME - PARGING		

Masonry Work — Foundation Damp-proofing

Brush-type asphalt coating covers approximately 30-35 square feet of wall to the gallon. Figure the square-foot area to be covered and divide this by 30. If the specifications call for two coats, the second coat will require only about half as much. Asphalt coating is sold in 5-gallon cans, so divide the number of gallons required by 5.

Pieces	Feet	Description	Cost	Total
		5-GALLON CANS ASPHALT COATING - FOUNDATION		

Masonry Work — Drain Tile

To figure the amount of plastic drain tile or pipe needed, find the distance around the perimeter of the house and divide by 10. This is the number of 10-foot pieces needed. A 90-degree elbow is required at every corner, and where the drain goes under the footings to a sump a tee is required. Crushed stone is sold by the ton and approximately one ton of stone is required for every 18 linear feet of drain pipe.

Pieces	Feet	Description	Cost	Total
		10-FOOT PERFORATED PLASTIC DRAIN TILE		
		90-DEGREE ELBOWS - DRAIN TILE		
		PLASTIC TEES		
		TONS NO. CRUSHED STONE - DRAINAGE		

CARPENTRY ESTIMATE: FRAMING AND EXTERIOR WORK

Carpentry Work — Girders

The girder and its supporting posts support the main bearing partitions as well as part of the weight of the floors and their contents. These are usually built up using three or more members. The lengths of the members should be determined so the ends of the wood members making up the girder should be joined only at the support columns. If possible use pieces that can run from wall to wall. Determine the length of the members and list them according to length and size.

Pieces	Feet	Description	Cost	Total
		2 X 8 (OR 10 OR 12) X LENGTH - BUILT UP GIRDER		

Carpentry Work — Termite Shield

If the specifications and working drawings call for a termite shield, determine the size and length needed and list by either the linear foot or the number of pieces.

Pieces	Feet	Description	Cost	Total
		TERMITE SHIELD		

Carpentry Work — Sill Seal and Sill

Enough sill seal and sill must be figured to cover the perimeter of the building. If a steel supporting beam is used for the main bearing beam and the joists rest on top of the beam, then enough sill must be added to cover the length of the girder.

Pieces	Feet	Description	Cost	Total
		LINEAR FEET 6-INCH SILL SEAL		
		2 X 6 X (LENGTH) SILL		

Carpentry Work — Box Sill

The box sill is the header joists placed at right angles to the ends of the floor joists. It is the same width as the floor joists and rests on the top of the sill. The box sill should be the same as the perimeter required for the sill.

Pieces	Feet	Description	Cost	Total
		2 X (8 OR 10 OR 12) X LENGTH BOX SILL		

Carpentry Work — Floor Joists

To determine the number of floor joists needed, multiply the length of the foundation wall on which they rest by 3/4 then subtract 1. This provides three joists for every four feet. Also add one joist for every partition that runs in the same direction as the joists.

Pieces	Feet	Description	Cost	Total
		2 X (8 OR 10 OR 12) X LENGTH- FLOOR JOISTS		

Carpentry Work — Bridging

To determine the amount of bridging needed, multiply the number of floor joists by 3. This is the number of linear feet of bridging needed for one row. If the span of the joists is over 14 feet, figure two rows of bridging for that span.

Pieces	Feet	Description	Cost	Total
		LINEAR FEET 1 X 3 SPRUCE - BRIDGING		

Carpentry Work — Headers and Trimmers

Headers are used to support the cut ends of the joists at such places as stairs, chimneys, or fireplaces. The joists at the ends of the headers are called trimmers. Trimmers should be doubled to support the extra load. Headers and trimmers must be the same width as the floor joists and trimmers should be the same length as the floor joists. Measure the length of the headers from the working drawings and list the headers and trimmers by the piece.

Pieces	Feet	Description	Cost	Total
		2 x (8 OR 10 OR 12) X LENGTH-HEADERS		
		2 x (8 OR 10 OR 12) X LENGTH-TRIMMERS		

Carpentry Work — Subfloor

Estimating the subflooring is done by the square-foot area to be covered. If the subflooring is plywood, take the square-foot area to be covered and divide by 32 for the number of 4' x 8' pieces. If shiplap or T&G is used for the subflooring, take the square-foot area to be covered and add 25 percent to determine the amount needed.

Pieces	Feet	Description	Cost	Total
		(THICKNESS) X 4' X 8' PLYWOOD-SUBFLOOR		

Carpentry Work — Plates and Shoe

This is estimated by the linear foot. Take the perimeter of the outside walls. Figure the linear feet of interior walls. Figure all walls solid. Do not deduct for door or window openings. Add the exterior linear feet of wall and the interior linear feet of interior partitions together and then multiply this total by 3 for the total linear feet of 2 x 4's needed for the plates and shoe. Check to see if exterior walls are 2 x 4 or 2 x 6.

Pieces	Feet	Description	Cost	Total
		LINEAR FEET 2 X 4 PLATES AND SHOE		

Carpentry Work — Studs

In estimating for studs, allow one stud for every linear foot of interior and exterior walls and partitions, and two extra for every corner. If exterior wall is 2 x 6, exterior wall studs will be 2 x 6.

Pieces	Feet	Description	Cost	Total
		2 X 4 X 8' STUDS		

Carpentry Work — Headers

Headers are required where openings for windows or doors are cut out of the walls to carry the vertical loads to other studs. The size of headers varies. They must be of sufficient size to carry the load. To find the length of headers, measure each opening for doors or windows. Allow 3 feet for single openings. Multiply the number of single openings by three and then double this (headers are double). Then determine the standard lengths that can be divided into this total.

Pieces	Feet	Description	Cost	Total
		2 X 4 X (LENGTH) DOOR AND WINDOW HEADERS		

Carpentry Work — Second Floor Framing

If the working drawings show a second floor, you estimate the materials in the same manner in which you did the first floor. List your box sill first, then the second floor joists. The headers and trimmers should be figured, the bridging for the second floor joists, and the subflooring. You then list the plates and shoe for the exterior walls and the interior partitions on the second floor. Next list the studs and headers for all door and window openings.

Pieces	Feet	Description	Cost	Total
		LINEAR FEET 2 X SIZE BOX SILL - SECOND FLOOR		
		2 X (8 -10 -12) X LENGTH JOISTS - SECOND FLOOR		
		2 X (8 -10 -12) X LENGTH HEADERS AND TRIMMERS - SECOND FLOOR		
		LINEAR FEET 1 X 3 SPRUCE - BRIDGING, SECOND FLOOR		
		SIZE X 4' X 8' PLYWOOD SUBFLOOR - SECOND FLOOR		
		LINEAR FEET 2 X 4 PLATE AND SHOE - SECOND FLOOR		

Pieces	Feet	Description	Cost	Total
		2 X 4 X 8' STUDS - SECOND FLOOR		
		2X SIZE X LENGTH - HEADERS FOR DOORS AND WINDOWS, SECOND FLOOR		

Carpentry Work — Ceiling Joists

Ceiling joists support the ceiling of the top floor of the house. They also support the rafters and help secure them to the house. The ceiling joists must span from the outside wall of the house to an interior bearing wall or to the opposite exterior wall. To estimate ceiling joists you find the direction in which they are to run. This is usually on the second floor plan. Multiply that distance by 3, divide by 4, and add 1. The length is the span that they have to cover.

Pieces	Feet	Description	Cost	Total
		2 X SIZE X LENGTH - CEILING JOISTS		

Carpentry Work — Rafters

To determine the number of rafters 16 inches o.c. required on each side of a gable roof, multiply the length of the ridge by 3, divide by 4, and add 1. Assume jack rafters the same size and the same length as the common rafter. To find the number of jack rafters needed, take 1 1/2 times the width of the building and add 2. Allow one hip rafter for each hip and one valley rafter for each valley. Hip and valley rafters should be 2 inches wider than the common rafters to allow for a solid nailing surface.

Pieces	Feet	Description	Cost	Total
		2 X SIZE X LENGTH - RAFTERS		
		2 X SIZE X LENGTH - RIDGE BOARD		
		2 X SIZE X LENGTH - HIP AND VALLEY RAFTERS		

Carpentry Work — Trusses

If the plans and specifications call for pre-assembled roof trusses instead of rafters, these are usually spaced two feet on center. To determine the number required for a straight roof, divide the length of the building in feet by 2 and add 1.

Pieces	Feet	Description	Cost	Total
		ROOF TRUSSES - GIVE PITCH AND TAIL (OVERHANG)		

Carpentry Work — Collar Beams

Occasionally one collar beam is used for each pair of rafters, but most buildings have one for every third pair of rafters. If collar beams are spaced 16 inches on center, divide the number of rafters by 2 and add 1. If they are spaced 32 inches on center, divide the number of rafters by 4 and add 1.

Pieces	Feet	Description	Cost	Total
		1 X 8 X LENGTH COLLAR BEAMS		

Carpentry Work — Gable Studs

When a gable roof is framed, gable studs are necessary between the end rafters of the gable roof and the top plate of each end wall. On a straight roof with two gable ends, measure one end and figure one stud for each linear foot. The height of the gable from the top of the wall plate is the length of the studs.

Pieces	Feet	Description	Cost	Total
		2 X 4 X LENGTH GABLE STUDS		

Carpentry Work — Roof Sheathing

On a plain gable roof, multiply the length of the ridge times the length of a common rafter to find the area of one side. Multiply this by 2 to find

the total roof area. If the roof area is to be shiplap or T&G, add 25 percent to this figure to find the number of board feet needed. If plywood is used then take the total area to be covered and divide by 32 to find the number of sheets of plywood.

Pieces	Feet	Description	Cost	Total
		SIZE X 4' X 8' PLYWOOD - ROOF SHEATHING		

Carpentry Work — Roofing

Roofing material is estimated by the square. A square is enough material to cover 100 square feet. Divide the square foot area of the roof by 100 to find the number of squares required. For asphalt shingles, add 1 1/2 square feet for every linear foot of ridge, eaves, hip, and valley. When asphalt roofing is used, building felt is placed between the roof sheathing and the finished roof. Building felt is sold in 500 square foot rolls. Add 10 percent for overlapping the edges. Take the square-foot area of the roof, add 10 percent, and divide by 500 to determine the number of rolls of felt.

Pieces	Feet	Description	Cost	Total
		SQUARE ASPHALT ROOFING		
		ROLLS 15-POUND ASPHALT FELT - ROOF		

Carpentry Work — Drip Edge

When asphalt shingles are used for the finished roofing, a metal drip edge is installed along the eaves and the rake of the house. Drip edge is manufactured in 10-foot lengths. It is nailed over the top of the roofing felt and the asphalt shingles are installed over the top of the drip edge. To determine the amount of drip edge needed, measure the length of the eaves and the rake and add them together, then divide by 10 for the number of pieces required.

Pieces	Feet	Description	Cost	Total
		PIECES OF GALVANIZED DRIP EDGE - 10-FOOT LENGTHS		

Carpentry Work — Cornices

All cornices are made up of finished millwork consisting of boards and moldings used in combination with each other to finish the ends of the rafters which extend beyond the face of the outside walls. Cornices are listed by the linear feet. The parts are named according to the position they are placed in. Usually for a tight cornice the rake is made using 5/4 stock material. Start with your rakes and list the cornice material.

Pieces	Feet	Description	Cost	Total
		LINEAR FEET 5/4 X SIZE RAKE		
		LINEAR FEET 1 X SIZE FASCIA		
		LINEAR FEET 1 X SIZE SOFFIT AND FRIEZE		
		LINEAR FEET CROWN MOLD (SIZE) - CORNICE		
		LINEAR FEET BED MOLD (SIZE) - CORNICE		

Carpentry Work — Exterior Sheathing

To determine the amount of sheathing needed, take the distance (perimeter) around the house and multiply this by the height. If you have a gable roof, do not forget the gable ends. This will give you the total square-foot area to be covered. If shiplap is used, add 25 percent to this total. If plywood is used, take the total square-foot area to be covered and divide by 32 to give you the number of sheets of plywood required.

Pieces	Feet	Description	Cost	Total
		THICKNESS X 4' X 8' PLYWOOD - SHEATHING		

Carpentry Work — Siding

Siding is estimated according to the material used. If it is shingles and sold by the square, then you figure the number of squares required. If it is sold by the board foot, then you find the number

of square feet, add for cutting and waste and list it as board feet. All sheathing should be covered with a building paper before the finished siding is installed. Kraft building paper is sold in 500 square foot rolls. Take the square-foot area to be covered, add 10 percent for lapping, and divide by 500.

Pieces	Feet	Description	Cost	Total
		(THICKNESS X SIZE) BEVEL SIDING		
		BUNDLES OF WOOD SHINGLES-SIDING		
		ROLLS KRAFT BUILDING PAPER - SIDING		

Carpentry Work — Exterior Finish Work

If you have any special work to be done to finish the outside of the building, you would list this work now. This would cover such items as corner boards, porches, special overhangs, entrance porch railings, etc.

Pieces	Feet	Description	Cost	Total
		LINEAR FEET 5/4 X SIZE CORNER BOARDS		
		THICKNESS X 4' X 8' EXTERIOR PLYWOOD - OVERHANG, ETC.		
		WOOD LOUVERS NO. 600-L - GABLE ENDS		
		ANY FRAMING NECESSARY FOR ENTRANCE ROOFS, PORCHES, ETC.		
		FINISH WORK FOR ANY ENTRANCES, PORCHES, ETC.		
		WROUGHT IRON RAILINGS, ETC.		

CARPENTRY ESTIMATE: INTERIOR WORK AND FINISHES

Carpentry Work — Insulation

In estimating ceiling insulation, figure the square-foot area to be covered. Fiberglass insulation is sold in batt form. The batts are 15" x 48"

and come ten pieces to the bag. Each bag covers 50 square feet. Determine the square-foot area to be covered and divide this by 50 to get the number of bags required.

Pieces	Feet	Description	Cost	Total
		BAGS (THICKNESS) FIBERGLASS INSULATION - CEILINGS		

In estimating insulation for the sidewalls, do not figure the gable ends. Multiply the perimeter of the house by the height of the walls to find the total area in feet; 3 1/2-inch thick fiberglass insulation is sold in 70 square foot rolls so divide 70 into the total square-foot area to get the number of rolls required.

Pieces	Feet	Description	Cost	Total
		BAGS (THICKNESS) INSULATION - SIDEWALLS		

Carpentry Work — Wallboard

Estimating wallboard can be done in two ways. One is to lay out a sketch of each wall and ceiling, showing the dimensions. From this the number of lengths and pieces can be determined. The second system is to figure the square foot area to be covered. The amount of wallboard needed for the ceiling is the same as our floor area. If you have two floors, double this amount. In figuring wallboard, no deductions are made for openings, so take the length of the outside walls and multiply them by the height. The length of our interior partitions has been established in our framing. Add the total length of all interior partitions and multiply this by 8 (8-foot ceiling heights) for the square-foot area. Interior walls have wallboard on both sides, so double this figure. Add the ceiling areas, the outside wall areas, and the total interior wall areas for the total amount of wallboard required.

Pieces	Feet	Description	Cost	Total
		SQUARE FEET (SIZE) WALLBOARD - CEILING AND SIDEWALLS		

Carpentry Work — Joint System

Joint compound is a premixed vinyl-based cement that is used directly from the container. It is used to apply the paper to the joints of the wallboard and for concealing the nail heads. The tape that goes over the seams comes in 250- and 500-foot rolls of tape. Joint compound is sold in 5- and 1-gallon cans. Estimate one 250-foot roll of tape and six gallons of compound for every 1,000 square feet of wallboard.

Pieces	Feet	Description	Cost	Total
		ROLLS PERFA-TAPE 250 FEET - WALLBOARD		
		5 GALLON CANS JOINT CEMENT - WALLBOARD		

Carpentry Work — Molding

Cove molding is estimated by the linear foot. The distance around the outside of the house is known. The total length of the interior partitions is already known. The cove mold goes on each side of the interior partitions. Double the length of the interior partitions and add this to the perimeter of the exterior walls.

Pieces	Feet	Description	Cost	Total
		LINEAR FEET (SIZE) COVE MOLD - CEILING		

Carpentry Work — Baseboard and Carpet Strip

Baseboard and carpet strip can be estimated in two ways. One is to find the perimeter of each room and deduct the width of the door openings. The other is to deduct twice the width of the interior door openings and the width of the exterior door openings and subtract from the total length of the ceiling moldings.

Pieces	Feet	Description	Cost	Total
		LINEAR FEET BASE (SIZE) AND CARPET STRIP (SIZE)		

Carpentry Work — Underlayment

To estimate the underlayment, take the square-foot area where the underlayment is required. Divide this area by 32 to get the number of pieces of 4' x 8' needed.

Pieces	Feet	Description	Cost	Total
		(THICKNESS) X 4' X 8' UNDER-LAYMENT - GIVE LOCATION		

Interior Finish — Linoleum or Resilient Flooring

To estimate the quantity, find the square-foot area to be covered and then divide by 9 for square yards. If flooring is foot square, then list by the piece.

Pieces	Feet	Description	Cost	Total
		SQUARE YARDS INLAID LINOLEUM - LOCATION		
		SQUARE FEET RESILIENT FLOORING - LOCATION		

Interior Finish — Slate Flooring

Slate is sold by the square foot. Find the square-foot area to be covered. You will also need one gallon of special adhesive for every 100 square feet of slate, and five pounds of grout for every 100 square feet.

Pieces	Feet	Description	Cost	Total
		SQUARE FEET MULTI-COLORED SLATE - LOCATION		
		QUARTS SPECIAL SLATE ADHESIVE		
		POUNDS GROUT - FOR SLATE JOINTS		

Carpentry Work — Hardwood Flooring

Find the square-foot area where the hard-

wood flooring is required. Allow one-third for matching. Add the total square-foot area and the third for the total board feet. Deadening felt is laid between the subfloor and the underlayment and hardwood flooring. Felt is sold in 500 square-foot rolls. Divide 500 into the square-foot area where the deadening felt is required.

Pieces	Feet	Description	Cost	Total
		1 X 3 SELECT (GRADE) OAK (TYPE) - FLOORING		
		ROLLS DEADENING FELT		

Interior Finish — Ceramic Wall Tile

Find the square-foot area to be covered. Tile is sold by the square foot. Cap tile is sold by the linear foot. Inside and outside corners are sold by the piece. Ceramic fixtures such as soap and grab, grab bars, towel bars, etc., are sold by the piece. One gallon of ceramic adhesive will cover approximately 100 square feet; five pounds white cement grout will cover about 60 square feet of wall tile.

Pieces	Feet	Description	Cost	Total
		SQUARE $4\frac{1}{4}$" x $4\frac{1}{4}$" CERAMIC WALL TILE		
		LINEAR FEET CERAMIC TILE CAP		
		INSIDE CORNERS CAP		
		OUTSIDE CORNERS CAP		
		CERAMIC FIXTURES (SOAP AND GRAB, ETC.)		
		QUARTS CERAMIC TILE ADHESIVE		
		POUNDS WHITE CERAMIC TILE GROUT		

Interior Finish — Ceramic Floor Tile

Floor tile is estimated by the square foot. One gallon of adhesive is required for every 100 square feet and five pounds of white grout is needed for every 100 square feet. A marble threshold is needed between doors where ceramic tile and other type floors meet.

Pieces	Feet	Description	Cost	Total
		SQUARE FEET CERAMIC FLOOR TILE		
		QUARTS CERAMIC FLOOR TILE ADHESIVE		
		POUNDS WHITE GROUT FOR CERAMIC FLOOR TILE		
		6-INCH MARBLE THRESHOLD X LENGTH		

Carpentry Work — Exterior Doors

List exterior doors first. Door frames and sills are listed according to size. Doors are always listed first width and then height and then thickness. Any special entrance doors with sidelights or special heads should be listed first.

Pieces	Feet	Description	Cost	Total
		EXTERIOR DOOR FRAME WITH SILL 3'0" X 6'8" - FRONT DOOR		
		EXTERIOR DOOR FRAME WITH SILL 2'8" X 6'8" - SIDE DOOR		
		EXTERIOR DOOR 3'0" X 6'8" X 1¾" M-108 - FRONT DOOR		
		EXTERIOR DOOR 2'8" X 6'8" X 1¾" NO. -SIDE DOOR		
		COMBINATION STORM AND SCREEN DOOR 3'0" X 6'8" NO. -FRONT		
		COMBINATION STORM AND SCREEN DOOR 2'8" X 6'8" NO. -SIDE		
		LINEAR FEET 1¾" DOOR STOP- EXTERIOR DOORS		
		LINEAR FEET CASING - EXTERIOR DOORS		

Carpentry Work — Interior Doors

Interior doors are listed the same as the exterior doors, according to size and style. You list any cased opening first.

Pieces	Feet	Description	Cost	Total
		DOOR JAMBS 3'0" X 6'8"		
		DOOR JAMBS 2'8" X 6'8"		
		DOOR JAMBS 2'6" X 6'8"		
		DOOR JAMBS 2'4" X 6'8"		
		DOOR JAMBS 2'0" X 6'8"		
		LOUVER DOORS (SIZE X HEIGHT X THICKNESS)		
		DOORS (SAMPLE) 2'6" X 6'8" X 1⅜" BIRCH FLUSH		

Carpentry Work — Door Stop and Casing

This is listed by the linear foot. It requires approximately 17 linear feet of stop for each door and 34 linear feet of casing for both sides of a door opening. Count the number of doors and multiply by 17 for the stop and by 34 for the linear feet of casing.

Pieces	Feet	Description	Cost	Total
		LINEAR FEET DOOR STOP - INTERIOR DOORS		
		LINEAR FEET CASING - INTERIOR DOORS		

Carpentry Work — Closet Shelves and Rods

List this by the linear feet. Closet shelves are usually 1 x 12 pine boards and the supports are 1 x 3 pine. Closet rods are metal chrome plated and are usually adjustable.

Pieces	Feet	Description	Cost	Total
		LINEAR FEET 1 X 3 PINE - CLOSET SHELVES SUPPORTS		
		LINEAR FEET 1 X 12 PINE - CLOSET SHELVES		
		ADJUSTABLE CLOSET RODS		

Carpentry Work — Windows

Windows are listed according to size and style. Starting at the first floor, list each window on the first floor. Do the same for the second floor. Group them and list according to size.

Pieces	Feet	Description	Cost	Total
		SINGLE DOUBLE HUNG 3'0" X 4'6" (STYLE) LOCATION		
		SINGLE DOUBLE HUNG 3'0" X 3'2" (STYLE) LOCATION		
		MULLION DOUBLE HUNG 2'8" X 4'6" (STYLE) LOCATION		
		AWNING WINDOW - SIZE AND STYLE		
		PICTURE WINDOW (SAMPLE) 2'0" X 4'6" X 5'0" X 4'6" X 2'0" X 4'6"		
		STORM SASH AND SCREEN 3'0" X 4'6" - LOCATION		
		STORM SASH AND SCREEN 3'0" X 3'2" - LOCATION		
		ETC.		

Carpentry Work — Window Trim

Window trim is listed by the linear foot. Estimate trim required for each window separately and then group together. List according to part and style.

Pieces	Feet	Description	Cost	Total
		LINEAR FEET WINDOW CASING (STYLE)		
		LINEAR FEET WINDOW STOP (STYLE)		
		LINEAR FEET WINDOW STOOL (STYLE)		
		LINEAR FEET WINDOW APRON (STYLE)		
		LINEAR FEET WINDOW MULLION CASING (STYLE)		

Carpentry Work — Fireplace Mantel

List according to style and manufacturer. This is found on the working drawings and also in the specifications.

Pieces	Feet	Description	Cost	Total
		MANTEL M-1448 – FIREPLACE		

Carpentry Work — Blinds or Shutters

List according to size and style.

Pieces	Feet	Description	Cost	Total
		PAIR BLINDS (SIZE) AND (STYLE) – LOCATION		

Carpentry Work — Stairs

List by size and style. List each part separately.

Pieces	Feet	Description	Cost	Total
		SET HOUSED STRINGERS WITH RISERS AND TREADS		
		NEWEL POSTS (STYLE)		
		BALUSTERS (STYLE AND LENGTH)		

		Description		
		LINEAR FEET HANDRAIL (STYLE)		
		OAK TREADS (SIZE AND LENGTH)		
		(MATERIAL) RISERS (SIZE AND LENGTH)		
		(MATERIAL) LANDING TREAD (SIZE AND LENGTH)		

Interior Finish — Hardware

This is usually made as an allowance and is found in the specifications. You list the hardware allowance on the list. Nails should have an allowance because this is not covered with the hardware.

Pieces	Feet	Description	Cost	Total
		FINISH HARDWARE ALLOWANCE		
		NAILS ALLOWANCE		

Carpentry Work — Kitchen Cabinets and Bath Vanities

List by units. List the wall units first, the base units second, and then the linear feet of counter tops. List giving size and manufacturer's finish. Width first then height.

Pieces	Feet	Description	Cost	Total
		W3018 WALL CABINET – OAKMONT		
		W1830 WALL CABINET – OAKMONT		
		W3330 WALL CABINET – OAKMONT		
		W2430 WALL WITH SPIN SHELVES – OAKMONT		
		W3315 WALL CABINET – OAKMONT		

Pieces	Feet	Description	Cost	Total
		B-18" BASE CABINET - OAKMONT		
		SF-42" SINK FRONT - OAKMONT		
		CB-45 SQUARE CORNER BASE - OAKMONT		
		LINEAR FEET LAMINATED PLASTIC COUNTER TOP - MOLDED EDGES		
		BATH VANITY RV-36" - LOCATION		
		BATH VANITY RV-30" - LOCATION		
		LAMINATED PLASTIC COUNTER TOPS FOR ABOVE		
		1/4" SILVER-PLATED GLASS MIRRORS (SIZE) - LOCATION		
		ANY MEDICINE CABINETS REQUIRED (SIZE AND MODEL - LOCATION)		

PAINTING

Exterior Painting — Siding

Figure the square-foot area to be painted. Allow 450 square feet to the gallon for one coat of primer. House paint should cover 600 square feet per gallon over the prime coat. Take the square-foot area to be covered and divide by 450 for the primer and by 600 for the house paint.

Pieces	Feet	Description	Cost	Total
		GALLONS EXTERIOR PRIMER		
		GALLONS EXTERIOR HOUSE PAINT		

Exterior Painting — Cornices and Porches

Figure two square feet for every linear foot of a box cornice and 1/2 square foot for every linear foot of a tight cornice such as the rake. The sash and frame of one side of an average window is equal to 25 square feet. Most doors are estimated at 25 square feet for one side.

Pieces	Feet	Description	Cost	Total
		GALLONS TRIM AND SHUTTER PAINT - EXTERIOR		
		LINEAR FEET (SIZE AND MATERIAL) COVE MOLDING		
		PINTS WHITE GLUE		
		WEDGES		

Interior Painting — Walls and Ceilings

This is figured by the square-foot area. Allow 450 square feet to the gallon for wall and ceiling primer and 600 square feet to the gallon for the finish coat. Divide 450 into the square-foot area to be covered by the primer and 600 into the square-foot area for the second coat.

Pieces	Feet	Description	Cost	Total
		GALLONS INTERIOR WALL PRIMER		
		GALLONS INTERIOR FLAT WALL PAINT		
		GALLONS INTERIOR SATIN WALL PAINT		

Interior Painting — Trim and Doors

Paint for the interior trim is the same as that for the outside windows. Figure 25 square feet for each window and 25 square feet for each door. For each cased opening, figure 20 square feet. For the baseboard, figure one square foot for every three linear feet. Figure 500 square feet for each gallon of undercoat enamel and 500 feet for the finish coat.

Pieces	Feet	Description	Cost	Total
		GALLONS UNDERCOAT ENAMEL		
		GALLONS ENAMEL		
		GALLONS VARNISH FOR DOORS (NATURAL)		
		GALLONS SHELLAC - FILLER FOR DOORS		

Interior Painting — Finishing Hardwood Floors

Wood filler covers 600 square feet to the gallon. Varnish covers 600 square feet to the gallon. Take the square-foot area of the hardwood floors and divide by 600 to determine the number of gallons of filler and varnish. Varnish usually requires two-coat work.

Pieces	Feet	Description	Cost	Total
		GALLONS PASTE WOOD FILLER - FLOORS		
		GALLONS VARNISH - FLOORS		

Painting — Miscellaneous Supplies

You should make an allowance for miscellaneous paint supplies such as drop cloths, brushes, paint thinners, etc.

Pieces	Feet	Description	Cost	Total
		MISCELLANEOUS PAINT SUPPLIES ALLOWANCE		

LABOR ESTIMATING GUIDELINE

In estimating labor, it is necessary to establish guidelines as a starting point. Conditions, labor market, and availability of competent workers will affect the amount of work performed. Just as material estimating has factors beyond the estimator's control, such as how the workmen will use certain materials on the job, also in estimating labor, you will find that there are factors beyond your control. This checklist is an estimate based upon the experience of many successful contractors. You will have to adjust your estimate, taking into consideration the ability of your workmen and their methods of getting a job done.

We will follow the same sequence in estimating labor, as we did in establishing the material list. In this manner, with experience, you will be able to estimate the materials required and the labor necessary to put those materials in place according to the working drawings and the specifications.

The first thing you should do in any estimating either for materials or labor, is to study and fully understand the working drawings and the specifications for that particular project. After you fully understand the size, scope, and requirements of the project by studying the working drawings and specifications, then you should start your estimate with *site work.*

The site should always be visited to ascertain the existing conditions. This is important in understanding what preparatory work must be done before excavation can be started. If there is a large amount of brush or any large trees to be removed, this must be figured in the cost. Large trees can be expensive to cut down and haul away. Also the site should be visited to ascertain if any excess earth has to be removed or additional fill brought in.

We will start our labor estimate with site work.

SITE WORK

1. Trees to be removed from premises _____   @ _____ = _____

2. Excess earth to be removed from premises _____ yards   @ _____ = _____

3. Fill to be brought to premises _____ yards   @ _____ = _____

4. Brush to be removed — hours unskilled _____   @ _____ = _____

MASONRY WORK

Foundation Work

Layout, stakes, batter boards, levels, etc. _____ hours unskilled @ _____ = _____

_____ hours skilled @ _____ = _____

Excavation — With a bulldozer, 100 cubic yards of earth can be removed in three (3) hours.

_____ hours skilled @ _____ = _____

Backfilling — With a bulldozer, 100 cubic yards of earth can be backfilled in one (1) hours.

_____ hours skilled @ _____ = _____

Digging footings — 100 linear feet based on 8" x 16" requires 5 manhours of unskilled labor.

_____ linear feet _____ hours unskilled @ _____ = _____

Footing formwork — 100 linear feet setting forms to level grade requires 2 hours skilled and 2 hours unskilled labor.

_____ linear feet _____ hours skilled @ _____ = _____

_____ linear feet _____ hours unskilled @ _____ = _____

Pouring concrete footings — 100 linear feet, placing ready mix concrete, average conditions and wheeling distance to forms, figure 1/2 hour skilled and 4 hours unskilled.

_____ linear feet _____ hours skilled @ _____ = _____

_____ hours unskilled @ _____ = _____

Concrete floors — base preparation — For every 100 square feet, placing, grading, tamping base material, figure 1 1/4 hours unskilled. Setting forms to level grade for screeding concrete, figure 1/2 hour skilled.

_____ square feet _____ hours skilled @ _____ = _____

_____ hours unskilled @ _____ = _____

Placing vapor barrier, polyethylene, sisal kraft, etc. requires 1/2 hour unskilled for every 100 square feet.

_____ square feet _____ hours unskilled @ _____ = _____

Placing reinforcing rods or wire mesh, for every 100 square feet figure 1/2 hour unskilled labor.

_____ square feet _____ hours unskilled @ _____ = _____

Pouring and finishing, for every 100 square feet of concrete 4-inch thick, figure 1 1/2 hours of skilled and 1 1/2 hours of unskilled

_____ square feet _____ hours skilled @ _____ = _____

_____ hours unskilled @ _____ = _____

Porches and Sidewalks — For every 100 square feet of porches and walks for grading, leveling, and forming, figure 3/4 hour skilled and 1/2 hour unskilled. Pouring and finishing concrete, for every 100 square feet figure 1 1/2 hours skilled and 1 1/2 hours unskilled. Removing forms after the concrete has set, figure 1/2 hour unskilled. For every 100 square feet of porches and sidewalks, figure 2 1/4 hours of skilled and 2 1/2 hours of unskilled.

_____ square feet _____ hours skilled @ _____ = _____

_____ hours unskilled @ _____ = _____

Porch steps — For every 10 square feet of tread with an 8-inch rise and 12-inch tread to form requires 2 hours skilled and 1 hour unskilled. Pouring concrete and finishing requires 1 hour skilled and 1/2 hour unskilled. Removing forms and finishing requires 1 hour skilled and 1/2 hour unskilled. For every 10 square feet of tread figure 4 hours skilled and 2 hours unskilled in all for steps of poured concrete.

_____ square feet _____ hours skilled @ _____ = _____

_____ hours unskilled @ _____ = _____

Brick treads and risers, tooled joints — For every 10 square feet of tread area, figure 5 hours skilled and 2 1/2 hours unskilled.

_____ square feet _____ hours skilled @ _____ = _____

_____ hours unskilled @ _____ = _____

Rough slate or stone treads to 4-inches thick — For every 10 square feet of tread area, figure 2 hours skilled and 2 hours unskilled.

_____ square feet _____ hours skilled @ _____ = _____

_____ hours unskilled @ _____ = _____

Concrete block — Based on 100 square feet of wall area with average conditions, struck or tooled joints, common bond, including openings:
8 x 8 x 16 cement block requires 6 hours skilled and 6 hours unskilled. 10 x 8 x 16 cement block requires 6 1/2 hours skilled and 6 1/2 hours unskilled. 12 x 8 x 16 cement block requires 7 hours skilled and 7 hours unskilled. Placing reinforcing rods requires 1/2 hour unskilled.

Block Foundation

_____ square feet 8 x 8 x 16 _____ hours skilled @ _____ = _____

_____ hours unskilled @ _____ = _____

_____ square feet 10 x 8 x 16 _____ hours skilled @ _____ = _____

_____ hours unskilled @ _____ = _____

_____ square feet 12 x 8 x 16 _____ hours skilled @ _____ = _____

_____ hours unskilled @ _____ = _____

Wall Reinforcing

_____ square feet _____ hours unskilled @ _____ = _____

Concrete foundation wall — Figuring a wall 8 inches thick with plywood forms, 1/2-inch reinforcing, pouring concrete, removing forms and hand rubbing walls, for every 100 square feet of wall figure 3 hours skilled and 6 1/4 hours unskilled. Poured Concrete Wall with 1/2-inch reinforcing rods 8 inches thick.

_____ square feet _____ hours skilled @ _____ = _____

_____ hours unskilled @ _____ = _____

Foundation windows — For each unit either wood or metal, figure 1/4 hour skilled. Setting poured-in-place basement frames and sash figure 1/2 hour skilled.

Basement windows _____ each _____ hours skilled @ _____ = _____

Fireplace — Estimating average fireplace for base, firebrick hearth, setting prefabricated unit, and brick facing, figure 13 1/2 hours of skilled and 14 hours of unskilled. For exterior face brick on chimney with one (1) 8 x 8 flue and one (1) 8 x 12 flue liner, figure 2 hours skilled and 2 hours unskilled for every foot in height. Scaffolding, figure 3 hours of unskilled labor for metal scaffolding erection and dismantling for an outside fireplace to 36-feet in height.

Fireplace — Interior work — brick, tile, and unit

_____ hours skilled @ _____ = _____

_____ hours unskilled @ _____ = _____

Exterior work _____ height _____ hours skilled @ _____ = _____

_____ hours unskilled @ _____ = _____

Scaffolding _____ height _____ hours unskilled @ _____ = _____

Brick Veneer — 100 square feet of facebrick based on 3/8-inch flush joint, normal openings, sills and headers to 16-foot height, 4-inch brick veneer, stretcher bond, including wall ties, figure 12 hours skilled and 8 hours unskilled. Cleaning brick using muriatic acid and water, figure 1 hour skilled and 1/2 hour unskilled.

Brick veneer _____ square feet

_____ hours skilled @ _____ = _____

_____ hours unskilled @ _____ = _____

Cleaning w/muriatic acid _____ square feet

_____ hours skilled @ _____ = _____

_____ hours unskilled @ _____ = _____

Masonry Dampproofing — 100 square feet of cement plaster on block, one coat work with two (2) coat brush coats asphalt or tar, figure 2 hours skilled and 2 1/2 hours unskilled.

Dampproofing _____ square feet

_____ hours skilled @ _____ = _____

_____ hours unskilled @ _____ = _____

Drain tile — 100 linear feet of digging and installing 6-inch perforated drain pipe, 10-foot lengths, and covered with No. 2 crushed stone, figure 14 3/4 hours unskilled labor.

Drain Tile _____ linear feet

_____ hours unskilled @ _____ = _____

CARPENTRY WORK

Girders — Built-up girders nailed together, set in place, leveled and support posts in place, for every 10 linear feet, figure 1 hour skilled and 1/2 hour unskilled.

Girder 3-2 x __ x __ linear feet _____

_____ hours skilled @ _____ = _____

_____ hours unskilled @ _____ = _____

Basement stairs — Frame per stairway, average type straight flight to 4 feet wide and 12 feet long, rough cutting, framing, placing, figure 5 hours skilled and 2 hours unskilled.

Basement stairs 2 x 10s and 2 x 8s _____ hours skilled @ _____ = _____

_____ hours unskilled @ _____ = _____

Floor joists — first floor — 100 square feet including sill, plate, box sill, bridging and plywood subflooring, using 2 x 8 joists 16 inches o.c., up to 12-foot span, figure 5 1/2 hours skilled and 2 1/4 hours unskilled.

First Floor Platform with 2 x 8s-16 inch o.c. with 5/8 x 4 x 8 plywood.

_____ square feet _____ hours skilled @ _____ = _____

_____ hours unskilled @ _____ = _____

First floor studding — exterior walls — For every 100 square feet, including normal openings, average type outside walls, frame or veneer construction with plates, headers, fillers, bracing, firestops, girts with 2 x 4s-16 inches o.c. from 8-foot to 12-foot heights, figure 2 1/2 hours skilled and 1/2 hour unskilled labor. With 4' x 8' plywood sheathing to 25/32 inch thick figure 1 hour skilled and 1/2 hour unskilled. For 100 square feet of exterior walls with 2 x 4 framing and 1/2" x 4' x 8' plywood sheathing, figure 3 1/2 hours skilled and 1 hour unskilled.

First floor exterior walls _____ square feet

_____ hours skilled @ _____ = _____

_____ hours unskilled @ _____ = _____

First floor studding — interior walls — For every 100 square feet of wall including normal openings, plates, headers and studs 2 x 4s-16 inches o.c., figure 2 hours skilled and 1/4 hour unskilled.

First floor interior walls _____ square feet

_____ hours skilled @ _____ = _____

_____ hours unskilled @ _____ = _____

Second floor framing — For every 100 square feet including box sill, joists, bridging and 5/8-inch plywood subfloor on 2 x 8s-16 inches o.c., figure 7 1/2 hours skilled and 2 1/2 hours unskilled.

Second floor platform 2 x 8s-16 inches o.c. with 5/8" x 4 x 8 plywood.

_____ square feet _____ hours skilled @ _____ = _____

_____ hours unskilled @ _____ = _____

Second floor exterior walls — For every 100 square feet 2 x 4 framing with 1/2-inch plywood sheathing, figure 4 1/2 hours skilled and 1 1/2 hours unskilled.

_____ square feet _____ hours skilled @ _____ = _____

_____ hours unskilled @ _____ = _____

Second floor interior walls — For every 100 square feet 2 x 4 framing, figure 2 1/2 hours skilled and 1/2 hour unskilled.

_____ square feet _____ hours skilled @ _____ = _____

_____ hours unskilled @ _____ = _____

Overlays or ceiling joists — For every 100 square feet, normal construction first and second floor levels, including bridging, trimmers to 16-foot spans using 2 x 6s-16 inches o.c., figure 5 hours skilled and 2 hours unskilled labor.

_____ square feet _____ hours skilled @ _____ = _____

_____ hours unskilled @ _____ = _____

Rafters — For every 100 square feet, average-type construction to 22 foot lengths, using 2 x 6s-16 inches o.c. and 1/2" x 4' x 8' plywood roof sheathing, figure 3 3/4 hours skilled and 1 1/2 hours unskilled.

_____ square feet _____ hours skilled @ _____ = _____

_____ hours unskilled @ _____ = _____

Gable studs — For every 100 square feet using 2 x 4s, 1/2" x 4' x 8' sheathing, figure 4 1/2 hours skilled and 1-1/4 hours unskilled labor.

_____ square feet _____ hours skilled @ _____ = _____

_____ hours unskilled @ _____ = _____

Roofing — Asphalt Shingles — For every 100 square feet of strip shingles 10" x 36" with 15-pound asphalt felt underlayment, figure 2 hours skilled and 1 1/4 hours unskilled.

Asphalt shingles _____ square feet

_____ hours skilled @ _____ = _____

_____ hours unskilled @ _____ = _____

Wood shingle roof — For every 100 square feet using 16-inch wood shingle with 5-inch exposures, figure 4 hours skilled and 1 1/2 hours unskilled.

Wood shingles _____ square feet

_____ hours skilled @ _____ = _____

_____ hours unskilled @ _____ = _____

Slate roofing — For every 100 square feet using 10" x 20", figure 3 1/2 hours skilled and 2 hours unskilled.

_____ square feet _____ hours skilled @ _____ = _____

_____ hours unskilled @ _____ = _____

CARPENTRY FINISH WORK — EXTERIOR

Cornice — For every 100 linear feet, including fascia, crown mold, soffit, bed mold, and drip edge, figure 12 hours skilled and 3 hours unskilled labor.

_____ linear feet _____ hours skilled @ _____ = _____

_____ hours unskilled @ _____ = _____

Wood siding — For every 100 square feet of 1/2″ x 8″ bevel siding with kraft building paper applied underneath, figure 2 3/4 hours skilled and 1 hour unskilled.

_____ square feet _____ hours skilled @ _____ = _____

_____ hours unskilled @ _____ = _____

Corner Boards — For every 100 linear feet, figure 3 hours skilled and 2 hours unskilled.

_____ linear feet _____ hours skilled @ _____ = _____

_____ hours unskilled @ _____ = _____

Front overhang soffit — For every 100 linear feet exterior plywood, 1 x 8 pine fascia, wood drip cap 24–30 inches wide, figure 6 hours skilled and 1 hour unskilled labor.

_____ linear feet _____ hours skilled @ _____ = _____

_____ hours unskilled @ _____ = _____

Louvers or vents — Each unit, figure 1/2 hour skilled labor for each.

Louvers _____ each _____ hours skilled @ _____ = _____

Porch railings — Wrought iron set into masonry steps and porches and attached to the framing, figure 1 hour skilled and 1/4 hour unskilled for each set.

Porch railing _____ sets _____ hours skilled @ _____ = _____

_____ hours unskilled @ _____ = _____

CARPENTRY — INTERIOR WORK

Insulation — ceilings — For every 100 square feet of nonrigid type 15″ x 48″ batts, figure 1/2 hour skilled and 1/2 hour unskilled labor.

Ceiling insulation _____ square feet

_____ hours skilled @ _____ = _____

_____ hours unskilled @ _____ = _____

Insulation — sidewall — For every 100 square feet of nonrigid-type blanket 16-inch to 8-foot lengths, figure 3/4 hour skilled and 1/2 hour unskilled.

Sidewall insulation _____ square feet

_____ hours skilled @ _____ = _____

_____ hours unskilled @ _____ = _____

Wallboard — ceilings — Gypsum board ceilings, for every 100 square feet 3/8″ x 4′-12′ lengths with joints finished ready for paint, figure 2 1/2 hours skilled and 1 1/2 hours unskilled.

Ceilings _____ square feet _____ hours skilled @ _____ = _____

_____ hours unskilled @ _____ = _____

Wallboard — sidewalls — For gypsum board walls, for every 100 square feet 3/8″ x 4′-10′ lengths with joints finished ready for paint, figure 2 hours skilled and 1 hour unskilled.

Sidewalls _____ square feet _____ hours skilled @ _____ = _____

_____ hours unskilled @ _____ = _____

Cove molding 1 3/4-inch — For every 100 linear feet of ceiling cove mold, figure 3 hours skilled and 1 hour unskilled.

Cove mold _____ linear feet _____ hours skilled @ _____ = _____

_____ hours unskilled @ _____ = _____

Baseboard and carpet strip — For every 100 linear feet of baseboard and carpet strip, figure 4 hours skilled and 1 hour unskilled labor.

Base and carpet strip _____ linear feet

_____ hours skilled @ _____ = _____

_____ hours unskilled @ _____ = _____

Underlayment — For every 100 square feet of 5/8″ x 4′ x 8′ plywood, figure 1 hour skilled labor.

Underlayment _____ square feet _____ hours skilled @ _____ = _____

Resilient-type flooring — For every 100 square feet resilient flooring, figure 3 hours skilled and 1/2 hour unskilled labor.

Resilient flooring _____ square feet

_____ hours skilled @ _____ = _____

_____ hours unskilled @ _____ = _____

Slate flooring — For every 100 square feet of random cut sizes to 12″ x 12″ including grouting or pointing of joints, figure 2 hours skilled and 2 hours unskilled.

Slate flooring _____ square feet _____ hours skilled @ _____ = _____

_____ hours unskilled @ _____ = _____

Ceramic floor tile — For every 100 square feet 1/2-inch to 2 inches with paper backing, figure 4 hours skilled and 4 hours unskilled labor.

Ceramic floor tile _____ square feet

_____ hours skilled @ _____ = _____

_____ hours unskilled @ _____ = _____

Ceramic wall tile — For every 100 square feet 4 1/4″ x 4 1/4″ including installing waterproof gypsum wallboard backing, adhesive, grouting and installing ceramic fixtures, figure 8 hours skilled and 8 hours unskilled labor.

Ceramic wall tile _____ square feet

_____ hours skilled @ _____ = _____

_____ hours unskilled @ _____ = _____

Hardwood flooring — For every square foot 25/32″ x 2 1/4″ face, including the placing of the deadening felt, figure 3-1/2 hours skilled and 1/2 hour unskilled.

Hardwood flooring _____ square feet

_____ hours skilled @ _____ = _____

_____ hours unskilled @ _____ = _____

CARPENTRY — MILLWORK AND TRIM

Exterior doors — Setting exterior door frames, wood residential type, setting wood threshold, hanging standard size door 1-3/4-inch with 3 butts, fancy outside lockset, and door trim, figure 5 hours skilled and 3/4 hour unskilled.

Exterior doors _____ each _____ hours skilled @ _____ = _____

_____ hours unskilled @ _____ = _____

Exterior combination storm/screen wood door complete with hardware and closer, standard size, figure 2 hours skilled and 1/4 hour unskilled.

Combination doors _____ each _____ hours skilled @ _____ = _____

_____ hours unskilled @ _____ = _____

Interior doors — Interior door jambs and heads, assembly from stock sections, setting interior door frames, standard sizes, hanging interior wood doors 1-3/8", door trim both sides, installing hardware, figure 5 1/4 hours skilled and 1 hour unskilled labor.

Interior doors _____ each _____ hours skilled @ _____ = _____

_____ hours unskilled @ _____ = _____

Closet shelving — For every 100 linear feet of shelving including 1 x 3 and 1 x 12 stock sizes and installing the adjustable closet rods, figure 3 hours skilled and 1 hour unskilled.

Closet shelving _____ linear feet _____ hours skilled @ _____ = _____

_____ hours unskilled @ _____ = _____

Window units — Each unit, setting complete window units, all types glazed to prepared opening including casing, stool, stop, apron, and hardware for windows 3' x 3' to 3' x 5'6", figure 2-1/4 hours skilled and 1/2 hour unskilled labor.

Window units _____ each _____ hours skilled @ _____ = _____

_____ hours unskilled @ _____ = _____

Storm/sash and screen (prefitted wood combination), for each unit, figure 1/2 hour skilled and 1/2 hour unskilled.

Storm and screen _____ each _____ hours skilled @ _____ = _____

_____ hours unskilled @ _____ = _____

Fireplace mantel — Each unit, setting average type factory-built mantel unit to prepared wall, figure 2 1/2 hours skilled and 1 1/2 hours unskilled labor.

Mantel unit _____ each _____ hours skilled @ _____ = _____

_____ hours unskilled @ _____ = _____

Blinds or shutters — Each pair, wood, small, figure 1/2 hour skilled labor. Medium, figure 3/4 hour skilled labor. Large, figure 1 hour skilled labor.

Blinds Small _____ pair _____ hours skilled @ _____ = _____

Medium _____ pair _____ hours skilled @ _____ = _____

Large _____ pair _____ hours skilled @ _____ = _____

Main stairs — Each flight, setting housed stringers, installing risers and treads, setting newel post, balusters, and handrail, figure 8 hours skilled and 3 hours unskilled.

Main stair, flight _____ each _____ hours skilled @ _____ = _____

_____ hours unskilled @ _____ = _____

Kitchen cabinets — For each 100 square feet of face area, average-type work base cabinets, figure 4 hours skilled and 2 hours unskilled labor. For each 100 square feet of face area for wall cabinets, figure 3 hours skilled and 2 hours unskilled.

Kitchen cabinets (factory built)

Base units _____ square feet _____ hours skilled @ _____ = _____

_____ hours unskilled @ _____ = _____

Wall units _____ square feet _____ hours skilled @ _____ = _____

_____ hours unskilled @ _____ = _____

Counter tops — For every 10-square-foot area, placing factory-built top in place, figure 1/2 hour skilled and 1/4 hour unskilled labor.

Counter top _____ square feet _____ hours skilled @ _____ = _____

_____ hours unskilled @ _____ = _____

Bathroom vanities — For each unit, factory built, placing, scribing and anchoring in place with laminated plastic top, figure 2 hours skilled labor.

Bathroom vanities _____ each

_____ hours skilled   @ _____ = _____

Medicine cabinets or mirrors over vanities — For each unit, figure 3/4 hour skilled and 1/2 hour unskilled.

Mirrors over vanity _____ each

_____ hours skilled   @ _____ = _____

_____ hours unskilled @ _____ = _____

PAINTING: EXTERIOR WORK

Exterior walls — For every 100 square feet with oil, stain, or water-based paint, doing 1 coat primer and 1 coat finish on wood siding, figure 2 hours skilled labor.

Wood siding _____ square feet _____ hours skilled   @ _____ = _____

Exterior cornice — For every 25 linear feet of standard box cornice, figure 1 hour skilled.

Wood cornice _____ linear feet _____ hours skilled   @ _____ = _____

Exterior window trim — For each window medium size 2 coats, figure 3/4 hour skilled labor for each unit.

Window trim _____ windows _____ hours skilled   @ _____ = _____

Combination storm and screen windows, figure 3/4 hour skilled for each.

Combination storm sash _____ each

_____ hours skilled   @ _____ = _____

Exterior doors — For each exterior door finished two sides with 2 coat work, figure 1 hour skilled labor.

Exterior doors _____ each _____ hours skilled   @ _____ = _____

Combination storm doors — For each door finished 2 sides with 2 coat work, figure 1 1/2 hours skilled.

Combination storm doors _____ each

_____ hours skilled   @ _____ = _____

Shutters or blinds — For each pair with 2 coat work, figure 1 hour skilled labor.

Blinds _____ pairs _____ hours skilled   @ _____ = _____

PAINTING: INTERIOR WORK

Interior ceilings — For every 100 square feet of primer, flat, casein, or water-based paint, 2 coat work, figure 2 hours skilled labor.

Interior ceilings _____ square feet _____ hours skilled   @ _____ = _____

Interior walls — For every 100 square feet of primer, flat, casein, or water-based paints for 2 coat work, figure 1 1/2 hours skilled labor.

Interior walls _____ square feet _____ hours skilled   @ _____ = _____

Interior window trim — For each window, figure 1/2 hour skilled labor.

Window trim _____ each _____ hours skilled   @ _____ = _____

Interior doors – For each door, casing, jambs, stops for 2 coat work on two sides, figure 1 1/2 hours for each door.

Interior doors _____ each _____ hours skilled @ _____ = _____

Baseboard and carpet strip – For every 25 linear feet for 2 coat work, figure 1 hour skilled.

Base and carpet strip _____ linear feet

_____ hours skilled @ _____ = _____

Hardwood floors – For every 100 square feet of hardwood for filling, wiping and 2 coats of varnish, figure 1 3/4 hours skilled labor.

Hardwood floors_____ square feet _____ hours skilled @ _____ = _____

Stairs – For each complete open type stairs for each flight, figure 4 hours skilled for either stain, varnish, or paint.

Main Stairs _____ flights _____ hours skilled @ _____ = _____

Fireplace mantel – For each unit, figure 1/2 hour skilled labor for finishing with either stain or paint with 2 coat work.

Fireplace mantel _____ each _____ hours skilled @ _____ = _____

HARDWARE

This is usually made as an allowance and is found in the specifications. Just list this total on the material list. You should also list nails and make an allowance for them on your material list.

Pieces	Feet	Description	Cost	Total
		FINISH HARDWARE ALLOW		
		NAILS ALLOW (MAKE AN ALLOWANCE)		

You have now completed the material list for the construction of the building and have estimated the labor to put that material in place. If you are going to have sub-contractors do the painting, electrical, and the heating and plumbing work, you would get the complete installation price from them including all fixtures. These figures together with your material and labor figures will give you the total cost of the building. If you are a contractor, you must add your overhead (cost of insurance, equipment, bonds, sales tax, fees for permits, etc.) and also your profit for the job.

SUMMARY

The working drawings and the specifications work together to spell out in detail how a building is to be constructed, the material and workmanship required, and all placed in a workable order. We have determined the materials and labor necessary to build the structure according to the plans and the specifications starting with the site work and the excavation. We then figured the material required for the footings, foundation wall, chimney, concrete floors and porches, sidewalks, all masonry materials, all carpentry materials and labor for the exterior and interior finish including floors, walls, cabinets, insulation, finish floors, floor tile, wall tile, doors, and windows.

We should check continually the working drawings and the specifications while making our material list to make sure we are figuring what is called for and required.

REVIEW QUESTIONS

Select the letter preceding the best answer.

1. Excavation is figured by the _____ .
   (a) Square foot   (b) Square yard   (c) Cubic feet   (d) Cubic yard

2. Concrete footings are figured by the _____ .
   (a) Square foot   (b) Square yard   (c) Cubic feet   (d) Cubic yard

3. Concrete floors are estimated by the _____ .
    (a) Square foot   (b) Square yard   (c) Cubic feet   (d) Cubic yard

4. Cement block is figured by the _____ .
    (a) Square foot   (b) Square yard   (c) Piece   (d) Cubic yard

5. Flue liners are estimated by the _____ .
    (a) Piece   (b) Linear foot   (c) Square foot   (d) Cubic foot

6. Brick veneer is figured by the _____ .
    (a) Square foot   (b) Cubic foot   (c) Square yard (d) Cubic yard

7. Drain tile is figured by the _____ .
    (a) Square foot   (b) Linear foot   (c) Square yard   (d) Cubic yard

8. Sills are estimated by the _____ .
    (a) Linear foot (b) Piece   (c) Square foot   (d) Cubic foot

9. Floor joists are listed by the _____ .
    (a) Square foot   (b) Cubic foot   (c) Piece   (d) Yard

10. Bridging is estimated by the _____ .
    (a) Linear foot (b) Cubic foot (c) Square foot   (d) Square yard

11. Sub-flooring is figured by the _____ .
    (a) Piece (b) Square foot   (c) Cubic foot   (d) Square yard

12. Plate and shoe is figured by the _____ .
    (a) Square foot   (b) Square yard   (c) Cubic foot   (d) Linear foot

13. Studs are figured by the _____ .
    (a) Piece   (b) Square foot   (c) Cubic foot   (d) Linear foot

14. Rafters are listed by the _____ .
    (a) Piece   (b) Square foot   (c) Cubic foot   (d) Linear foot

15. Roofing is figured by the _____ .
    (a) Piece   (b) Square foot   (c) Square   (d) Bundle

16. Cornices are usually listed by the _____ .
    (a) Piece   (b) Linear foot   (c) Square foot   (d) Cubic foot

17. Insulation is figured by the _____ .
    (a) Piece   (b) Square foot   (c) Cubic foot   (d) bag

18. Wallboard is figured by the _____ .
    (a) Square foot   (b) Cubic foot   (c) Square yard   (d) Piece

19. Cove mold and base and carpet strip is estimated by the _____ .
    (a) Linear foot (b) Square foot   (c) Cubic foot   (d) Piece

20. Ceramic tile is listed by the _____ .
    (a) Square foot   (b) Linear foot   (c) Cubic foot   (d) Square yard

21. Windows and doors are usually listed by the _____ .
    (a) Unit   (b) Piece   (c) Square foot   (d) Cubic foot

22. Kitchen cabinets are listed by the _____ .
    (a) Piece   (b) Unit   (c) Square foot   (d) Cubic foot

SECTION 5 MECHANICAL
unit 40 plumbing

OBJECTIVES

After studying this unit the student will be able to

- explain the specifications for plumbing.
- list the quantity of materials required for the plumbing in a house by reading the specifications and drawings.

As with some of the other highly-specialized phases in the construction of a house, the plumbing work is usually performed by a subcontractor. The plumbing contractor will make an accurate list of materials before beginning the job. However, in order for the general contractor to estimate the overall cost of the project and act as the coordinator of all of the work, general knowledge of the basics in plumbing is necessary.

ROUGH PLUMBING

The plumbing is installed in two operations. First the *rough plumbing* is installed. This involves the installation of piping for fresh water and waste, up to the wall or floor surface where the fixtures are attached. When the fresh water *(supply)* pipes are installed, all openings are capped or plugged and pressure is applied to them. In this manner, the piping is tested for leaks before it is concealed in the structure.

Bathtubs and shower bases are fixtures, but they are installed when the rough plumbing is done. This is because the wall and floor covering must be installed up to the edge of the tub or shower. Shower bases are available in a wide variety of sizes, so the drawings and specifications must be consulted to determine the size required. Bathtubs may be ordered in several sizes, but 5 feet long by 2 feet 6 inches wide is considered a standard size. Bathtubs are available with a finished return at either end or with no returns at the ends, figure 40-1, so the tub must be enclosed between two

walls. They are available with either a left-hand drain or a right-hand drain. There are also steel and fiberglass assemblies, which include the tub and the walls in the tub area, figure 40-2.

Fig. 40-1 Bathtub with a return on the left end

Fig. 40-2 One piece fiberglass bathtub enclosure

The waste plumbing for a house includes the pipes and fittings that carry the waste water from the fixtures to the house sewer, which runs from the building to the municipal sewer or the house septic system, and the pipes and fittings that make up the vent system. Figure 40-3 shows several of the most common types of fittings. The piping from a fixture to the point in the plumbing system where it joins piping from other fixtures is called a *branch*. The large vertical pipe in the main house drain, into which the branches run, is called a *stack*. The *vent* runs vertically up through the roof to allow atmospheric pressure to enter the system and prevent vacuum from building up as the waste water is discharged.

Traps are included at each fixture and at the point where the sewer leaves the house. A trap is a U-shaped fitting, figure 40-4, which prevents sewer gas from entering the house. As water passes through the system, the trap remains filled, so that gasses cannot pass backward through it.

Normally, each branch of the supply includes a valve to shut off the water. This valve may be a *stop valve* (a simple valve to turn the flow on and

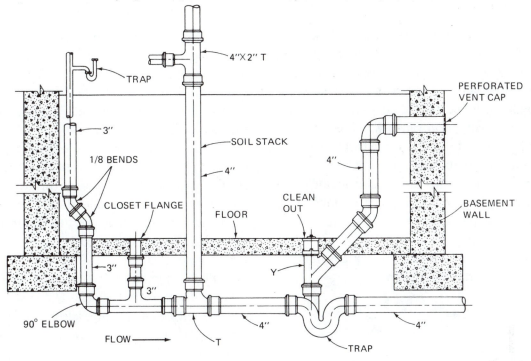

(A) COMMON FITTINGS FOR WASTE PLUMBING

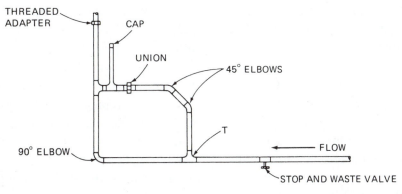

(B) COMMON FITTINGS FOR SUPPLY PLUMBING

Fig. 40-3 Waste fittings (A) are generally available in cast iron and plastic, supply fittings (B) are available in copper, galvanized iron, and plastic.

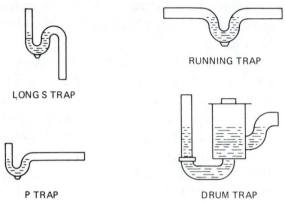

Fig. 40-4 Common types of traps

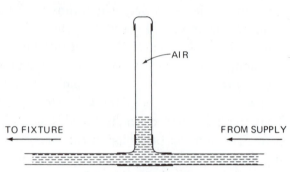

Fig. 40-5 Air chamber

off) or it may be a *stop and waste valve* (turns flow on and off and includes a means to drain the system). Usually, a stop and waste valve is used at a low point in the system, such as the main shut off for the building. The individual shut offs at the fixtures are usually stop valves which come with chrome-plated supply set, installed between the wall or floor and the fixture.

Each branch of the supply also includes an air chamber, figure 40-5. The air chamber is a short section of pipe, capped at the end to trap air. When a sudden surge of pressure occurs in the line, the trapped air acts like a shock absorber to prevent the pipes from hammering.

PLASTIC PIPING

In construction plastic piping is being more widely used because of its ease in handling. It is easy to fasten together at joints, and its cost is less than with copper and cast iron, which require oakum and lead at the joints.

Plastic pipe is made from polyvinyl chloride (PVC) which is used for cold water installation and chlorinated polyvinyl chloride (CPVC) which can be used for hot or cold water installation. You should check the label on the pipe to be sure if it pertains to hot and cold water installation or only for cold water installation.

Most brands of CPVC and PVC piping are sold in the same size as copper tubing. Just like copper tubing fittings, plastic piping 45 and 90 degree

elbows, straight couplings, tees, etc., are used to assemble the pipe. The pipe is recessed into these fittings approximately 1/2 to 5/8 inch. Most fittings have a shoulder inside the fitting, the same as copper, and the pipe should fit tightly against this shoulder when assembled.

Plastic pipe expands and contracts, so when installing, this should be taken into consideration. If the plastic piping goes through a framing member or a wall, the hole should be a little larger than the pipe to allow for this expansion. Plastic piping should be supported by hangers every three feet of pipe run.

Plastic piping is fastened together at all fittings with a solvent-weld. The pipe should be cleaned with a plastic pipe cleaner before the pipe cement is applied. Care must be taken when applying the cement and fastening it to the fitting. Once a weld is made, it cannot be changed. If necessary to make a change, the part must be cut out and discarded.

DVW stands for Drain, Vent, and Waste. This is the larger diameter plastic pipe, usually three to four inches. It is the part of the plumbing system that removes waste from the house and vents, at the roof, the odors created by the waste.

Plastic piping and fittings are accepted by most National Plumbing Codes, but some local codes will not accept it. Plumbing codes, sometimes called sanitary codes, are necessary to protect the health of the homeowners. Be sure to check the plumbing department of your local government about your local plumbing code and what materials are required.

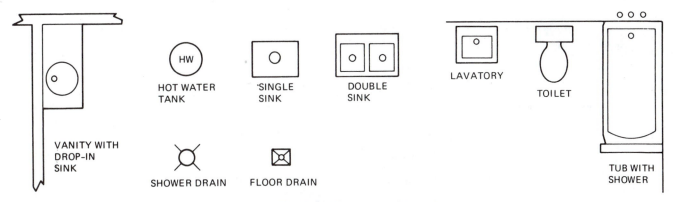

Fig. 40-6 Plumbing fixtures on a floor plan

FINISHED PLUMBING

Once the rough plumbing is installed, and the walls or floors are covered with whatever finish is to be used, the finished plumbing is installed. *Finished plumbing* includes fixtures, such as lavatories, sinks, water closets (toilets), and trim, such as faucets. Figure 40-6 shows how fixtures are indicated on a floor plan.

Water closets are available in a variety of types, and most manufacturers make more than one type; so the specifications must be consulted to determine which type is required.

Lavatories (wash basins) may be either wall hung or mounted in a built-in vanity cabinet. Many wall-hung lavatories require supporting legs. These legs may be included with the lavatory, or it may be necessary to order them separately. Lavatories vary in size from 12 inches by 12 inches, to 21 inches by 24 inches.

The water heater is basically a tank for storing water. It includes a device for heating the water with electricity, oil, or gas. Once the water is heated to the desired temperature, a thermostat shuts off the heating device. A relief valve is installed on the water heater, to release the pressure in the event that either the temperature or the pressure becomes dangerously high.

In addition to the above fixtures, the finished plumbing includes *sill cocks* (or *hose bibbs*), shower heads, faucets for all fixtures requiring them, tub and lavatory drains, and any other exposed hardware which is connected to the plumbing. Us-ually one or more sill cocks are included. A sill cock is a faucet mounted on the outside of the building to which a hose may be attached.

There are several materials which are commonly used for pipes and fittings. At one time, most fresh water plumbing was done with galvanized iron, and waste plumbing was done with cast iron soil pipe. Copper has replaced galvanized iron as the most common supply piping material. Plastic pipe and fittings are also available for supply plumbing, but copper is more generally used. When the plumbing is done with copper, fittings made of copper or brass are soldered to the pipe. The method of soldering used in plumbing is called *sweat* soldering and the fittings are called *sweat fittings*.

Waste plumbing may be done with any one of three materials. Cast iron pipe, called *soil pipe*, is still used frequently, but plastic and copper are also common. Cast iron soil pipe is manufactured in 5-foot lengths and is joined with *oakum* (a fibrous rope-like material) and a molten lead or a man-made material, called plastic lead. Plastic pipe is available in 10- or 20-foot lengths and is joined with a special adhesive. Copper waste plumbing is joined by sweat soldering, the same as copper supply plumbing. Usually all underground waste plumbing such as the house sewer is cast iron.

Many areas have plumbing codes which specify what material can be used for the plumbing. The estimator should be familiar with local and state plumbing codes.

The following are the specifications for the plumbing in the sample house:

Specifications

DIVISION 11: MECHANICAL – PLUMBING

A. WORK INCLUDED

Plumbing work includes, but shall not be limited by the following:

1. Complete systems of waste drainage, and vent piping, including all connections.

2. Complete systems of hot and cold water piping and all connections.

3. Furnishing, delivering, installing, and setting of all plumbing fixtures.

4. All excavation and backfilling for the plumbing.

B. GENERAL REQUIREMENTS

All work shall be in accordance with local, state or federal requirements, and shall comply with all applicable codes. The contractor shall obtain and pay for all permits and certificates of inspection if any are required for the plumbing.

C. SANITARY DRAINAGE SYSTEM

1. Sanitary drainage piping buried, shall be extra heavy asphalt coated hub and spigot soil pipe, of uniform weight and thickness, with corresponding cast iron soil pipe fittings. Above-ground piping shall be copper pipe with corresponding fittings.

2. Waste piping shall be uniformly pitched in the direction of the flow. Main drains shall be pitched approximately 1/8 inch per foot, with branch laterals pitched 1/4 inch per foot where possible.

3. Cleanouts shall be provided and installed as indicated.

D. VENT PIPING

1. Vent piping shall be installed, connected to fixtures and drainage piping as indicated.

2. Vents shall be extended through the roof to a height of approximately 12 inches above the roof deck. Vents shall be provided with approved flashing hub or cap.

E. INTERIOR WATER PIPING

1. All inside water piping shall be of standard hard drawn Type L copper except that exposed branches and exposed piping to fixtures shall be chrome brass.

2. All joints in copper tubing shall be sweat type. Soldered joints shall be thoroughly cleaned with steel wool and approved, oil-free, soldering paste.

3. All water piping shall be properly and adequately supported on hangers spaced not over eight feet apart.

F. FIXTURES

Kessler plumbing fixtures have been used as a guide. An approved equal will be accepted. Colors will be selected by the owner.

1. *Bath tub* shall be Royal K-786-SA recessed tub 5'-0" long, 1'-4" high. K-7004-T Benson shower and bath faucet with valet units. K-7370 shower head with volume regulator. K-7150-R1 1/2" pop-up drain.

2. *Water closets* shall be Johnson K-3512-PBA siphon jet with K-4662 Glimmer seat and cover. K-7638 supply with stop.

3. *Lavatories* shall be Continental K-2208-F round lavatory with K-7346-T Fisher supply faucet with valet units, aerator, and pop-up drain. K-7606 3/8" supplies with stops. K-900 1 1/4" cast brass P trap.

4. *Kitchen sink* shall be Titan, double bowl 21 x 32, Model SA-410; stainless steel with conventional trim. K-7665-T Marnell sink supply faucet.

5. *Water heater* shall be Williams brand, glass lined standard electric hot water heater; forty gallon; Model WG-40.

Estimating Plumbing

Some architects include elevations and plans of the plumbing with the working drawings. If these are not included with the drawings, the estimator will want to make a sketch of the plumbing system. The easiest way is to sketch the supply plumbing on one sheet, and the waste plumbing on another sheet. These sketches, in figures 40-7 and 40-8, were drawn from the Floor Plans of the sample house.

The estimate discussed here is for the plumbing within the house itself. Normally, the materials necessary to connect houses to the municipal sewer are included in the estimate, but these vary depending on the setback of the house and the location of the municipal sewer.

With the sketches drawn, the estimator can measure the length of the pipes, count the necessary fittings, and list the finished plumbing. The specifications indicate the material and size of the piping, and the manufacturer and model of each item of finished plumbing. Brand names and model numbers may be included to indicate the type and grade; substitutions are permitted.

SUMMARY

The plumbing in a house can be categorized in one of two ways: Waste and supply plumbing or rough and finished plumbing. Waste plumbing includes all of the pipes and fittings necessary to carry the waste water away from the house, while supply plumbing includes the pipes and fittings necessary to supply the fixtures with water. Rough plumbing is all of that which is installed before the walls and floors are covered, whereas finished plumbing includes the fixtures and all exposed hardware.

Plastic, cast iron, and copper are all used in modern plumbing, so the estimator must consult the specifications to determine which is used. Many localities have plumbing codes which control what materials may be used, as well as other features. The estimator should be familiar with these codes.

To estimate the materials for the plumbing in a house, first draw a sketch of the supply plumbing and one of the waste plumbing. The lengths of pipes and number of fittings required can easily be taken from these sketches. The type and number of fixtures can be listed directly from the specifications.

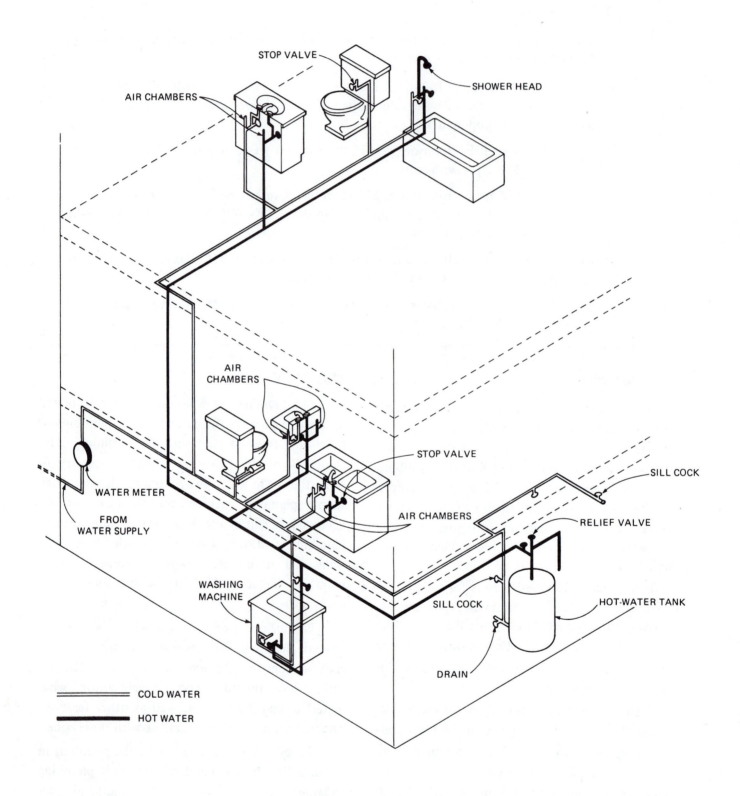

STOP VALVE

AIR CHAMBERS

SHOWER HEAD

AIR
CHAMBERS

STOP VALVE

SILL COCK

WATER METER

AIR CHAMBERS

RELIEF VALVE

FROM
WATER SUPPLY

WASHING
MACHINE

SILL COCK

HOT-WATER TANK

DRAIN

══════ COLD WATER

────── HOT WATER

Fig. 40-7 Hot and Cold Water piping layout

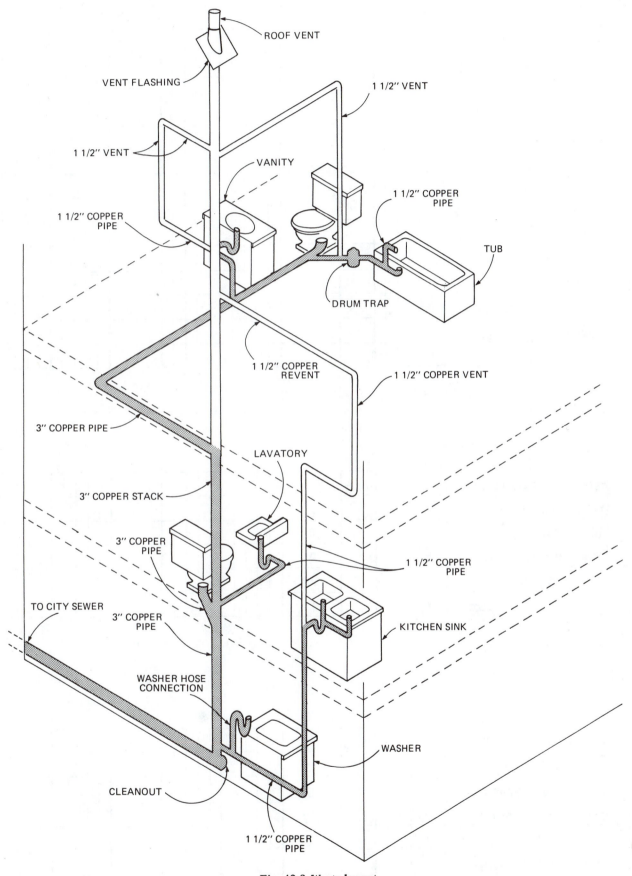

Fig. 40-8 Waste layout

A

BATHTUB			1	
BATH FAUCET & SHOWER SET			1	
BATHTUB DRAIN SET			1	
WATER CLOSETS W/ SEATS & COVERS			2	
ROUND LAVATORIES W/ FAUCETS AND DRAINS			2	
DOUBLE BASIN, STAINLESS STEEL KITCHEN SINK			1	
KITCHEN SINK FAUCET SET			1	
ELECTRIC WATER HEATER	40 GAL.		1	
CHROME SUPPLY SETS			7	

B

TYPE L COPPER PIPE	½"	LIN. FT.	132	
T- FITTINGS- BRASS	½"		14	
90° ELBOWS- BRASS	½"		19	
STOP & WASTE VALVES	½"		9	
CAPS — BRASS	½"		5	
FROSTPROOF SILL COCKS	½"		1	

C

TYPE L COPPER PIPE	½"	LIN. FT.	124	
T FITTINGS- BRASS	½"		10	
90° ELBOWS- BRASS	½"		13	
STOP & WASTE VALVES	½"		7	
CAPS — BRASS	½"		5	
RELIEF VALVE — W. H.			1	

D

COPPER PIPE - WASTE	3"				IIN.FT.	60																							
CLOSET BENDS - BRASS	3"					2																							
CLOSET FLANGES - WATER																													
CLOSET	3"					2																							
90° ELBOWS - BRASS	3"					2																							
45° Y BRANCH - BRASS	3" x 1½"					1																							
SANITARY T - BRASS	3"					4																							
CLEAN OUT - BRASS	3"					1																							
VENT FLASHING - ROOF	3"					1																							
COPPER PIPE - WASTE	1½"				IIN.FT.	82																							
TRAPS W/CLEANOUTS - BRASS	1½"					4																							
T FITTINGS - BRASS	3" x 1½"					5																							
T FITTINGS - BRASS	1½"					3																							
DRUM TRAP	1½"					1																							
90° ELBOWS - BRASS	1½"					7																							
ADAPTER 3" COPPER TO																													
4" CAST IRON						1																							

REVIEW QUESTIONS

A. Select the letter preceding the best answer.

1. Which of the following is usually installed with the rough plumbing?

   a. Bathtub

   b. Water heater

   c. Lavatories

   d. Water closets

2. What is the standard size of the bathtub?

   a. 5'-0" x 2'-0"

   b. 5'-6" x 2'-6"

   c. 5'-6" x 2'-0"

   d. 5'-0" x 2'-6"

3. What is the purpose of the vent in the waste plumbing?

   a. To discharge sewer gas from the building

   b. To allow atmospheric pressure to enter the system

   c. To prevent excess pressure from building up in the system

   d. To provide an overflow in case the system becomes blocked

4. What is the purpose of a trap?

   a. To discharge sewer gas from the building

   b. To prevent sewer gas from entering the building

   c. To prevent excess pressure from building up in the system

   d. To trap solid waste material

5. Which of the following is usually installed with the finished plumbing?

    a. Water closets
    b. Vents
    c. Bathtubs
    d. Air chambers

6. What is an air chamber?

    a. An air bubble which has accidentally become trapped in the system
    b. A special device attached to the fixtures to prevent air from entering the system
    c. A short, capped length of pipe, which holds air to act as a cushion and prevent the pipes from hammering
    d. The part of a trap which holds air

7. Which of the following materials is not used for supply piping?

    a. Galvanized iron
    b. Cast iron
    c. Copper
    d. Plastic

8. Where should the estimator look to find a complete list of the fixtures for a house?

    a. Plumbing code
    b. Working drawings
    c. Specifications
    d. Plumbing elevation

B. Using the following given information, construct a material list for installing the rough plumbing in the house in figure 40-9. Assume that all supply pipes are 1/2″ copper and waste pipes are 3″ plastic, with 1 1/2″ plastic branches. It is not necessary to list any fixtures for this estimate. Do not estimate the materials required for any plumbing which is outside the building. Remember to include shut offs for the tub and both the hot and cold main supplies. Also remember to include air chambers. It may be helpful to plan a separate vent stack for the kitchen sink in this house. The water supply and sewer enter the house near the front door.

Include a plumbing sketch with the list.

Cold Water

	Quantity		Material
a.	_____		1/2″ type-L copper pipe
b.	_____		1/2″ T fittings
c.	_____		1/2″ 90-degree elbows

d. _____   1/2″ stop and waste valves

e. _____   1/2″ caps

Hot Water

f. _____   1/2″ type-L copper pipe

g. _____   1/2″ T fittings

h. _____   1/2″ 90-degree elbows

i. _____   1/2″ stop and waste valves

j. _____   1/2″ caps

Waste

k. _____   3″ plastic pipe

l. _____   3″ closet flange

m. _____   3″ 90-degree elbow

n. _____   3″ plastic T's

o. _____   3″ plastic clean out

p. _____   Flashing for roof vent

q. _____   1 1/2″ plastic pipe

r. _____   1 1/2″ plastic 90-degree elbows

s. _____   1 1/2″ plastic T's

t. _____   3″ x 1 1/2″ plastic T's

u. _____   1 1/2″ plastic clean out

v. _____   3″ plastic to 4″ cast iron adapter

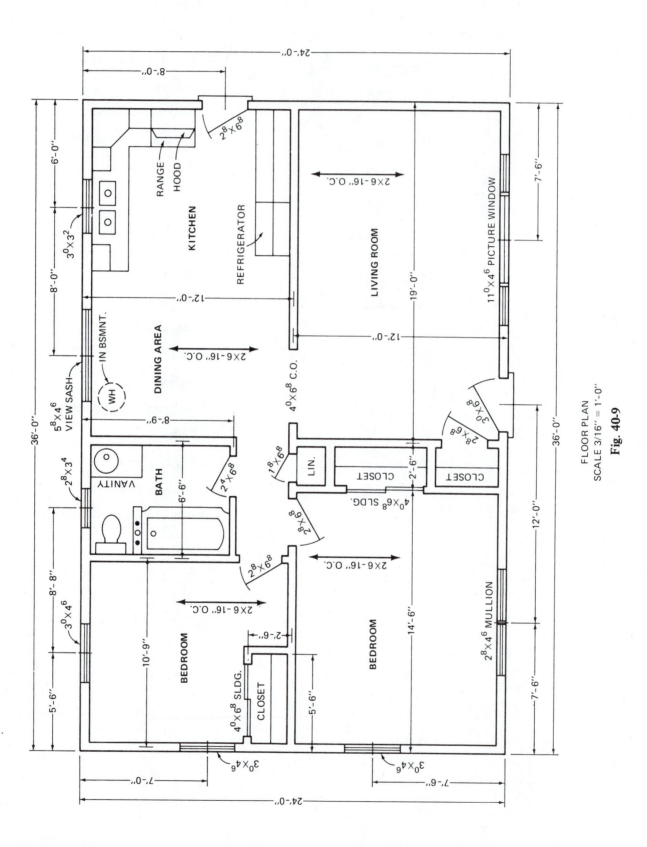

FLOOR PLAN
SCALE 3/16" = 1'-0"

Fig. 40-9

unit 41 electrical wiring

OBJECTIVES

After studying this unit the student will be able to

- identify the electrical symbols on working drawings.
- explain the specifications for the electrical work in a house.
- determine the number of outlets and switches for a house.
- list the equipment for the electrical service to a house.

The estimator for a general contractor is not usually called upon to prepare an itemized list of materials for the electrical work in a house. Electrical wiring is a highly specialized trade and the specific equipment required is usually estimated by the electrical contractor. However, the general estimator should know how the material list is prepared, so that the estimates received from the electrical contractor can be checked.

In most localities, electrical wiring is governed by the *National Electrical Code.* Many communities have local codes in addition to the *National Electrical Code.* In addition to the requirements set forth in the various codes, special building considerations may affect the design of the electrical installation. Because of all these variables, the information presented here should be considered as a guide only; the appropriate codes should always be consulted.

CONDUCTORS

An electrical wire is frequently referred to as a conductor, because it conducts electricity. Conductor sizes are listed by *American Wire Gauge* (AWG) numbers. The higher the AWG number is the smaller the diameter of the wire is. The size, or gauge, determines the current capacity of a wire. According to the National Electrical Code, a 14-gauge copper conductor can be used for circuits carrying up to 15 amperes; a 12-gauge conductor can be used for up to 20 amperes; and 4/0-gauge conductors can be used for up to 200 amperes. Because aluminum has a slightly higher electrical resistance than copper, a larger size conductor must be used for aluminum wiring.

SERVICE

The wiring and equipment required to deliver electricity from the utility company's pole to the building constitutes the electrical service. This may be either an underground service, with special wiring buried in the ground; or overhead service, with the wiring run overhead from the pole to the building, figure 41-1. The *weatherhead* may be attached directly to the building or mounted on a *mast* made of 2-inch steel conduit (pipe). From the weatherhead, the entrance cable runs through the utility company's meter, to the service panel. This service panel normally includes the main disconnect, which is required by the code, and a means of distributing the electricity to the various circuits within the building, figure 41-2. The entrance cable from the service head to the service panel must be at least 4/0 gauge for a 200 ampere service.

BRANCH CIRCUITS

In the service (or distribution) panel, the entrance cable feeds several individual branch circuits. These are the circuits that carry the current to the various parts of the building. Each branch circuit must be protected against overloading by a circuit breaker or fuse. The size of the conductors in the circuit and the size of the overcurrent protection are governed by the code. In general. lighting circuits have 15 ampere circuit breakers and 14-gauge copper conductors; many outlet

circuits are required to have 20 ampere circuit breakers and 12-gauge conductors. The estimator can base all general lighting and convenience outlet circuits on the 20 amp, 12-gauge figures. Most houses are wired with at least eight to ten branch circuits. Of these, 2 are lighting circuits, 2 are kitchen outlet circuits (these are required), 2 are general outlet circuits, 1 is a laundry outlet circuit, 1 is an electric range circuit, and 1 is an electric clothes dryer circuit.

Some circuits are required to have additional protection provided by a *ground-fault device*. This is an electronic device that interrupts the flow of current in an extremely short amount of time in the event of excessively high current flow, such as occurs when the circuit is shorted. Ground-fault protection greatly reduces the hazard of electrical shock resulting from short circuits. Residences are required to have at least one waterproof outlet with ground-fault protection on the exterior of the house.

The material in a branch circuit includes: the cable, which runs from the distribution panel (circuit breaker) to the nearest outlet, and then to each of the other outlets in turn; a steel or plastic box, figure 41-3, to house the receptacle, switch, or other device; and the device (duplex receptacle,

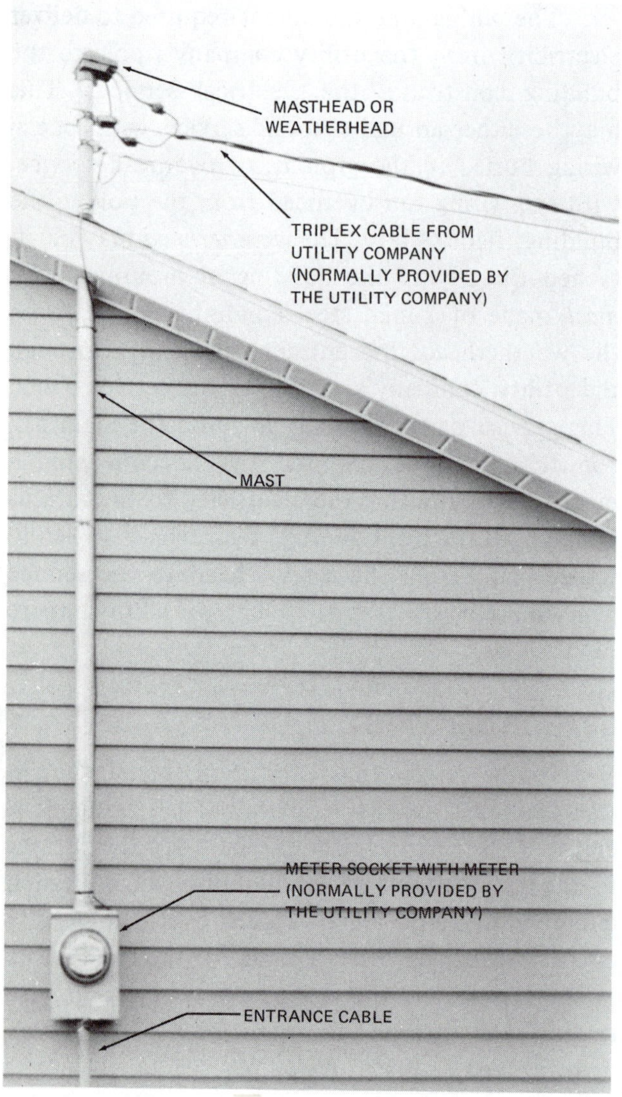

MASTHEAD OR WEATHERHEAD

TRIPLEX CABLE FROM UTILITY COMPANY (NORMALLY PROVIDED BY THE UTILITY COMPANY)

MAST

METER SOCKET WITH METER (NORMALLY PROVIDED BY THE UTILITY COMPANY)

ENTRANCE CABLE

Fig. 41-1 Overhead electrical service

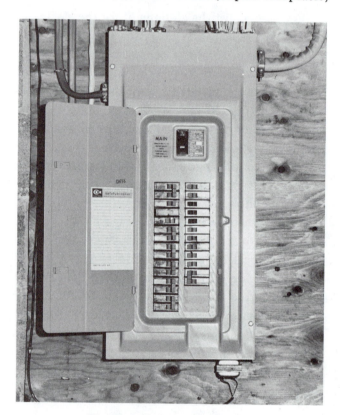

Fig. 41-2 Circuit breaker panel

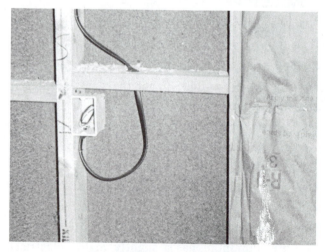

Fig. 41-3 Steel electrical box for a wall switch

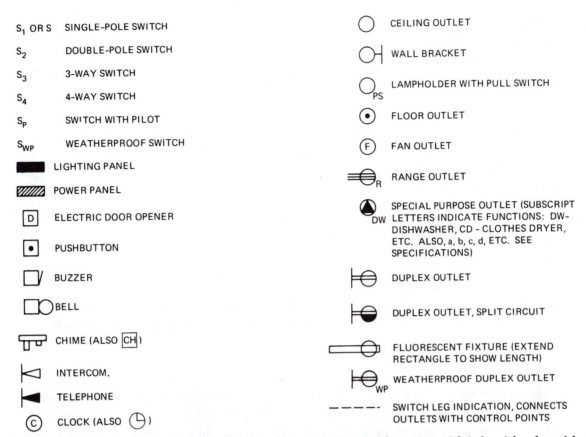

S₁ OR S SINGLE-POLE SWITCH

S₂ DOUBLE-POLE SWITCH

S₃ 3-WAY SWITCH

S₄ 4-WAY SWITCH

Sₚ SWITCH WITH PILOT

S_WP WEATHERPROOF SWITCH

LIGHTING PANEL

POWER PANEL

D ELECTRIC DOOR OPENER

PUSHBUTTON

BUZZER

BELL

CHIME (ALSO CH)

INTERCOM.

TELEPHONE

C CLOCK (ALSO)

CEILING OUTLET

WALL BRACKET

LAMPHOLDER WITH PULL SWITCH

FLOOR OUTLET

F FAN OUTLET

R RANGE OUTLET

DW SPECIAL PURPOSE OUTLET (SUBSCRIPT LETTERS INDICATE FUNCTIONS: DW- DISHWASHER, CD - CLOTHES DRYER, ETC. ALSO, a, b, c, d, ETC. SEE SPECIFICATIONS)

DUPLEX OUTLET

DUPLEX OUTLET, SPLIT CIRCUIT

FLUORESCENT FIXTURE (EXTEND RECTANGLE TO SHOW LENGTH)

WP WEATHERPROOF DUPLEX OUTLET

SWITCH LEG INDICATION, CONNECTS OUTLETS WITH CONTROL POINTS

Fig. 41-4 Common electrical symbols. There are many other symbols in use for commercial, industrial and special applications; those shown here are only the most common symbols for residential construction.

switch, light fixture, etc.) All splices in the conductors must be made inside an approved box and must not be concealed in a wall or partition. Usually all splices not made at a device are made in junction boxes located in the cellar or attic.

In addition to switches, light fixtures, and convenience outlets, a variety of special electrical devices may be installed in a residence. These are described in the specifications and their location is indicated on the working drawings. Some sets of working drawings include the location of electrical equipment on the floor plan; others include a separate electrical plan. In the sample house, the location of the electrical equipment is located on the floor plans. Figure 41-4 shows some of the most common electrical symbols.

SPECIFICATIONS

The specifications may suggest brand names of electrical fixtures, as a guide to the quality expected. Usually the specifications include a statement that the contractor may substitute brands as long as the quality is the same. Because of the great variations in the cost of fixtures, a certain amount of money is normally indicated as a fixture allowance. The owner selects the fixtures to be purchased by the contractor; the contract price of the building is then adjusted to make up any difference between the actual fixture cost and the fixture allowance.

The following are the specifications for the electrical work in the sample house.

Specifications

DIVISION 12: ELECTRICAL

A. WORK INCLUDED

In general, the electrical work includes, but shall not be limited by the following:
1. Service entrance
2. Service panel

3. Wiring
4. Fixture allowance
5. Special outlets
6. All cutting and patching for installation of the work

B. GENERAL REQUIREMENTS

The complete installation shall be made in a neat, workmanlike manner in conformance with best modern trade practices, by competent, experienced mechanics and to the full satisfaction and approval of the architect. All work shall be in accordance with local, state or federal requirements, and shall comply with all applicable codes.

C. GUARANTEE

The contractor guarantees that:

1. All work executed under this contract will be free from defects of material and workmanship for a period of one year from the date of acceptance by the owner.
2. The contractor, at his or her own expense, will repair and replace all such electrical work and all other work damaged thereby, which becomes defective during the period of the guarantee.

D. GUIDE SPECIFICATIONS

1. *Service* supplied to the structure shall be a wire, 116/230 volts, 60-cycle, single phase.
2. *Service Panel* shall have a 200-ampere capacity, with automatic circuit breakers. The service panel shall be flush mounting with a flush door and shall accommodate 30 circuits.
3. *Wiring:* All circuit wiring is to be 12-gauge or larger, type TW for general use.
4. *Boxes:* Outlet boxes and junction boxes are to be galvanized steel approved for purposed use and of a suitable size to accommodate the requirements of the fixture, wiring device or equipment, and the wiring connections.
5. *Switches* for general lighting shall be of the quiet AC-rated toggle type. Switches shall be installed 48 inches to center above the finished floor, unless otherwise specified.
6. *Receptacles* shall be standard duplex grounding type.
7. *Locations of outlets:* Outlets, as shown, are in approximate locations. These must be checked on the job for possible conflicts with other trades or built-ins. Convenience outlets shall be 16 inches to center from the finished floor, unless otherwise indicated.
8. *Grounding:* The complete electrical system will maintain a solid ground, in accordance with the *National Electrical Code.*

E. SPECIAL OUTLETS

The contractor shall furnish and install receptacles, and switches as required, for the following special outlets:

1. Exhaust fan over the kitchen range
2. Electric range in the kitchen
3. Electric hot water heater in the basement
4. Furnace control switch at the head of the basement stairs
5. Waterproof outlet on side porch

F. TELEPHONE OUTLET

Furnish and install telephone wiring with one outlet in the living room-hall area and one in the master bedroom.

G. EXHAUST FAN

Furnish and install exhaust hood and fan in the kitchen as indicated on drawings. The fan shall have at least 300 cfm free delivery.

H. SIGNAL CHIMES

Furnish and install chimes with circuit connection to approved transformer and light circuit. Provide an outside push button at each door.

I. LIGHTING FIXTURES

The contractor shall furnish and install all electrical fixtures and shall allow the sum of Four Hundred Dollars ($400.00) for the purchase of these fixtures. This allowance covers the net cost to the contractor and does not include any labor, overhead, or profit. Fixtures shall be selected by the owner, but will be purchased by the contractor. The net differences in cost, if any, shall be added to or deducted from the contract.

Estimating Electrical Work

The electrical subcontractor will probably prepare an itemized list of the materials required to do the electrical work before commencing work on a job. However, due to the highly technical nature of the work, and the strict requirements imposed by electrical codes, the general estimator does not usually prepare such a list. The estimate is frequently based on the following:

- The type of electrical service to the building. (underground or overhead)

- The *ampacity* (capacity in amperes) of the electrical service.

- Special equipment, such as garbage disposals and exhaust fans, that are to be included.

- The number of outlets for lighting, receptacles, and switches.

- The amount of money to be included as a fixture allowance.

The type of service to be provided is indicated in the specifications. The length of the entrance cable can be determined by reading the plot plan. Most estimators calculate the cost of burying underground cable on the basis of what it costs to bury one foot. If the service is to be overhead, the cost may be absorbed by the utility company.

·The ampacity of the service should be given in the specifications. The estimator usually allows a fixed amount for the service equipment depending on the ampacity of the service. If a more accurate estimate is desired for this phase of the job, the material for the service can be itemized.

Any special electrical equipment which is to be included in the construction of the building is listed in the specifications. Some specifications do not indicate manufacturer's names, but sufficient information is always included to determine the quality and type of special equipment required.

The National Electrical Code specifies that outlets must be installed as follows:

- So that no point measured horizontally along a wall is more than 6 feet from an outlet

- In any wall which is 2 feet or more in width

- For each counterspace wider than 12 inches

- At least one outlet in each bathroom

- At least one outlet in the basement

- At least one outlet in the garage

- At least one outlet in laundry rooms

- At least one outdoor outlet with ground-fault protection

- Every room, hallway, stairway, garage, and outdoor entrance must have a lighting outlet, which is controlled by a wall switch.

The location of these outlets should be indicated on the floor plans or a separate electrical plan. If they are not shown, the estimator should include them in the estimate for electrical work. It is common practice to base the estimate on the cost of a typical outlet. When this is done the estimator determines the cost of installing one typical outlet, then multiplies this figure by the number of outlets of that type. In this manner the estimator does not need to determine the number

of circuits or the quantity of cable required for the entire construction project.

The following is a list from which an electrical estimate can be prepared for the sample house:

Piece	Feet	Material	Cost	Amount
1		200-ampere service with 30-circuit panel		
1		Exhaust hood with 300 cfm fan		
1		Set of door chimes		
1		Door-chime transformer		
3		Push buttons for door chimes		
29		Lighting outlets		
19		Switch Outlets		
1		Range outlet		
1		Clothes dryer outlet		
2		Weatherproof outlets with ground-fault protection		
33		Convenience Outlets		
2		Telephone outlets		

SUMMARY

Electrical work can generally be viewed as one of three parts of a system. Electricity enters a building through the service, which may be either underground or overhead. A service panel provides some means of disconnecting all power to the building, terminals to distribute the power to several branch circuits, and circuit breakers or fuses to protect against excessive current.

Individual branch circuits carry the electricity through conductors, made of copper or aluminum, to steel or plastic boxes, which enclose all electrical connections. Separate branch circuits supply electricity throughout the house for lighting, convenience outlets, and appliances. When all of the wiring is installed electrical devices are installed. These include switches, receptacles for general use and special equipment, light fixtures, and equipment.

The National Electrical Code specifies the design of most electrical work. In addition, many communities have local electrical codes. The estimator should be familiar with all existing codes in the area, as they constitute a major part of the specifications for electrical work on most jobs.

The building specifications for electrical work primarily describe the electrical equipment to be used. The location of all electrical devices can be found on either the floor plans or special electrical plans. The architectural drafter uses standard symbols to indicate electrical devices.

General estimators are not normally called upon to prepare itemized lists of the electrical supplies for a construction job. The estimator counts the number of outlets of each type shown on the working drawings, and allows a certain price for each complete outlet. The size and type of electrical service are also considered in estimating the total electrical work.

REFERENCE

National Fire Protection Association. *National Electrical Code.* Bloomfield: Connecticut Printers Inc.

REVIEW QUESTIONS

A. Select the letter preceding the best answer.

1. Who normally prepares an itemized list of materials for electrical work?
   a. The architect
   b. The estimator
   c. The general contractor
   d. The electrical contractor

2. What does the abbreviation AWG stand for?
   a. Architectural Wiring Guide
   b. American Wire Gauge

c. Associated Wirer's Guild

d. None of these

3. Which of the following is considered a part of the electrical service?

   a. The mast

   b. The wiring from a circuit breaker to a light fixture

   c. A wall switch to control an overhead light fixture

   d. An electric range or cook stove

4. What is a fixture allowance?

   a. The number of light fixtures in a building

   b. The capacity of a lighting circuit

   c. A list of suggested manufacturers of light fixtures

   d. The amount of money that has been included in a contract for the purchase of light fixtures

5. In the event of a conflict, which of the following should determine the design of the wiring for a building?

   a. The working drawings

   b. The National Electric Code

   c. The construction specifications for the job

   d. The desires of the owner

B. What does each of the following symbols indicate?

   1. $\bigcirc$

   2. $\bigcirc_{PS}$

   3. ⊢⊖

   4. s_1

   5. s_3

   6. ⊢⊖$_{WP}$

   7. Ⓕ

   8. ◬

   9. ▨

   10. ⎯ ⎯ ⎯ ⎯ ⎯-

C. For each type of outlet, indicate the quantity shown on the Floor plan in figure 41-5.

   1. Single pole switch

   2. Three way switch

   3. Convenience outlet

   4. Ceiling

   5. Weatherproof outlet

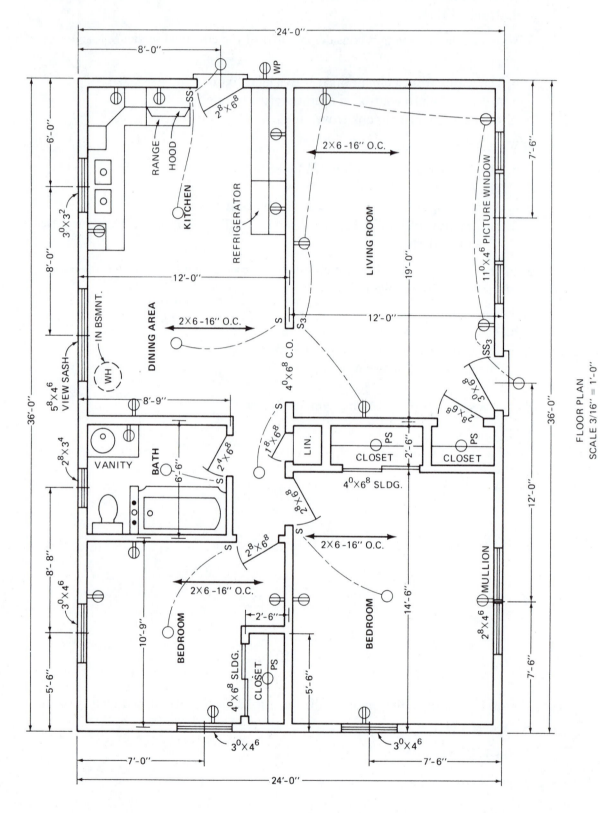

FLOOR PLAN
SCALE 3/16'' = 1'-0''

Fig. 41-5 Floor Plan

unit 42 heating

OBJECTIVES

After studying this unit the student will be able to

- describe the major types of heating systems found in residential construction.
- define K factor, U factor, R factor, and Btu.
- estimate the size furnace required for a residence.

PRINCIPLES OF AIR HANDLING

The design of an efficient heating system involves more than an elementary knowledge of air handling. The heating system is usually designed by a specialized heating contractor or engineer. However, the estimator should understand the principles involved in order to interpret the information provided by the engineer.

All heating systems are based on the warming of air in the living space. This warming may be accomplished by passing the air through a chamber heated by the combustion of oil, natural gas, air passing through radiators and convectors to which hot water has been piped, or by the use of electricity to heat the air.

Heat flows from warm objects to colder ones. As a house loses its heat to the colder outdoors, persons inside the house lose their body heat to the colder air. The basic function of a heating system is to supply heat to the building as quickly as it is lost to the outside.

Warm air is lighter than cold air. For this reason, the heated air rises to the ceiling and slowly descends to the floor as it becomes cooler. As the air passes cold windows, it is chilled rapidly and becomes heavier, causing it to descend faster. This causes a draft and a layer of cold air on the floor. The heating system can be designed to overcome this by positioning heating outlets on the outside walls and under windows, figure 42-1. As the warm air leaves the outlet, it forms a curtain of warm air in front of the windows and exterior walls. This reduces the rapid cooling of air and eliminates drafts.

There are three basic types of heating systems in common usage: central furnace, which heats air to be directed into the living space; baseboard units, which are located in the living spaces; and radiant heat, which is provided by heat lamps and electrical units imbedded in the building structure. The source of energy for these may be electricity, natural gas, or fuel oil. The type of heating system selected depends on the local climate, the size of the house, the availability of fuel, the cost of installation, and the owner's preference.

HOT AIR HEATING

In hot air heating, the cold air is passed over a surface heated by electricity or the combustion of fuel oil or gas. This heated air is carried by ducts to outlets at various points throughout the house, figure 42-2. The heated air is generally forced

Fig. 42-1 Heating outlets are positioned under windows to prevent drafts.

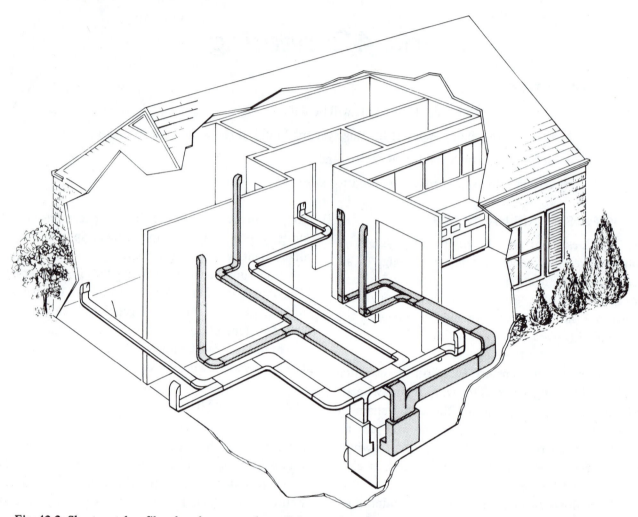

Fig. 42-2 Sheet metal or fiberglass ducts carry heated air to each room of the house, and return cool air to the furnace.

through the ducts by a fan in the furnace, but it may rely on the force of gravity acting on cold air. As the air in the room cools, it is drawn into a cold air return, which carries it back to the furnace to be reheated.

Hot air heating systems frequently incorporate a humidifier to replace some of the moisture which has been removed from the air in the heating process. Humidifying the air provides a more comfortable atmosphere for the occupants of the building and protects the woodwork from excessive drying. Central air conditioning is also easily incorporated in a hot air heating system. The cooled air is forced through the same duct work as the heated air.

HOT WATER HEATING

In hot water heating a boiler uses either gas or fuel oil to heat the water, which is then circulated

through pipes to radiators or convectors. A *convector* is a type of radiator with metal fins attached, figure 42-3. These fins are heated by the hot water and they, in turn, transfer the heat, by a process called convection, to the surrounding air. In a radiator, the heat is simply transferred from the part containing the water to the surrounding air by radiation.

The heated water is circulated through the system by an electrically operated circulating pump. This pump is usually on the return side of the system. That is, it draws the water from the radiators or convectors back to the boiler to be reheated.

Hot water heating may use either a one-pipe or two-pipe system. In a one-pipe system a main line carries the heated water throughout the house and returns it to the boiler, figure 42-4. Branch lines at each heating outlet carry the hot water

to each heating outlet and back to the main. A heat control valve is installed at the inlet side of each outlet. This allows the heat to be controlled at each outlet without affecting others in the system.

In a two-pipe system one pipe carries heated water to all of the outlets and a second pipe returns the cooled water to the boiler, figure 42-5. The outlets are fed by a branch pipe from the main supply, and the cooled water is returned to the return line by another pipe.

Hot water systems frequently have zoned heat. In zoned heating, the temperature in various parts of the house can be controlled separately. In residential construction it is common for the bedrooms to be zoned separately from the other living areas, so that the temperature can be kept at a lower setting in the bedrooms. Each zone of a zoned system has a separate thermostat which regulates electrically operated valves at the outlets.

ELECTRIC HEATING

A variety of electric heating devices is available for residential construction. The most common is electric baseboard heat, figure 42-6. This system uses an electrical resistance element, similar to that found on an electric cooking range or toaster, to heat aluminum fins which transfer the heat to the surrounding air. Some electric heating outlets include a fan to circulate the air.

Another type of electric heat uses special resistance cable imbedded in the floor or ceiling. This type provides uniform heat from the floor to the ceiling without the need for additional equipment on the walls.

THERMOSTATS

A thermostat is a switch which is activated by changes in temperature. As the temperature drops below that for which the thermostat is set, the switch closes and activates the heating device to which it is connected. As the temperature rises, the switch opens.

All heating systems use thermostats. With electric heat, the thermostat simply turns on the power to the heating outlet.

Fig. 42-3 Hot-water convector with the cover removed to show the fins.

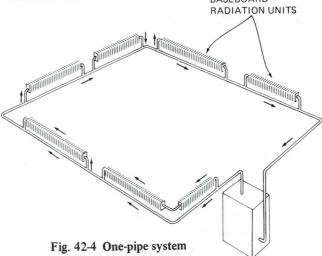

Fig. 42-4 One-pipe system

Fig. 42-5 Two-pipe system

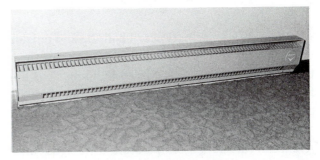

Fig. 42-6 Electric baseboard heating unit

With hot air heating, thermostats are used to control the flow of fuel to the furnace. As the temperature drops, the thermostat opens a valve allowing fuel to enter the burner. When the air inside the furnace reaches a preset temperature, an automatic switch activates the blower.

In a hot water heating system the thermostat opens and closes valves to control the flow of hot water. In a zoned system each thermostat controls the valves on the radiators in their respective zones.

THERMAL CONDUCTIVITY AND RESISTANCE

Architects and heating engineers measure heat by Btu's (British thermal units). A *Btu* is the amount of heat required to raise the temperature of one pound of water, one degree Fahrenheit. The rate of heat transfer is expressed as Btuh. (British thermal unit per hour).

All materials conduct heat. The rate at which a material conducts heat is its *K factor,* This is the number of Btuh that is conducted by one square foot of the material one inch thick with a difference in temperature of one degree Fahrenheit at each side. For example, consider a one inch thick piece of gypsum wallboard with one side being one degree colder than the other. If one Btu passes through a one square-foot area every hour, the K factor is 1.

Buildings are composed of an assortment of materials, so the K factor is not a practical way to measure the amount of heat that is lost through a wall, ceiling, or floor. The combination of all of the K factors in a section of a building is the *U factor.* Figure 42-7 lists the approximate U factors for some common types of construction.

The purpose of thermal insulation is to resist the flow of heat. The resistance of a material to the flow of heat is the *R value* of that material. The R value is the reciprocal of the U value (R = 1/U).

The difference between the inside and outside temperature greatly affects the amount of heat that is lost through the shell of a building. The difference between the lowest probable outside temperature and the desired inside temperature *(design temperature)* is called the *design temperature difference.* This design temperature difference must

Type of Building Section	U Value
Wood frame with plywood sheathing, wood siding, and 1/2 inch drywall, no insulation.	0.24
Wood frame with plywood sheathing, wood siding, and 1/2 inch drywall, 3 1/2 inch (R-11) insulation.	0.07
8-inch concrete block	0.25
Single-glazed window	1.1
Double-glazed window	0.6

Fig. 42-7 U factors for some typical building sections

be known in order to determine the heating requirements of a building.

DETERMINING THE THERMAL RESISTANCE OF A BUILDING

Properly installed insulation is vital to comfortable, economical heating. Although it is impossible to completely stop the flow of heat through a building section, insulation can greatly reduce it. Insulating materials vary in their ability to restrict the flow of heat depending on their type, density, and other characteristics. For this reason insulation is specified according to its R value, rather than its thickness.

A building section consists of the materials provided to support the structure, the inside and outside surface materials, and whatever material is used to minimize heat transfer and air infiltration. Figure 42-8 shows a building section and the thermal resistance of each component. Notice that the total R value for the section is 4.59. Fiberglass insulation has a thermal resistance of approximately 3.7 resistance units per inch of thickness. If all 3 1/2 inches of available stud space were filled with such insulation, 12.95 resistance units would be substituted for the .95 units provided by the air space. This would increase the total resistance of the building section to an R value of 16.59.

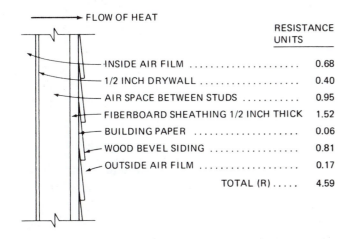

	RESISTANCE UNITS
INSIDE AIR FILM	0.68
1/2 INCH DRYWALL	0.40
AIR SPACE BETWEEN STUDS	0.95
FIBERBOARD SHEATHING 1/2 INCH THICK	1.52
BUILDING PAPER	0.06
WOOD BEVEL SIDING	0.81
OUTSIDE AIR FILM	0.17
TOTAL (R)	4.59

FLOW OF HEAT

Fig. 42-8 A typical building section, showing thermal resistance, without insulation.

Type of Building Section	R value
Wood Frame Walls	3 plus the R value of the insulation used.
Floors above unheated spaces	2 plus the R value of the insulation used.
Single-glazed windows	1 1/2 plus the R value of the insulation.
Double-glazed windows	0.88
Doors with glass	Use R value of the glass for the entire door.
Doors without glass Less than 1″ thick or metal Over 1″ thick nonmetal	 0.88 1.67

Fig. 42-9 R values for common building sections

It can readily be seen that the original resistance is less than one quarter of the insulated resistance. With substantial insulation, slight variations in the resistance of the structural and finish components of the building have a minor effect on the overall resistance. In wall sections, the total resistance can be assumed to be the resistance of the insulation plus three units for the structural and finish components. Similar reasoning can be applied to floor and ceiling construction to arrive at values for these sections. Uninsulated floors can be assumed to have a resistance of approximately 2 resistance units and uninsulated ceilings can be assumed to have a resistance of approximately 1 1/2 units.

Windows and doors offer much less resistance to the flow of heat than do other building materials, so the values mentioned above do not apply to them. A single layer of window glass has an R value of approximately 0.88. However, trapped air offers substantial resistance to the flow of heat, so double glazing increases the R value to approximately 1.67. The resistance values of several common building sections is shown in figure 42-9.

The amount of heat lost through the various sections of the entire building can be found from the R value for each section and the area of each exposed section. By dividing the resistance (R) into the area, the heat transmission load (the heat that is lost through the building materials) is found in Btuh per degree Fahrenheit of temperature difference. The following three steps are used to find the building transmission load:

1. Find the square-foot area of each exposed outside section.

2. Divide the R value listed in figure 42-9 into the section area to find the heat transmission for that section.

3. Add all section transmission load per degree Fahrenheit.

INFILTRATION LOSSES

In addition to the heat that is lost by transmission through the various building sections, some heat is also lost through *infiltration*. That is the heat that is lost as air enters the building through cracks and small openings. To replace the heat lost in this manner, an infiltration factor is used. The infiltration factor is based on changing all of the air in the building periodically. The number of Btuh required to reheat the infiltrated air in one cubic foot of space is the infiltration factor. For example, if the air is changed once

in two hours, the infiltration factor is 0.0088 Btuh. To find the total infiltration heat loss at this rate, multiply 0.0088 times the number of cubic feet in the building.

> Example: Find the infiltration heat loss for a building 40 feet long by 24 feet wide, with 8-foot ceilings. Use an infiltration factor of 0.0123 Btuh (.7 air change per hour). Volume = 40' x 8' = 7680 cubic feet. Heat loss per degree F = 7680 x 0.0123 = 101 Btuh.

The sum of all transmission losses and the infiltration loss is the building load per degree Fahrenheit. Multiply the building load per degree Fahrenheit by the design temperature difference to find the total building load. This total building load is the number of Btu per hour that the heating system must be capable of replacing or the required output of the heating system.

Estimating Heating Loads

The floor plan to be estimated in this unit is shown in figure 42-10. This house is used instead of the two-story house found in other units, to avoid time-consuming calculations. All calculations for this house are based on .7 air change per hour infiltration rate (0.0123 Btuh) and a design temperature difference of 80 degrees Fahrenheit. This difference is appropriate for an outside temperature of minus 10 degrees Fahrenheit and an inside (or design) temperature of 70 degrees Fahrenheit. For the purposes of this estimate, assume that the house is of standard frame construction with R-19 insulation in the ceiling, R-11 insulation in the walls, and R-11 insulation in the floors. All windows are to be double glazed.

First find the square-foot area of the outside walls of the building. The house is 36 feet long and 24 feet wide, so the perimeter is 120 feet. The ceilings are 8 feet high, so the wall area is 960 square feet.

The sizes of the windows are indicated on the floor plan. The window areas are as follows:

Living room picture window	40 sq. ft.
Bedroom mullion window	27 sq. ft.
Three windows 3/0 x 4/6	40 sq. ft.
Bathroom window	8 sq. ft.
Dining area window	23 sq. ft.
Kitchen window	9 sq. ft.
Front door	21 sq. ft.
Kitchen door	20 sq. ft.
Total	**188 sq. ft.**

The area of the ceiling is found by multiplying the width of the house by the length of the house. The house is 36 feet by 24 feet, so the ceiling area is 864 square feet. The area of the floor is the same as that of the ceiling.

To find the infiltration load, it is necessary to know the cubic-volume of the building. This is found by multiplying the floor area by the ceiling height. (864 x 8 = 6912 cubic feet)

Using this information, the heat loss can be found as follows:

SECTION	DESIGN DATA	LENGTH AREA OR	UNIT HEAT LOSS	HEAT LOSS Btuh per DEGREE
Glass and Doors	Double Glass	188 sq. ft.	R = 1.67	188/1.67 = 113
Walls	R 11 Insulation	960 sq. ft.	R = 11 + 3 = 14	960/14 = 69
Ceilings	R 18.5 Insulation	864 sq. ft.	R = 18.5 + 1.5	864/20 = 43
Floors	R 7.4 Insulation	864 sq. ft.	R = 7.4 + 2	864/9.4 = 92
Infiltration	.7 air change/hr.	6,912 cu. ft.	0.0123	6,912 x 0.0123 = 115
			Total	432

Heat Load Btuh (80°F design temperature difference) 432 x 80 = 34,560 Btuh. The heating system for this house must be capable of supplying 34,560 Btuh.

SUMMARY

Nearly all heating systems use either fuel oil, gas, or electricity to replace the heat lost in cold weather. This heat is lost by one of two processes; transmission through the building materials, and infiltration of unheated air.

To determine the amount of heat lost through transmission, the thermal resistance of the materials and the area they cover must be known.

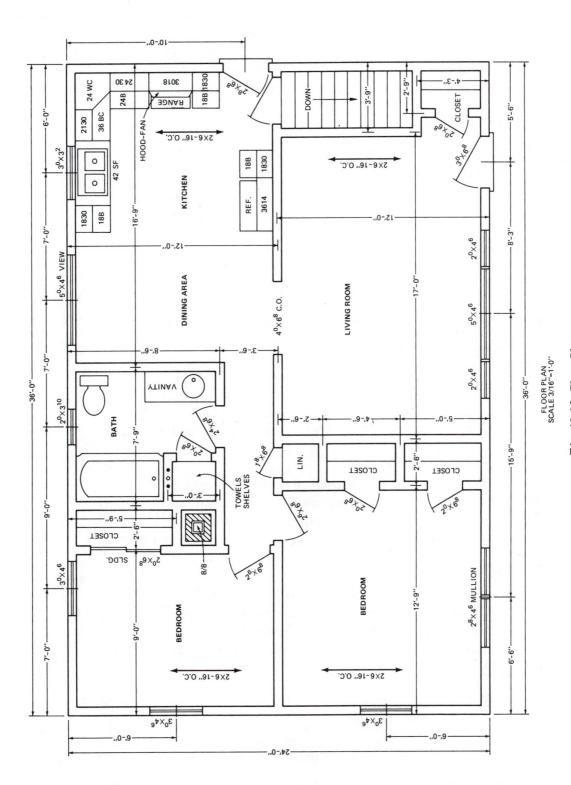

Fig. 42-10 Floor Plan

FLOOR PLAN
SCALE 3/16"=1'-0"

Dividing the area involved by the resistance unit (R value) of the material indicates the number of Btuh's lost for each degree Fahrenheit of temperature difference. This must be calculated for all exposed surfaces, including walls, floors, and ceilings.

The infiltration heat loss is based on the rate at which the air in the building is changed. This is normally determined by an engineer or architect. An infiltration factor in Btuh is established according to the air change rate. To find the total infiltration load, the infiltration factor is multiplied by the cubic-foot volume of the building; this product is then multiplied by the number of degrees of temperature difference on the outside and the inside of the building.

REVIEW QUESTIONS

A. Select the letter preceding the best answer.

1. The design and layout of a heating system is usually done by

a. The architect   c. A heating contractor
b. The contractor   d. The owner

2. Which of the following is a correct statement?

a. Warm air is heavier than cold air.
b. Cold air is heavier than warm air.
c. Cold air rises.
d. Air temperature does not affect the weight of the air.

3. What is a U factor?

a. The resistance of a material to the flow of heat
b. The resistance of a combination of materials to the flow of heat
c. The rate at which a given material conducts heat
d. The rate at which a combination of materials conducts heat

4. What is a K factor?

a. The resistance of a material to the flow of heat
b. The resistance of a combination of materials to the flow of heat
c. The rate at which a given material conducts heat
d. The rate at which a combination of materials conducts heat

5. What is a Btu?

a. A unit of measurement of the quantity of heat
b. A unit of measurement of the flow of heat
c. A unit of measurement of the flow of heat
d. A unit of measurement of the rate at which an object heats up

6. What is an R value?

a. The resistance of a material to the flow of heat
b. The reciprocal of Btuh
c. The infiltration factor for a building
d. The rate at which a given material conducts heat

7. What is the R value of a building section with a U value of 0.13?

a. Approximately 13   c. Approximately 0.076
b. Approximately 7.7   d. Approximately 1.3

8. What is considered an advantage of hot air heating systems?

   a. They are very adaptable to zoned heating.

   b. The heat can be controlled independently in each room.

   c. They are the easiest to install.

   d. Central air conditioning and humidification can easily be included with them.

9. What is considered an advantage of electric heat?

   a. It is the least expensive to operate.

   b. The heat can be controlled independently in each room.

   c. It requires less insulation.

   d. Central air conditioning and humidification can easily be included.

10. What is considered an advantage of hot water heating?

   a. It is the least expensive to install.

   b. The heat can be controlled independently in each room.

   c. It is very adaptable to zoned heating.

   d. Central air conditioning and humidification can easily be included.

B. Find the total heating load for the house shown in the floor plan in figure 42-11. The house is to be of frame construction, with 1/2-inch plywood sheathing. The siding is to be 16-inch wood shingles. All inside walls are covered with 3/8-inch gypsum wallboard. The ceilings are 8 feet high. The insulation is to be R-21 in the ceilings, R-13 in the walls, and R-9 in the floors. All windows are double glazed. The design temperature difference is 50 degrees and the infiltration is based on .7 air change per hour.

Window and door area = _____ R value = _____ Btuh per degree _____

Wall area   = _____ R value = _____ Btuh per degree _____

Ceiling area   = _____ R value = _____ Btuh per degree _____

Floor area   = _____ R value = _____ Btuh per degree _____

Total transmission loss in Btuh per degree _____

Cubic-foot volume of building = _____

Infiltration factor = _____

Infiltration loss = _____ Btuh per degree

Total heat loss in Btuh per degree F = _____

Total building heat load = _____

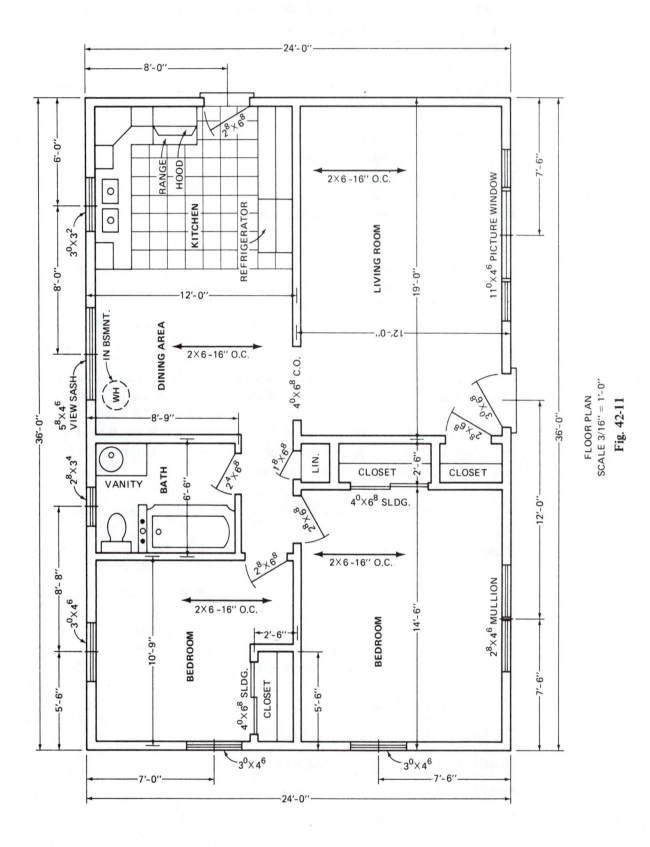

FLOOR PLAN
SCALE 3/16" = 1'-0"

Fig. 42-11

unit 43 heat pumps

OBJECTIVES

After studying this unit the student will be able to

- explain what a heat pump is.
- why defrosting is essential for operation.
- tell the different types of heat pumps used.

WHAT'S A HEAT PUMP?

In simplest terms, a heat pump extracts heat from the outside air and pumps it into the house in winter, while removing the heat from the house in summer. It does this by using a compressor (or pump) in conjunction with a series of controls, along with coils (tubes and fins) that function much like an automobile radiator. There is a fluid called refrigerant that circulates in the sealed system to pick up and transfer the heat.

Until the energy crisis gained prominence, everyone accepted the fact that home heating was best accomplished by burning gas or oil. Heating with these fuels is accomplished by transferring the heat from flames to hot water piping. Or, more commonly, directly to the air inside the home through a device within the furnace called a heat exchanger.

A heat exchanger is usually made of steel or cast iron. It "exchanges" the heat from the flames to the air in the home without allowing the flames and combustion products to mix with the air in your home.

Home heating with a heat pump uses an entirely different concept. The most pronounced difference between a heat pump and all other commonly used heating methods is the source of heat. Heat pump heat is obtained from the outside air, a virtually unlimited source and one that is basically free. There is always heat in the outside air, generated mainly by the sun's energy. Even in winter time, the sun heats the air, which is then captured by the heat pump. Even at -10 degrees below zero, the air contains 84% of the heat that is normally available at 75 degrees above zero. The temperature would have to be -460 degrees below zero to eliminate all heat from the outside air.

DEFROSTING

When it's heating, a heat pump's outdoor coil is colder than the outside air. This allows the coil to "extract" the heat from the outside air. This occurs because heat always "flows" toward a colder surface, just like the heat from your stove always "flows" into a cold pan of water to make it boil. The cold outside coil on a heat pump (when heating) not only attracts the heat in the air, it also attracts moisture.

To understand why, think of a glass of iced tea. When you fill the glass with ice cubes, you make it colder than the air in the room. Thus, the "warm" air in the room flows to the cold glass, warming it (which is why the ice cubes eventually melt) but the cold glass also attracts the humidity in the air which condenses out in the form of drop-

lets of water. You soon find the glass dripping wet, so you need a coaster to protect your furniture. For another example, put a glass mug in the freezer and make it ice cold. Take it out and, in a matter of minutes, it's frosted. Why? Because the cold glass immediately attracts the warm air in the room, along with the humidity or moisture in the air which condenses on the glass creating the "frosting."

When this happens to the cold outside coil on a heat pump, the moisture collects on the coil. When the outdoor temperature is near or below freezing, the moisture in the air can freeze on the coil, requiring defrosting (just as is required by your refrigerator). The reason why defrosting is so necessary is that the more ice that builds up on the coil, the less heat the coil can extract from the air. This is because the ice tends to "insulate" the coil surfaces. Thus, defrosting is a vital part of efficient heat pump operation. Moisture, in varying amounts, is always present in the outdoor air. Defrosting becomes most necessary during periods of high humidity. In the wintertime this means periods of freezing rain, heavy fog, snow, etc. This is where the moisture comes from and its constant attraction to cold surfaces. Defrosting is generally accomplished by reversing the system so that, during defrost, it operates as an air conditioner. To overcome the possibility of cold air drafts, electric heat is operated to counteract the air conditioning effect during defrost.

Most heat pumps defrost on a time-temperature basis. That is, whenever the outdoor temperature falls below a preset level (usually around 40 degrees F) the unit defrosts. And, as long as the outdoor temperature remains at (or below) that temperature, the unit will defrost on a timed cycle every 30, 60, or 90 minutes of unit operation, depending upon the field adjusted time period.

Early versions of the heat pump did not allow for balance-point control. Also, many had problems with defrosting and other mechanical functions. Today modern heat pumps have overcome these problems as the state of the art has advanced with new engineering technology. Today modules have been developed that in a real sense are a self-contained computer containing, among

its many functions the vital balance-point and defrosting controls. Solid-state technology has opened up a whole new world. It's being applied to everything from television sets to automobiles. Because of the extraordinary reliability and accuracy of solid state controls, today's TV sets, cars, hand-held calculators, and other products last longer and they perform more efficiently. This electronic brain helps ensure comfort, cuts electric bills, saves energy and eliminates traditional sources of heat pump problems. One of its major advantages is that it is not affected by sunlight, random winds, or similar conditions that can give false information to systems operated by more traditional, mechanical controls.

WHAT ABOUT COOLING?

A built-in benefit of the heat pump is that the system can reverse itself. When it is heating, the outside coil enables the system to pick up heat from the outdoor air and transfer it into the house through another coil. By making the heat pump system work in the opposite manner, the inside coil extracts the heat from the inside of the house and discharges it outdoors. Hence, the name "heat pump." The device literally pumps heat indoors in the winter (thereby heating) and outside in the summer (thereby cooling). It's a complete year-round system that offers exceptional efficiencies and it conserves both oil, gas, and other fossil fuel energy.

With the advent of more sophisticated and reliable control systems on several versions of the heat pump, these products can now be applied to any home or virtually any building.

SPLIT SYSTEM

A split-system version of a heat pump means that the outdoor section is separated from the indoor section. In other words there are two "boxes" connected by piping; one outside and one inside. Split systems are available with complete, self-contained indoor sections that provide everything necessary to heat, cool, filter, and circulate the air in the house.

ADD-ON SYSTEM

Also a split system, but the indoor section is only a heating/cooling coil added to an existing gas, oil, or electric forced-air furnace, allowing the furnace blower to circulate and filter the air. Thus the designation "add-on." This heat pump is often the most economical that can be used in a home.

SINGLE PACKAGE SYSTEM

Here the entire system is in one "box" complete with all necessary coils and blowers (similar to room air conditioners). These units are especially applicable in areas requiring roof-mounted heating/cooling systems, such as in the West and Southwest. Single package units are also ideal for any home which has a crawl space (space under the floor) that is used to contain the ductwork necessary to heat and cool the structure. There is no specific advantage in one type system over another, but as with all heating and cooling systems, they should be designed by a heating and air conditioning engineer.

SUMMARY

A heat pump takes heat from the outside air and pumps it into the house in winter and reverses itself and removes the heat from the house in the summer; thus providing both a heating system and an air conditioning system all in one unit. All outside air contains heat. At -10 degrees below zero the air has 84 percent of the heat normally available at 75 degrees F. Defrosting is a vital part of heat pump operation. With today's solid state technology the heat pump has been developed to be an efficient, economical method of heating and cooling homes. There are three heat pump systems available: the split system, which has two units, one located outdoors and one located in the house; the add-on system, where the unit is placed in the existing furnace, using the existing furnace ductwork and blower system to circulate the air; the single package system, where the unit is mounted on the roof or in a crawl space under the floor. These last units are used more in the West and Southwest where most houses do not have basements. There is no specific advantage of one type system over another.

REVIEW QUESTIONS

Select the letter preceding the best answer.

1. A heat pump uses _____ as its main source of heat.

    (a) Oil   (b) Gas   (c) Electricity   (d) Outside air   (e) All of these

2. Heat flows _____ a colder surface.

    (a) Towards   (b) Away from   (c) Neither

3. A vital part of efficient heat pump operation is _____ .

    (a) Cold air   (b) Air conditioning   (c) Defrosting   (d) All of these

4. A built in benefit of a heat pump is _____ .

    (a) It's efficient   (b) Cools air   (c) Heats air   (d) Reverses itself

5. The greatest improvement to heat pumps is _____ .

    (a) Modern technology   (b) Heating & cooling   (c) Defrosting

6. A _____ is both a heating and cooling system in one unit.

    (a) Heat pump   (b) Furnace   (c) Air conditioner

7. The design of a heat pump should be done by a/an _____ .

    (a) Contractor   (b) Heating engineer   (c) Architect

SECTION 6 TECHNOLOGY
unit 44 energy conservation

OBJECTIVES

After studying this unit the student will be able to

- describe solar energy.
- explain "active" and "passive" solar systems.
- determine which system is most efficient according to geographical location.

Solar energy is not something new. The earliest man was aware of the value of the sun and its warming rays. In 1774, Joseph Priestley experimented with the sun's rays and found that by concentrating the rays of the sun onto mercuric oxide a gas was formed that caused a candle to burn brighter than in the regular air. Thus, oxygen was discovered. In 1881, a French scientist experimented with the theory of using the ocean as a source of energy. He used the temperature difference between the warm upper layer of the ocean heated by the sun and the deeper cold layers. Today, engineers are experimenting with the ocean tide, to develop a new source of energy and power.

The major focus of Solar Energy Technology is broken into six general areas. *Solar climate control* is a term used to describe solar radiation collection systems utilizing energy-absorbent materials located on roofs or walls of buildings or homes. *Solar energy focusing systems* use the solar energy received by a large area (in terms of acres rather than in square feet as with solar climate control systems) which is focused upon energy-absorbent materials for transfer to a fluid medium, such as water. *Direct solar energy conversion systems* employ thermoelectric devices such as solar cells made with silicon or other materials and thermionic devices; they receive a solar energy input and provide an electric power output.

Indirect solar energy conversion systems capture solar energy which is not immediately converted to a heat-fluid medium or electricity for immediate transmission, but rather converted to some other form of energy that is easy to store.

Ocean thermal energy conversion takes advantage of the temperature difference between the upper layer of the ocean that is heated by the sun and the deeper cold layers.

Wind power is also classified with other solar approaches because the winds arise from atmospheric thermal differences.

Solar radiation is measured by calorie (heat) of radiation energy per square centimeter. Converting this to Btu's (British Thermal Units) means that solar radiation per minute (one calorie per square centimeter per minute) is equivalent in Btu's to 221 Btu's per square foot per hour.

Scientists have determined that approximately 43 percent of the radiation that reaches the earth from the sun is changed to heat. Clouds have a strong influence on the amount of energy that reaches the earth's surface. It is estimated that a typical cloud reflects approximately 75 percent of the sunlight that strikes it back into space. This means that on a cloudy overcast day only 25 percent of the sun's energy reaches the earth's surface.

The solar energy reaching the earth's surface is absorbed and reflected in varying degrees depending upon the surface it strikes. Bright objects such as shining aluminum, glass mirrors, bright steel, white snow, and to some degree water will reflect about 75 percent of the sunlight it receives and green grass and forest absorb anywhere from 80 to 95 percent of the solar energy it receives and changes it to heat.

The earth's axis and its rotation around the sun along with the earth's atmosphere determines

the amount of solar energy reaching the earth's surface and this in turn accounts in part for the difference in temperature and climate of various sections of the earth's surface.

Scientists and engineers have been working for over half a century to find a way to absorb, store, and distribute in an economical system this vast source of energy. Solar energy is not yet suitable for use in cold cloudy regions, or in large cities, where the acres of sunlight is not enough to supply the needs. Solar energy is being used and is suitable for use in sunny rural areas.

Solar radiation is so low in intensity that collectors of large areas are required to get enough energy concentrated in small workable units.

Solar energy used in residential construction can be classified into two categories: Active and passive. An active solar system entails the use of solar collectors or panels attached to the outside of a structure which collects solar radiation and transfers it into the building by using either a fluid or air; pumps transfer the fluid to a storage tank; controls operate the heat exchanger.

A passive solar system uses solar energy by collecting it through glass on the side of the building which receives the most sunlight. In its purest definition, a passive solar house would use no energy to circulate air, relying instead on the simple principle of convection to regulate the flow of heat within the building.

An active solar heating system is a very simple thing to understand. The sun's rays heat up a blackened absorber plate that is usually covered with a sheet or two of glass or plastic and is insulated on the back and sides. A liquid, circulated in tubes is attached to the absorber; or the liquid is allowed to trickle down the absorber surface; or air is ducted to the absorber. The rays of the sun heat the liquid or the air and it is carried where it is needed; either to a hot water tank, to the living space, or to a heat storage space, where it will be used at a later time.

There are many types of systems used to store the energy. Water, sand and pebbles, rocks, antifreeze liquids, etc. The size of the storage unit depends upon the number and size of the collectors. Most manufacturers of solar systems recommend that with pebbles and rocks that you use 1/2 cubic feet of rock for every square foot of collector.

A comparison of heat capacity on a volume basis between various storage media shows that water can store 62.5 Btu's per cubic foot per degree Fahrenheit, while rocks, brick, and gravel can store approximately 36 Btu's per cubic foot per degree.

ACTIVE SOLAR HOT WATER HEATING

There are many solar hot water heating systems in use today. The simplest system, with no

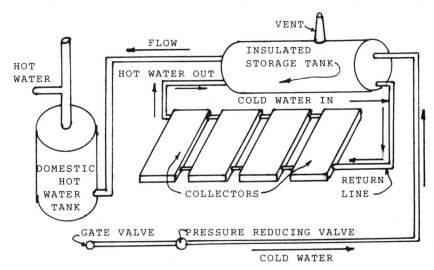

Fig. 44-1 Typical thermosiphon hot water system. The circulation between solar collector and the storage tank is affected by thermal gradients. The hot water, being less dense, rises to the storage tank at a higher level. If this is not possible, a pump can be used for forced circulation. Collectors are tilted at an angle to receive their optimum position at noon in March and September. In most climates an auxiliary heating unit is necessary.

energy input, is the thermosiphon. This system requires that the storage tank be located above the collectors. In thermosiphon or natural-circulation, the denser cold water—or anti-freeze liquid in freezing climates—falls and absorbs the heat in the collectors, and then rises to the storage tank.

It is estimated that each person uses approximately 25 gallons of hot water a day. The collectors are usually sized so that each collector can handle 25 gallons of hot water a day. The storage capacity also should be 25 gallons per person. A family of four would require four collectors and a 100 gallon storage tank. Figure 44-1 is a diagram of a simple thermosiphon system.

NOTE: Collectors must be tilted at an angle to be at their optimum position at noon in March and September.

The circulation between solar collector and the storage tank is affected by thermal gradients; the hot water being less dense rises to the storage tank at a higher level. If not possible, a pump can be used for forced circulation. (In most climates an auxiliary heating unit is necessary.)

SOLAR SWIMMING POOL HEATING

This system uses the same principle as the domestic hot water solar heating system. A pump is used to circulate the pool water through uninsulated collectors where it absorbs the heat. A special valve is placed in the pool's existing filtration plumbing and is controlled by a *differential thermostat*. This device compares two temperatures; the collector and the water. When heat can be added to the pool, the valve closes and the water is diverted to the collectors. This system can also be connected to an existing pool heater to lower the cost of heating the pool with conventional fuels.

SOLAR SPACE HEATING — ACTIVE

Active solar heating requires a storage unit. This is called a thermal storage unit. The basic elements of the active air heating system exclusive of pumps, valves, and controls are (a) a solar collector, (b) an auxiliary heating device, (c) hot storage system, and (d) heater element fan and air duct system. This system requires more solar

collectors, depending upon the requirements for heating the house (Btu loss), and it also requires a larger storage unit, usually one with the capacity to hold three to five days supply.

Put simply, air is heated in the following way: The cold air enters the bottom of the collectors, is heated by the rays of the sun, and rises to the top of the collectors. An air handler with blowers and dampers directs the air flow throughout the system. The heated air is forced into a storage bin, usually filled with pebbles and rocks. The size of the bin depends upon the required number of collectors. The storage bin itself is well insulated—usually made of concrete or concrete products.

When the heat is stored, air flows from the top of the bin to the bottom so the coldest air is returned to the collectors. When heating from the storage, the flow is from the bottom to the top so the hottest air goes to the living space.

SOLAR SPACE HEATING — PASSIVE

In passive solar heating systems the buildings themselves are the collectors, storage device, and heat distributer; and they must be carefully designed to take advantage of the changing position of the sun. In its purest definition, a passive solar house would use no energy to circulate air, relying instead on the simple principle of convection to regulate the flow of heat within the building.

Heat flows from hot areas to the cold areas. As sunlight warms the surface of a mass wall, heat will begin to flow to the cooler interior. When the room temperature is below the temperature of the wall surface, heat will flow from the mass into the room. Heat flow will stop when the temperature within the room equals the temperature within the walls.

The rate at which a mass conducts heat is directly related to its ability to absorb heat. Passive solar buildings utilize the facing of windows and skylights to bring in the warmth of the sun. Passive buildings must be carefully oriented in response to the seasonal and daily movements of the sun to maximize solar heat gain in the winter and to minimize solar heat gain in the summer.

Today the more effective passive solar houses utilize fans to increase and direct the air flow and

have very large thermal storage units. One of the most significant characteristics of a passive solar building is the use of the thermal mass; materials which have the ability to absorb and re-radiate large amounts of energy. Passive design measures also include use of good insulation, careful entrance location with regard to winter wind, use of air-lock vestibules, careful consideration of natural ventilation and natural light.

Passive measures such as cross ventilation, exhaustion of hot air by convection, evaporation, and absorption of heat by thermal mass can provide up to 100% of a building's cooling needs in summer.

A passive solar heating system for a house incorporates the passive solar design with a large internal thermal mass and interlocking air handling system. The most important feature of a passive solar house is its insulation envelope. The insulation envelope has sufficient thermal integrity to insure proper utilization of all available energies.

The greatest source of energy for a passive solar house comes naturally from the sun through glass. All windows are double-glazed with a minimal amount of windows on the opposite side where the sun does not shine upon it.

Excess energy from the sun and other sources should be stored within the structure, not only to increase its thermal performance but to prevent the house from experiencing large temperature swings. It is extremely important that the storage system be properly sized to transfer energy in and out of the storage medium in good fashion.

The thermal storage systems are usually underground sand, concrete, and crushed stone beds. The bed is completely enclosed in insulation and is an integral part of the building. These beds vary in size depending upon the size of the building. They average 150 to 200 tons under the first level of the building.

Within the mass storage bed are thermal transfer ducts connected to registers at the perimeter of the building from a central plenum. Air is continually drawn from the highest place down through the internal vertical mass (chimney). After passing through the storage bed it is circulated throughout the building. One or two small fans power this thermal loop. When the air in the building is warmer than the mass, the mass absorbs the excess heat, which acts to cool the air and increase the mass temperature. When the building begins to lose energy at night or during cloudy days, the warmed mass releases its stored energy to the air flowing through it back into the building. Figure 44-2 shows a cross section of how a passive system works.

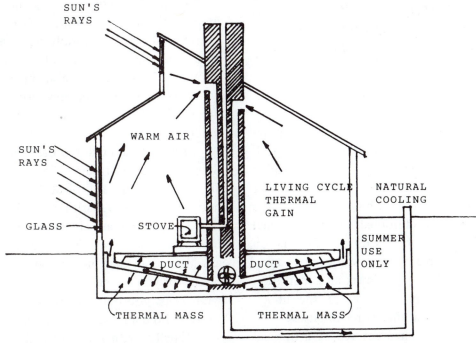

Fig. 44-2 Passive solar heating system. This interlocking thermal storage bed and air handling system is continually charged and discharged with Btu's. The storage bed can accept Btu's from woodburning, solar collectors, a wind generator, as well as the sun and the daily living cycle.

This interlocking thermal storage bed and air handling system is continually charged and discharged with Btu's. The storage bed can accept Btu's from wood burning, solar collectors, or a wind generator, as well as the sun and the daily living cycle.

Solar heating systems are sized to replace certain proportions of heat that is lost from the house. There are well established formulas for determining heat-loss characteristics. They are complicated and should be left to a solar engineer.

There are many factors to consider in solar heating. The size of the house, its location, the style (architecture), the cost of the solar system, the ability of the system to compensate for climate differences, etc. Most all solar heating systems will require an auxiliary heating system for backup.

With today's high technology, many changes have occurred in the building industry; this has caused the changing of local building codes to insure that the homeowner is getting the most and best built house for his money. As an estimator, you should be very familiar with your local building code and what is required.

Today in northern sections of the United States, where the temperature fluctuates from -20 degrees below zero in the winter to +90 degrees in the summer, many building codes require 2 x 6 framing for outside walls with 6'' insulation in the outside walls and 9 to 12'' inches of insulation in the ceilings. This is all being done for energy conservation.

ENERGY SHIELD

Energy shield sheathing is a rigid polyisocyanurate foam board with a triple face (foil-kraft-foil) on the weather side and either a triplex or single layer of pure aluminum foil on the other side. The energy-shield sheathing surrounds the entire house with an additional layer of thermal protection. This layer works with other insulation to achieve an even greater total wall performance. It reduces energy needs, helps lower air penetration and improves the comfort level of the home.

Energy shield sheathing is recommended for *concealed* applications when covered by a thermal barrier acceptable to local building codes. It can be used on exterior walls, interior walls/ceilings, cathedral ceilings, roofs, interior basement insulation, re-siding and crawl spaces.

WARNING: This product will burn. Do not leave exposed. The energy shield must be installed behind 1/2'' inch gypsum wallboard or other thermal barrier approved by local building codes.

HOUSEWRAP

Housewrap is building material that is manufactured in 1000 square foot rolls in either a nine foot width or 4.5 foot width. It is also manufactured in a 1755 square foot roll nine feet wide and a three foot wide roll that has 495 square feet. It is fabricated from spunbonded polypropylene or spunbonded olefin in a very fine, high-density fiber. It is not a film or a paper. It resists tears and punctures.

It reduces air infiltration and exfiltration by providing an air tight envelope around a house. It keeps the heated air from leaking out in the winter and the cool air from leaking out in the summer. It inhibits air movement in wall cavities by covering any construction gaps that occur normally between boards, sheathing, and insulation. Without housewrap, outside air can penetrate through these gaps and circulate around the insulation, reducing its "R-value" and its efficiency.

Insulation manufacturers measure "R-value" under still-air conditions (ZERO convection). Convection occurs naturally when air is allowed to enter the wall and can substantially reduce the R-value. Housewrap is also breathable to allow water vapor to escape, preventing moisture build-up inside the wall cavities which could cause mildew and rot.

Tests conducted by the National Bureau of Standards demonstrated a 35 percent reduction in air exchange in a typical house wrapped with housewrap. The Department of Energy suggests a rate around 0.6 air changes per hour. Less than that results in a "too-tight" house and means stale air. More than that means you are losing too much of the air you are paying to heat or to cool. Housewrap helps you build homes that are closer to the suggested 0.6 air change per hour.

Finally, with housewrap the convection loop is broken, the R-value is protected, and you get

maximum efficiency and savings from the insulation.

Vapor barriers, usually consisting of non-permeable plastic film normally are placed on the inside wall to keep moisture out of the wall cavity. Wall outlets and vent openings are cut through the plastic, creating draft points. With housewrap, the air barrier is placed on the exterior wall over the sheathing to keep the outside air out of the wall cavity. It is wrapped completely around the building covering window and door openings, plates and sills and corners. Cut an "X" from corner to corner at window and door openings and pull the fabric inside, fold over the excess and fasten. If installed properly, it will seal cracks and seams in sheathing and at sill and soleplates thereby minimizing draft points. It is estimated that two men can wrap a typical house in less than two hours so labor cost is minimal.

SUMMARY

Solar energy is not something new. Scientists and engineers have been working for over half a century on how to collect, store, and use the heat generated by the sun's rays. Solar energy technology is classified into six general areas. In residential construction, the two main areas are referred to as active solar heating and passive solar heating.

Active solar heating is very simple to understand. Either air or a liquid is circulated through an absorber plate built within a collector. The rays of the sun heat the liquid or air and it is carried to where it is needed, either to a hot water tank, or to the living space, or to a heat storage bin to be used at a later time.

Passive solar heating uses the buildings themselves as the collector, storage device, and heat distributer. Today the more effective passive solar houses utilize fans to increase and direct the air flow and they have very large thermal storage units.

Solar heating systems are sized to replace certain proportions of heat that is lost from the house. There are well established formulas for determining heat-loss characteristics. They are complicated and should be left to a solar engineer.

There are many factors to consider in solar heating: The size of the house, its location, the style (architecture), the cost of the solar system, the ability of the system to compensate for climate differences, etc.

REVIEW QUESTIONS

Select the letter preceding the best answer.

1. The main source of solar energy is (the) _____ .

   (a) Moon   (b) Sun   (c) Coal   (d) Oil   (e) Water

2. Solar radiation is measured by _____ .

   (a) Calorie   (b) Btu   (c) Degree   (d) Centigrade

3. Solar energy used in residential construction can be classified into _____ categories.

   (a) Two   (b) Three   (c) Four   (d) Six

4. In its purest definition, a _____ solar system would use no energy to circulate air.

   (a) Passive   (b) Active   (c) Indirect

5. A comparison of heat capacity on a volume basis between various storage media shows that _____ can store more Btu's per cubic foot per degree.

   (a) Rocks and gravel   (b) Brick   (c) Water   (d) Concrete

6. It is estimated that each person uses approximately _____ gallons of hot water a day.

   (a) 20   (b) 25   (c) 30   (d) 15

7. Heat flows from _____ to _____ .

   (a) Hot-cold   (b) Cold-hot   (c) Hot-hot   (d) Cold-cold

8. In solar heating, the storage units are usually referred to as _____ .

   (a) Bins   (b) Thermal mass   (c) Battery

9. It is estimated that a typical cloud reflects back into space approximately_____ percent of the sunlight that strikes it.

   (a) 50   (b) 75   (c) 90   (d) 85

unit 45 computer estimating

OBJECTIVES

After studying this unit the student will be able to

- explain what a microcomputer can do.
- tell the main functions of a computer.
- know how to use a computer in estimating.

A microcomputer is an electronic machine that can only do what it is told to do in instructions given to it by some human being. All the microcomputer does is execute the instructions it has been given. A computer is not human and cannot think as a human. The microcomputer is just a tool which the human uses for carrying out computing activities. In its basic form, all microcomputers functions are built up from a few very simple activities. It can add two numbers together, it can subtract one number from another number, and it can compare two numbers or symbols to see if they are the same.

Basically all microcomputers have the same fundamental parts regardless of their size: input device, central processor unit, and output device. The input device, such as a keyboard or microphone, contains a memory which can store information, including the instructions which have been given the microcomputer by the manufacturer; its program. The central processor processes the information in accordance with its program instruction. The output device receives messages from the microcomputer and does something useful with this information; it can be a screen, a printer, or both and capable of printing and/or storing information.

There are other parts used in computing such as tapes and discs but basically we have covered what a microcomputer consists of. Remember that it is only a tool to be used and is only capable of producing what it is programmed to produce.

In estimating for house construction, it would be necessary to have the computer programmed so that by input of certain information, we would get the answer on the printer. This would give you a list of materials and quantities needed in the same order that they would be used in the construction of the house. In doing this, we will use the same sequence we would do if we were going to estimate with our own facilities and not with the microcomputer. By using the computer, the computer screen, and the printer we can complete the estimate and have a completed material list much faster than if we listed the materials, determined the cost and then had to type the list on a typewriter. We would use the same sequence and the same formula for estimating regardless of the size and style of house.

We will start our estimate on the microcomputer the same as we would in making a regular material list. We start at the footings and work toward the completed house. You have to program the software to include the following:

1. **FOOTING & STAKES**

   Take the linear feet around the building and double it. Add for any support posts, fireplaces, and porches.

2. **FOOTINGS**

   Multiply the width of the footing in feet times the concrete's thickness in feet. Then multiply by total linear feet of footings. This

gives you the cubic feet of footing. Divide this product by 27 to find the cubic yards.

3. **FLOORS & PORCHES**

Take the total square foot area of all places to be covered by concrete. Multiply this area by the thickness in feet and divide by 27. This gives you the cubic yards required.

4. **SIDEWALKS**

Length in feet times width in feet times thickness in feet divided by 27 gives you the number of cubic yards.

5. **GRAVEL FILL**

Take the square foot area where the concrete is to be placed and divide by 2, (or the proportional part of foot the specifications call for in thickness). This gives you cubic feet. For cubic yards, divide by 27.

6. **CEMENT BLOCK**

Add perimeter of the building in feet. Add porches and fireplace base, if any. Multiply this by 3 and divide by 4. This is the number of block for one course. You figure three courses for every two feet in height. Usually ten courses in house construction. You multiply the number of block in one course by 10 for the number of full size block. List the number of block required for one course for the cap block. Corner block are listed separately. Count the number of corners and multiply by 10. Figure two half sash and two whole sash block for each basement window.

7. **POURED CONCRETE WALL**

Multiply the length of the foundation times the thickness in feet. Multiply this times the height in feet. This is the number of cubic feet in the wall. Divide this product by 27 to get the number of cubic yards.

8. **BASEMENT SASH**

Count the number of basement windows as shown on the Foundation and Basement Plan.

9. **METAL AREAWAYS**

The areaways are shown on the Foundation and Basement Plan. List by the number required. Should be the same as basement cellar sash.

10. **MORTAR FOR BLOCK**

Figure one bag regular cement and one bag masons lime for every 28 cement block.

11. **WASHED SAND**

Figure three cubic feet of sand for every bag of cement. Easiest method is figure one cubic yard of sand for every 10 bags of cement. This allows for waste.

12. **WALL REINFORCING**

Take the linear feet where reinforcing is required and multiply by the number of courses where it will be placed. This gives you linear feet required.

13. **CONCRETE FLOOR REINFORCING**

Take square foot area where concrete is to be placed. This gives you number of square feet of wire mesh. It is usually 6 x 6 wire mesh.

14. **STEEL SUPPORT POSTS**

Count number required on Foundation and Basement Plan. List according to size and number.

15. **CHIMNEY**

Determine total chimney height. Divide this by two for the number of flue liners. Using whole chimney block figure three block for every two feet in height. For each foot in height where brick is used, figure 27 brick for each foot for a single 8 x 8 flue chimney. Also list cement for mortar and flashing.

16. **FIREPLACE**

Determine the height of the chimney. Divide by two for the number of furnace flue liners. You will need three fewer fireplace flues, if fireplace is on first floor. Figure 30 firebrick for firebox base. An ash dump, a

12" x 15" cleanout door for ash pit and 8 x 8 cleanout door for furnace flue will be needed. An angle iron is needed for each fireplace opening. Estimate six and three quarter (6 3/4) brick for each square foot of surface. Add for fills. If you are using an 8 x 8 furnace flue and an 8 x 13 fireplace flue figure 55 brick for every foot in height.

17. **FIREPLACE HEARTH**

Figure by square foot area to be covered. Usually 4 x 4 quarry tile set in cement with cement joints.

18. **FIREPLACE MORTAR**

One bag regular cement, one bag masons lime and 48 shovels of sand will lay approximately 400 brick. Allow one third (1/3) for filling between brick and flue liners.

19. **BRICK VENEER**

Figure 675 brick for every 100 square feet of wall. Allow ten percent for waste. Figure 70 wall ties for every 100 square feet.

20. **BRICK VENEER MORTAR**

Figure one bag regular cement and one bag masons lime for every 400 brick. If using masonry cement figure eight bags for every 1,000 brick.

21. **FOUNDATION PLASTERING (PARGING)**

One bag regular cement and one bag masons lime with 48 shovels of sand will cover approximately 40 (forty) square feet. Determine the square foot area of the foundation walls. Divide this by 40 to get the number of regular cement and masons lime.

22. **DAMPPROOFING (FOUNDATION)**

Brush type asphalt covers 30-35 square feet to the gallon. Figure the square foot area to be covered and divide by 30. If two coats are called for the second coat will require half as much. Asphalt coating is sold in five gallon cans, so divide the number of gallons required by 5.

23. **DRAIN TILE**

Determine the linear distance around the house and divide by 10. This is the number of 10' pieces of pipe required. One 90 degree elbow for each corner and one TEE is needed. Crushed stone is sold by the ton. Figure one ton of stone for every 18 linear feet of tile.

24. **GIRDERS**

Determine the length of the members and list them according to length and size. Usually listed by the piece.

25. **SILL SEAL & SILL**

Listed by the linear feet required. Distance around the foundation. Sometimes listed by the number of pieces.

26. **BOX SILL**

Listed by linear feet required for perimeter. Same width as floor joists.

27. **FLOOR JOISTS**

Multiply length of foundation wall on which they rest by 3. Divide by 4 and deduct 1. Also add one joist for each partition that runs the same direction as the joists.

28. **BRIDGING**

Multiply the number of floor joists required by 3. This gives the linear feet of bridging needed for one row. If span of joists is over 14 feet, figure two rows of bridging for that span.

29. **HEADERS & TRIMMERS**

Headers support the cut ends of joists at such places as stairs, chimneys, or fireplaces. The joists at the end of the headers are called trimmers. They should be doubled and same width as joists. List by piece.

30. **SUB-FLOORING**

Estimated by the adjusted square foot area to be covered. Divide by 32 for number of 4' x 8' pieces of plywood or waferboard as specified.

31. **PLATE & SHOE**

    Estimated by the linear foot. Take the perimeter of outside wall; figure linear feet of interior wall. Do not deduct for openings. Add exterior and interior walls together and multiply by 3 for total linear feet needed.

32. **STUDS**

    Figure one stud for every linear foot of exterior and interior partition and add two for each corner.

33. **HEADERS**

    Estimated by the linear foot or by the piece depending upon the size and span they have to cover. Headers should be doubled.

34. **CEILING JOISTS**

    Determine the direction they run. Multiply that distance by 3; divide by 4 and add 1. The length is the span they have to cover.

35. **RAFTERS**

    Multiply the length of the ridge by 3, divide by 4 and add 1. Assume jack rafters the same size and length as common rafters. To estimate jack rafters, take 1 1/2 times width of building and add 2. Allow one hip for every hip and one valley for each valley.

36. **TRUSSES**

    To determine the number of trusses required in a straight roof, divide the length of the building in feet by 2 and add 1.

37. **COLLAR TIES**

    If spaced 16 inches on center, divide the number of rafters by 2 and add 1. If spaced 32 inches on center, divide number of rafters by 4 and add 1.

38. **GABLE STUDS**

    Measure one gable end and figure one stud for each linear foot of half width. Height of building from top of wall plate is length of studs.

39. **ROOF SHEATHING**

    Multiply length of adjusted ridge times the length of a common rafter. Multiply this by 2 to find the total roof area. Take this total area and divide by 32 to find the number of 4' x 8' pieces of plywood.

40. **ROOFING**

    Divide the square foot area of the roof by 100 to find the number of squares required. For asphalt shingles, add 1 1/2 square feet for every linear foot of ridge, eaves, hip, and valley.

41. **ASPHALT FELT**

    Take square foot area of the roof; divide by 400 to determine the number of rolls of felt required.

42. **DRIP EDGE**

    Measure the length of the eaves and the rake and add them together, then divide by 10 for the number of pieces required.

43. **CORNICES**

    List each part by the linear foot: rake, fascia, soffit, and frieze.

44. **EXTERIOR SHEATHING**

    Take the adjusted distance around the house in feet and multiply it by the height. Figure the gable ends. This is the total square foot area to be covered. If plywood or waferboard is used for sheathing, divide the total square foot area by 32 for the number of 4' x 8' required.

45. **SIDING**

    Estimate according to the material used. If sold by the square, divide by 100 for the number of squares. If sold by the board foot, find the number of square feet, add for cutting and waste and list as board feet.

46. **EXTERIOR FINISH WORK**

    Any special work to be done to finish the outside of the house should be listed. This would cover such items as corner boards, porches,

special overhangs, entrance porch, railings, etc.

47. INSULATION

Determine the square foot area to be covered. Insulation is sold by the bag. For 6″ fiberglass insulation in ceilings each bag covers 50 square feet. Three inch insulation, used in sidewalls, is sold in 70 square foot rolls. You can easily figure the number of bags and rolls required or list by the square foot area to be covered.

48. WALLBOARD

Figure by the square foot area to be covered. The ceiling is the same area as the floor. Take the length of the outside walls and multiply them by the height. Add the total length of the interior partitions and multiply by height. Double the interior wall area. Add ceiling area, the outside wall area and the total interior wall area for the total square foot area of wallboard. Do not deduct for any openings.

49. JOINT SYSTEM

Tape comes in 250′ and 500′ rolls. Joint compound is sold in five gallon and one gallon cans. Figure one 250′ roll of tape and six gallons of joint compound for every 1,000 square feet of wall.

50. MOLDING

This is estimated by the linear foot and by specific lengths. The distance around the house is already known. The total length of the interior partitions is already known. Double the length of the interior partitions and add the length of the outside walls for total molding.

51. BASEBOARD & CARPET STRIP

Deduct twice the width of the interior door openings and the width of the exterior doors and subtract from the total length of the ceiling molding.

52. UNDERLAYMENT

Take the square foot area where underlayment is required (kitchen, baths, entrance hall). Divide this area by 32 to get the number of 4′ x 8′ needed.

53. LINOLEUM OR RESILIENT FLOORING

Find the square foot area to be covered and divide by 9 for square yards. If flooring is square feet, then list by the piece.

54. SLATE FLOORING

Estimate by the square foot area to be covered. Figure one gallon special floor adhesive for every 100 square feet and five pounds of grout for every 100 square feet.

55. HARDWOOD FLOORING

Determine the square foot area to be covered and add one third for matching. This will give you total board feet required. Deadening felt (sold in 500 foot rools) is laid between the subfloor and finished floor. Divide the square foot area by 500 for number of rolls.

56. CERAMIC WALL TILE

Sold by the square foot. Cap tile is sold by the linear foot. Inside corners and outside corners are sold by the piece. Ceramic fixtures are sold by the piece. One gallon ceramic tile adhesive will cover approximately 100 square feet, 5 pounds white cement grout will cover about 60 square feet of wall tile.

57. CERAMIC FLOOR TILE

Sold by the square foot. One gallon of adhesive and five pounds of grout is needed for every 100 square feet. A marble threshold is needed between doors where ceramic tile and other type floors meet.

58. EXTERIOR DOORS

Doors are always listed width first, then height, and then thickness. Any special entrance doors with sidelights or special head should be listed first.

59. INTERIOR DOORS

Listed the same as exterior doors according to size and style. List any cased openings first.

60. DOOR STOP AND CASING

Listed by the specific lengths. It requires approximately 17 linear feet of stop for each door and 34 linear feet of casing for both sides of a door opening. Count the number of door and multiply by 17 for the door stop and by 34 for the casing.

61. CLOSET SHELVES & RODS

List by the linear foot. Closet shelves are usually 1 x 12 boards and supports are 1 x 3 pine. Closet rods are either wood or adjustable chrome-plated metal.

62. WINDOWS

List according to size and style. Here again the width is listed first and the height second.

63. WINDOW TRIM

Window trim is listed by the specific foot. Estimate the trim required for each window and then group them together. List according to part and style. Casing, stop, stool, apron, and mullion casing if required.

64. FIREPLACE MANTEL

List according to size and style and manufacturer's number.

65. BLINDS OR SHUTTERS

List according to size and style.

66. STAIRS

List by size and style. List each part separately. Stringers, newel posts, balusters, handrail, treads, risers, starting steps, landing treads, etc.

67. HARDWARE

Usually an allowance. Nails should have an allowance because this is not covered with hardware.

68. KITCHEN CABINETS & BATH VANITIES

List by units. List wall units first, the base cabinets second, and then the linear feet of countertop. List giving size and manufacturer's finish. List width first then height with wall cabinets. List vanities by manufacturer's style and size.

69. MISCELLANEOUS ITEMS

List any items necessary to complete the material list.

SUMMARY

A microcomputer is an electric machine that can do only what it is programmed to do. All the computer does is execute the instructions it has been given. All computers have the same basic parts. They have an input device such as a keyboard, a memory which stores information, a central processor which processes the information according to its programmed instruction, and an output device which receives messages from the computer and does something useful with this information such as printing it on paper. Using a computer for estimating requires that the computer be programmed for such use; and by following a certain sequence, we should be able to get a printed list of materials for the job in the same order they will be used in the construction of the house.

REVIEW QUESTIONS

Select the letter preceding the best answer.

1. All microcomputers have basically the following parts.

    (a) Input device   (b) Output device   (c) Memory   (d) All of these

2. Computers _____ as humans.

   (a) Can think   (b) Cannot think   (c) Do the same   (d) None of these

3. The important part of the computer for estimating is its _____ .

   (a) Program   (b) Output   (c) Processor   (d) Input device

4. Concrete is estimated by the _____ .

   (a) Square foot   (b) Cubic foot   (c) Cubic yard   (d) Square yard

5. Brick veneer work is estimated by _____ .

   (a) Square foot area   (b) Cubic foot area   (c) Linear foot   (d) None of these

6. Plate and shoe is estimated by the _____ .

   (a) Square foot   (b) Cubic foot   (c) Linear foot   (d) Piece

7. Studs are estimated by the _____ .

   (a) Piece   (b) Linear foot   (c) Square foot   (d) Cubic foot

8. Roofing is estimated by the _____ .

   (a) Cubic foot   (b) Cubic yard   (c) Square foot   (d) Square yard

9. Cornice material is estimated by the _____ .

   (a) Piece   (b) Linear foot   (c) Square foot   (d) Cubic foot

10. Linoleum is usually estimated by the _____ .

   (a) Square yard   (b) Square foot   (c) Cubic foot   (d) All of these

SECTION 7 COSTS
unit 46 labor and related costs

OBJECTIVES

After studying this unit the student will be able to

- list all of the areas of cost for the construction of a residential building.
- describe the basis for estimating labor for each phase of the construction of a residential building.
- determine a reasonable bidding price for the construction of a residential building when given the specifications, working drawings, and necessary cost factors, such as labor rates.

Most of the material in the preceding units has been concerned with listing the materials required to complete the construction of a building. This is called a *quantity takeoff*. To arrive at a bid price for the construction, the estimator must also determine all of the costs involved. These costs include such things as permit fees, the cost of required bonds, temporary facilities, labor costs, overhead, and profit.

The quantity takeoff provides the necessary information to determine how much labor is involved in each phase of the construction project. Some estimators prefer to estimate the labor involved as they prepare the quantity takeoff; others prefer to complete the list of materials, then go back over it and estimate the labor. In either case the result is the same.

The Appendix at the back of this textbook is a table showing the approximate amount of labor required for a unit of each type of construction. By keeping accurate records of the actual labor involved in each project, the estimator can prepare such a table.

Site Work

The contractor visits the site to learn what trees and shrubs are to be removed, what the slope of the land is, what the relationship of the new building is to any existing structures, and any other existing conditions that may affect the site work. Based on this first-hand knowledge of conditions, the specifications, and the working draw-

ings, the amount to be excavated is estimated in cubic yards. The cost of this excavation is normally based on a price per cubic yard. Work to be performed on remaining trees is based on a cost per tree. Finally, the cost of placing topsoil and seeding the lawn is based on a price per square foot.

Form Work

The labor required to set forms for footings and foundation walls is based on the number of linear feet involved. Form work for concrete piers and other special construction is generally estimated on a unit basis.

Concrete Footings

The contractor must determine how much labor will be involved in placing concrete footings. This is based on the cost to place one cubic yard of concrete. If the foundation walls are to be poured concrete, their cost is estimated in the same way.

Concrete Floors

The labor cost for concrete floors includes all labor to place and finish the concrete. This is usually estimated by the square foot, regardless of the depth of the concrete. This cost may vary depending on the location of the floor. It costs more, for example, to place 100 square feet of concrete on the third floor than it does in the basement.

Brick Work

The estimate for brick work is based on the cost of laying 1,000 bricks in a wall. This includes

erecting scaffolding and cleaning the brick work when the job is finished. For fireplaces and chimneys the estimate is normally based on a base price per fireplace or a price per linear foot of chimney height.

Concrete Block

The blockwork estimate is based on the cost per block. This figure includes all of the labor needed to install reinforcement, parge the wall, and apply dampproofing. The estimator may use varying price schedules for blockwork, depending on what additional work is to be performed.

Carpentry Work

Estimators base some of their estimate for carpentry work on a cost per piece or unit to be installed, and some of it on the cost of installing a specific area of the material. Girders, beams, sill seals, and sills are estimated by the linear foot. The labor for framing the floors and walls is usually based on the cost per square foot of area. This includes the box sill, joists, bridging, subflooring, plates, studs, headers, trimmers, and wall sheathing.

The labor for roof framing is based on the cost per rafter or truss. This includes the labor on the ridgeboard, valley rafters, hip rafters, gable studs, collar beams, and roof sheathing. The cost of installing the finished roof is based on the cost per square.

Wood trim is usually estimated on a cost-per-linear-foot basis. Naturally, the cost per foot is dependent on the type of trim; a foot of colonial cornice costs more to install than does a foot of window casing. The labor for siding is usually based on the cost per square. The siding estimate includes the labor for building paper and scaffolding necessary to install the siding.

Flooring is usually estimated on the basis of the labor to install one square foot. The number of square feet of hardwood flooring, ceramic tile, slate flooring, or carpeting can be determined from the quantity takeoff. If deadening felt is to be applied under a hardwood floor, this in included in the flooring labor.

Stairwork is estimated separately for every set of stairs. This is usually based on the style and extra finish work involved.

Insulation, Plaster, Wallboard, and Paneling

The labor involved in installing these materials is normally estimated on a square-foot basis. These estimates include the labor required to finish joints in wallboard, apply lath for plastering, and any other related work.

Cabinet Work

This includes the installation of kitchen cabinets, countertops, and bathroom vanities. The labor for this is normally based on the particular layout and design of cabinets involved. Any carpenter-built cabinets are estimated separately.

Tile Work

The application of ceramic or plastic tile is based on the cost of applying one square foot. This figure includes setting, grouting, cutting, and cleaning of the tile work, as well as the placing of ceramic accessories like soap dishes and grab bars. If the surface to be covered with tile requires special preparation, this is also included in the square-foot estimate for tile work.

Doors and Windows

The labor for windows and doors is usually based on the labor to install one unit. The labor for doors includes setting the door frame and hanging the door. For prehung doors much less labor is required. The labor for installing windows includes everything necessary to ready the window for trim.

Painting

Painting is frequently done by a subcontractor who furnishes an estimate for the entire job. However, the estimator must be able to determine approximately what the cost and time requirements of this part of the job are. Painting estimates are based on the labor for one square foot of the area to be painted. The cost per square foot varies considerably, depending on the location (interior or exterior), the kind of material to be used (varnish, water-based paint, or oil-based paint), the nature of the surface (siding, ceilings, or trim).

Plumbing

Plumbing is also normally performed by a subcontractor. The estimate for plumbing can be quite complex, because there are several ways in which the labor might be estimated. One typical method for estimating plumbing labor is to estimate the time for one typical fitting and multiply this by the number of fittings in the entire job. An allowance for labor on each major fixture, and an allowance for any required excavation must be added to the allowance for fittings.

Electrical

Electrical estimates may be similar to plumbing estimates. An allowance is made for a typical outlet, then this is multiplied by the number of outlets in the building. Additional allowances must be made for such things as special equipment and underground service.

Air Conditioning

There are two common methods by which the estimate for air conditioning may be done. A percentage of the overall equipment cost may be allowed as a labor estimate. This percentage depends on the type of heat being installed. Another method for estimating the labor for air conditioning, is to base the estimate on the total building heating load.

Related Costs

Many of the costs involved in the construction of a building do not fall within the categories of labor and materials, but are still directly related to the particular job. These things are often listed in the General Conditions and General Requirements of the specifications. These expenses include temporary power, temporary toilet facilities, pumping excess water from excavations, building permit fees, and insurance on the structure during construction. All such costs must be included on the estimate for the job.

Overhead and Profit

Some of the costs that any construction company must pay are not directly related to a particular construction project. These are, nonetheless, part of the expenses of the business and must be paid for out of the income from construction projects. These expenses include such things as rent for office spaces, bookkeeping costs, depreciation of equipment, and certain taxes and legal fees. These overhead expenses are normally allowed for as a percentage of the total construction project.

After all expenses have been paid, any remaining income is profit. One of the most important goals of most businesses is to make a profit, so this item must be included on the total construction estimate. Usually a percentage of the total construction cost is added for profit.

SUMMARY

The total building estimate includes cost allowances for materials, labor, legal fees, insurance, overhead, and profit. The most time-consuming part of the estimate is determining the quantity of construction involved. As the estimator prepares a detailed quantity takeoff, the amount of labor involved becomes apparent. Working from the quantity takeoff the estimator calculates the labor requirements for the job. In most cases this is done by determining the amount of labor involved in using a certain unit of materials; the number of units in the total project is then multiplied by the requirements for one unit to find the total project requirement. Other related costs, overhead, and profit are then added to the labor and material estimate to complete the total estimate.

REVIEW QUESTIONS

A. Matching

Indicate by letter, the unit of measurement from the column on the right that is the basis for estimating the labor for the items in the column on the left.

1. Excavation for the basement	a. Square foot
2. Concrete blockwork for the foundation	b. Linear foot
3. Concrete for a poured foundation	c. Cubic yard
4. Floor framing	d. Piece
5. Wall framing	e. 100 square feet
6. Siding	f. Truss or rafter
7. Roofing	g. 1,000 units

8. Painting

9. Interior finishing with gypsum wallboard

10. Tile work for a ceramic tile bathroom floor

11. Brickwork for brick veneer over frome construction

12. Kitchen cabinet work

13. Form work for concrete footings

14. Placing topsoil on a lawn

15. Repair work on a tree to remain on the site

B.   List five expenses in construction of a residence that are not included under a specific trade. Do not include overhead or profit.

COMPREHENSIVE REVIEW

The following pages include specifications and working drawings for a single-story and two-residence. This provides the student with an opportunity to review and apply the knowledge gained through the use of this textbook. The student is cautioned to review the specifications carefully and compare them with the appropriate drawings as the quantity takeoff is prepared. In cases of conflict, it should be remembered that the specifications take precedence over the drawings, and that detail drawings take precedence over large-scale drawings.

It should also be noted that construction specifications may be written in a variety of forms and styles. The specifications for this review use the Description of Materials, form 2005, from the United States Federal Housing Administration. This form of specifications has been specially selected for this review, because of their brevity in contrast with the extensive specifications used throughout the textbook. These specifications describe the material to be used, but do not provide additional information about the construction of the house. When a question arises about construction techniques, the student should rely on knowledge of accepted practices.

Partially completed quantity takeoff forms are included. The student need only indicate the quantities and sizes of materials required in the shaded spaces. The instructor may choose to provide separate forms, so that it is not necessary to write in this textbook.

SINGLE STORY RESIDENCE

FHA Form 2005
VA Form 26-1852
Rev. 2/75

U. S. DEPARTMENT OF HOUSING AND URBAN DEVELOPMENT
FEDERAL HOUSING ADMINISTRATION

For accurate register of carbon copies, form
may be separated along above fold. Staple
completed sheets together in original order.

Form Approved
OMB No. 63–R0055

DESCRIPTION OF MATERIALS

No. _____
(To be inserted by FHA or VA)

☐ **Proposed Construction**

☐ **Under Construction**

Property address ___ REVIEW HOUSE ___ **City** ___ Hometown ___ **State** ___ USA ___

Mortgagor or Sponsor ___ First Main Bank ___ ___ 101 Main Street ___
(Name) (Address)

Contractor or Builder ___ John Doe ___ ___ 225 Main Street ___
(Name) (Address)

INSTRUCTIONS

1. For additional information on how this form is to be submitted, number of copies, etc., see the instructions applicable to the FHA Application for Mortgage Insurance or VA Request for Determination of Reasonable Value, as the case may be.
2. Describe all materials and equipment to be used, whether or not shown on the drawings, by marking an X in each appropriate check-box and entering the information called for in each space. If space is inadequate, enter "See misc." and describe under item 27 or on an attached sheet. THE USE OF PAINT CONTAINING MORE THAN ONE HALF OF ONE PERCENT LEAD BY WEIGHT IS PROHIBITED.
3. Work not specifically described or shown will not be considered unless

required, then the minimum acceptable will be assumed. Work exceeding minimum requirements cannot be considered unless specifically described.
4. Include no alternates, "or equal" phrases, or contradictory items. (Consideration of a request for acceptance of substitute materials or equipment is not thereby precluded.)
5. Include signatures required at the end of this form.
6. The construction shall be completed in compliance with the related drawings and specifications, as amended during processing. The specifications include this Description of Materials and the applicable Minimum Property Standards.

---

1. EXCAVATION:

Bearing soil, type ___ Sand and Loam ___

2. FOUNDATIONS:

Footings: concrete mix ___ ; strength psi ___ 3000 ___ Reinforcing ___ none ___
Foundation wall: material ___ 8x8x16 & 8x12x16 Conc. Block ___ Reinforcing ___ every 3 courses ___
Interior foundation wall: material ___ none ___ Party foundation wall ___ none ___
Columns: material and sizes ___ 3" steel w/ base & cap ___ Piers: material and reinforcing ___ none ___
Girders: material and sizes ___ 6x8 wood, built-up ___ Sills: material ___ 2x6 fir ___
Basement entrance areaway ___ none ___ Window areaways ___ corrugated steel ___
Waterproofing ___ 1/2" cement plaster & 2 coats asphalt ___ Footing drains ___ 4" plastic ___
Termite protection ___ none ___
Basementless space: ground cover ___ none ___ ; insulation ___ none ___ ; foundation vents ___ none ___
Special foundations ___
Additional information: ___ Sisalkraft vapor barrier under basement floor. 6x6-10/10 ___
___ welded wire mesh on porches and garage ___

3. CHIMNEYS:

Material ___ brick & block ___ Prefabricated (make and size) ___
Flue lining: material ___ T.C. ___ Heater flue size ___ 8x8 ___ Fireplace flue size ___ none ___
Vents (material and size): gas or oil heater ___ ; water heater ___
Additional information: ___

4. FIREPLACES: ___ none ___

Type: ☐ solid fuel; ☐ gas-burning; ☐ circulator (make and size) ___ Ash dump and clean-out ___
Fireplace: facing ___ ; lining ___ ; hearth ___ ; mantel ___
Additional information: ___

5. EXTERIOR WALLS:

Wood frame: wood grade, and species ___ Fir ___ ☐ Corner bracing. Building paper or felt ___ doublekraft ___
Sheathing ___ plywood ___ ; thickness ___ 1/2" ___ ; width ___ ; ☐ solid; ☐ spaced ___ " o. c.; ☐ diagonal; ___
Siding ___ ; grade ___ ; type ___ ; size ___ ; exposure ___ "; fastening ___
Shingles ___ 16" Red Cedar ___ ; grade ___ #1 ___ ; type ___ 5x ___ ; size ___ 16" ___ ; exposure ___ 7" ___ "; fastening ___
Stucco ___ ; thickness ___ "; Lath ___ ; weight ___ lb.
Masonry veneer ___ Brick ___ Sills ___ Brick ___ Lintels ___ none ___ Base flashing ___
Masonry: ☐ solid ☒ faced ☐ stuccoed; total wall thickness ___ "; facing thickness ___ 4" ___ "; facing material ___ Brick ___
Backup material ___ ; thickness ___ "; bonding ___
Door sills ___ Oak ___ Window sills ___ Pine ___ Lintels ___ Base flashing ___
Interior surfaces: dampproofing, ___ none ___ coats of ___ ; furring ___
Additional information: ___
Exterior painting: material ___ Nonchalking Shake & Shingle paint ___ ; number of coats ___ 2 ___
Gable wall construction: ☒ same as main walls; ☐ other construction ___

301

6. FLOOR FRAMING:

Joists: wood, grade, and species ___2x8 fir___ ; other _____ ; bridging ___metal___ ; anchors _____

Concrete slab: ☒ basement floor; ☐ first floor; ☐ ground supported; ☐ self-supporting; mix _____ ; thickness ___4"___ ;

reinforcing ___6x6-10/10 welded wire___ ; insulation _____ ; membrane ___Sisalkraft___

Fill under slab: material ___sand___ ; thickness ___6"___. Additional information: _____

7. SUBFLOORING: (Describe underflooring for special floors under item 21.)

Material: grade and species ___APA CD Int. w/ Ext glue___ ; size ___5/8"___ ; type ___plywood___

Laid: ☐ first floor; ☐ second floor; ☐ attic _____ sq. ft.; ☐ diagonal; ☐ right angles. Additional information: _____

8. FINISH FLOORING: (Wood only. Describe other finish flooring under item 21.)

Location	Rooms	Grade	Species	Thickness	Width	Bldg. Paper	Finish
First floor	Liv. - BR	Select	Oak	1"	3"	deadening felt	2 coats varnish
Second floor							
Attic floor	sq. ft.						

Additional information: _____

9. PARTITION FRAMING:

Studs: wood, grade, and species ___Fir___ size and spacing ___2x4-16" OC___ Other _____

Additional information: _____

10. CEILING FRAMING:

Joists: wood, grade, and species ___Trusses___ Other _____ Bridging _____

Additional information: _____

11. ROOF FRAMING:

Rafters: wood, grade, and species _____ Roof trusses (see detail): grade and species ___See details___

Additional information: _____

12. ROOFING:

Sheathing: wood, grade, and species ___1/2" APA CD Int w/ Ext glue plywood___ ; ☐ solid; ☐ spaced ___" o.c.

Roofing ___Asphalt strip___ ; grade _____ ; size _____ ; type ___240 lb.___

Underlay ___Asphalt felt___ ; weight or thickness ___15 lb___ ; size ___36"___ ; fastening _____

Built-up roofing _____ ; number of plies _____ ; surfacing material _____

Flashing: material ___Aluminum___ ; gage or weight ___28 gauge___ ; ☐ gravel stops; ☐ snow guards

Additional information: _____

13. GUTTERS AND DOWNSPOUTS: none

Gutters: material _____ ; gage or weight _____ ; size _____ ; shape _____

Downspouts: material _____ ; gage or weight _____ ; size _____ ; shape _____ ; number _____

Downspouts connected to: ☐ Storm sewer; ☐ sanitary sewer; ☐ dry-well. ☐ Splash blocks: material and size _____

Additional information: _____

14. LATH AND PLASTER

Lath ☐ walls, ☐ ceilings: material _____ ; weight or thickness _____ Plaster: coats _____ ; finish _____

Dry-wall ☒ walls, ☒ ceilings: material ___gypsum___ ; thickness ___1/2"___ ; finish ___Tape & Joint system___

Joint treatment ___according to manufacturer's specifications___ ;

15. DECORATING: (Paint, wallpaper, etc.)

Rooms	Wall Finish Material and Application	Ceiling Finish Material and Application
Kitchen	2 semi-gloss enamel	2 semi-gloss enamel
Bath	2 semi-gloss enamel	2 semi-gloss enamel
Other	2 flat latex	2 flat latex

Additional information: _____

16. INTERIOR DOORS AND TRIM:

Doors: type ___Flush___ ; material ___Birch veneer___ ; thickness ___1 3/8"___

Door trim: type ___Colonial___ ; material ___pine___ Base: type ___colonial___ ; material ___pine___ ; size ___3"___

Finish: doors ___Varnish___ ; trim ___primed & painted___

Other trim (item, type and location) _____

Additional information: _____

17. WINDOWS:

Windows: type ___D.H. & Csmnt___ ; make ___Wilson___ ; material ___pine___ ; sash thickness ___1 3/8"___

Glass: grade ___insul.___ ; ☐ sash weights; ☐ balances, type _____ ; head flashing _____

Trim: type ___Colonial___ ; material ___pine___ Paint ___2 coats___ ; number coats _____

Weatherstripping: type ___factory-installed___ ; material ___vinyl___ Storm sash, number _____

Screens: ☒ full; ☐ half; type _____ ; number __10__ ; screen cloth material __aluminum__
Basement windows: type __2-light__ ; material __steel__ ; screens, number _____ ; Storm sash, number _____
Special windows _____
Additional information _____

18. ENTRANCES AND EXTERIOR DETAIL:
Main entrance door: material __pine__ ; width __3'-0"__ ; thickness __1 3/4"__ Frame: material __pine__ ; thickness __5/4"__
Other entrance doors: material __pine__ ; width __2'-8"__ ; thickness __1 3/4"__ Frame: material __pine__ ; thickness __5/4"__
Head flashing _____ Weatherstripping: type __interlocking__ ; saddles __oak__
Screen doors: thickness __none__ ; number _____ ; screen cloth material _____ Storm doors: thickness ____"; number ____
Combination storm and screen doors: thickness __1 1/8"__ ; number __2__ ; screen cloth material __aluminum__
Shutters: ☐ hinged; ☐ fixed. Railings _____ , Attic louvers _____
Exterior millwork: grade and species __#1 pine__ Paint __primed & painted__ ; number coats __2__
Additional information: _____

19. CABINETS AND INTERIOR DETAIL:
Kitchen cabinets, wall units: material __Kingswood Oakmont – Sutherland__ ; lineal feet of shelves _____ ; shelf width _____
Base units: material __Kingswood Oakmont__ ; counter top __Plastic laminate__ ; edging __Plastic Laminate__
Back and end splash __4"__ Finish of cabinets __Factory finished__ ; number coats _____
Medicine cabinets: make __none__ ; model _____
Other cabinets and built-in furniture __Bathroom vanity RV-48__
Additional information: _____

20. STAIRS:

STAIR	TREADS		RISERS		STRINGS		HANDRAIL		BALUSTERS	
	Material	Thickness	Material	Thickness	Material	Size	Material	Size	Material	Size
Basement	fir	5/4"	pine	3/4"	fir	3/4"	pine	1 1/2"	none	
Main										
Attic										

Disappearing: make and model number _____
Additional information: _____

21. SPECIAL FLOORS AND WAINSCOT:

	Location	Material, Color, Border, Sizes, Gage, Etc.	Threshold Material	Wall Base Material	Underfloor Material
FLOORS	Kitchen				
	Bath				

	Location	Material, Color, Border, Cap. Sizes, Gage, Etc.	Height	Height Over Tub	Height in Showers (From Floor)
WAINSCOT	Bath				

Bathroom accessories: ☐ Recessed; material _____ ; number _____ ; ☐ Attached; material _____ ; number _____
Additional information: _____

22. PLUMBING:

FIXTURE	NUMBER	LOCATION	MAKE	MFR'S FIXTURE IDENTIFICATION No.	SIZE	COLOR
Sink	1	Kitchen	Kessler	SA-410 Titan	21x32	ss
Lavatory	1	Bath	Kessler	K-2208 Continental	16" round	blue
Water closet	1	Bath	Kessler	K-3512—PBA		blue
Bathtub	1	Bath	Kessler	K-786-SA Royal	5'-0"	blue
Shower over tub △	1	Bath	Nisson	7004-T		chrome
Stall shower △						
Laundry trays						

△☒ Curtain rod △☐ Door ☐ Shower pan: material _____
Water supply: ☒ public; ☐ community system; ☐ individual (private) system. ★

Sewage disposal: ☒ public; ☐ community system; ☐ individual (private) system.★

★*Show and describe individual system in complete detail in separate drawings and specifications according to requirements.*

House drain (inside): ☐ cast iron; ☐ tile; ☐ other __copper__   House sewer (outside): ☒ cast iron; ☐ tile; ☐ other _____

Water piping: ☐ galvanized steel; ☒ copper tubing; ☐ other _____   Sill cocks, number __2__

Domestic water heater: type __electric__; make and model __Williams WG-40__; heating capacity __4000 Watts__ _____ gph. 100° rise. Storage tank: material _____; capacity __40__ gallons.

Gas service: ☐ utility company; ☐ liq. pet. gas; ☐ other _____   Gas piping: ☐ cooking; ☐ house heating.

Footing drains connected to: ☐ storm sewer; ☐ sanitary sewer; ☐ dry well. Sump pump; make and model _____ ; capacity _____ ; discharges into _____

23. HEATING:

☐ Hot water. ☐ Steam. ☐ Vapor. ☐ One-pipe system. ☐ Two-pipe system.

☐ Radiators. ☐ Convectors. ☐ Baseboard radiation. Make and model _____

Radiant panel: ☐ floor; ☐ wall; ☐ ceiling. Panel coil: material _____

☐ Circulator. ☐ Return pump. Make and model _____ ; capacity _____ gpm.

Boiler: make and model _____ Output _____ Btuh.; net rating _____ Btuh.

Additional information: _____

Warm air: ☐ Gravity. ☒ Forced. Type of system __Mitchell, upflow oil fired__

Duct material: supply __galv.__ ; return __galv.__   Insulation _____, thickness _____ ☐ Outside air intake.

Furnace: make and model __Mitchell__   Input __125,000__ Btuh.; output __100,000__ Btuh.

Additional information: __1500 cfm__

☐ Space heater; ☐ floor furnace; ☐ wall heater. Input _____ Btuh.; output _____ Btuh.; number units _____

Make, model _____   Additional information: _____

Controls: make and types __Sherwood T861A clock thermostat__

Additional information: _____

Fuel: ☐ Coal; ☐ oil; ☐ gas; ☐ liq. pet. gas; ☐ electric; ☐ other _____ ; storage capacity __1,000 gal.__

Additional information: _____

Firing equipment furnished separately: ☐ Gas burner, conversion type. ☐ Stoker: hopper feed ☐; bin feed ☐

Oil burner: ☐ pressure atomizing; ☐ vaporizing _____

Make and model _____   Control _____

Additional information: _____

Electric heating system: type _____   Input _____ watts; @ _____ volts; output _____ Btuh.

Additional information: _____

Ventilating equipment: attic fan, make and model _____ ; capacity _____ cfm.

kitchen exhaust fan, make and model _____

Other heating, ventilating, or cooling equipment _____

24. ELECTRIC WIRING:

Service: ☐ overhead; ☒ underground. Panel: ☐ fuse box; ☒ circuit-breaker; make __Cunningham__ AMP's __200__ No. circuits __30__

Wiring: ☐ conduit; ☐ armored cable; ☒ nonmetallic cable; ☐ knob and tube; ☐ other _____

Special outlets: ☒ range; ☒ water heater; ☐ other _____

☐ Doorbell. ☒ Chimes. Push-button locations __Front door__   Additional information: _____

25. LIGHTING FIXTURES:

Total number of fixtures __11__   Total allowance for fixtures, typical installation, $ __400.00__

Nontypical installation _____

Additional information: _____

26. INSULATION:

Location	Thickness	Material, Type, and Method of Installation	Vapor Barrier
Roof			
Ceiling	R-21	Fiberglass stapled to trusses	Kraft
Wall	R-11	Fiberglass stapled to studs	Kraft
Floor			

27. MISCELLANEOUS: (*Describe any main dwelling materials, equipment, or construction items not shown elsewhere; or use to provide additional information where the space provided was inadequate. Always reference by item number to correspond to numbering used on this form.*) _____

_____

_____

_____

_____

_____

HARDWARE: *(make, material, and finish.)* ___Welch, brass finish. Allowance of $500.00 for purchase of finish hardware, including bathroom mirrors.___

SPECIAL EQUIPMENT: *(State material or make, model and quantity. Include only equipment and appliances which are accept- able by local law, custom and applicable FHA standards. Do not include items which, by established custom, are supplied by occupant and removed when he vacates premises or chattles prohibited by law from becoming realty.)*_____

PORCHES: Porches of same construction as house.

TERRACES:

GARAGES: Attached — of same construction as house.

WALKS AND DRIVEWAYS:
Driveway: width __9'-0"__ ; base material __crushed stone__ ; thickness __3"__ "; surfacing material __Asphalt__ ; thickness __2"__ "
Front walk: width __3'-6"__ ; material __Conc.__ ; thickness __4"__ ". Service walk: width _____ ; material _____ ; thickness _____ "
Steps: material _____ ; treads _____ "; risers _____ ". Cheek walls _____

OTHER ONSITE IMPROVEMENTS:
(Specify all exterior onsite improvements not described elsewhere, including items such as unusual grading, drainage structures, retaining walls, fence, railings, and accessory structures.)

LANDSCAPING, PLANTING, AND FINISH GRADING:
Topsoil __6__ " thick: ☒ front yard; ☒ side yards; ☒ rear yard to __125__ feet behind main building.
Lawns *(seeded, sodded, or sprigged):* ☒ front yard __seeded__ ; ☒ side yards __seeded__ ; ☒ rear yard __seeded__
Planting: ☐ as specified and shown on drawings; ☐ as follows:
_____ Shade trees, deciduous, _____ " caliper.   _____ Evergreen trees. _____ ' to _____ ', B & B.
_____ Low flowering trees, deciduous, _____ ' to _____ '   _____ Evergreen shrubs. _____ ' to _____ ', B & B.
_____ High-growing shrubs, deciduous, _____ ' to _____ '   _____ Vines, 2-year _____
_____ Medium-growing shrubs, deciduous, _____ ' to _____ '
_____ Low-growing shrubs, deciduous, _____ ' to _____ '

IDENTIFICATION.—This exhibit shall be identified by the signature of the builder, or sponsor, and/or the proposed mortgagor if the latter is known at the time of application.

Date_____   Signature _____

Signature _____

FHA Form 2005
VA Form 26–1852

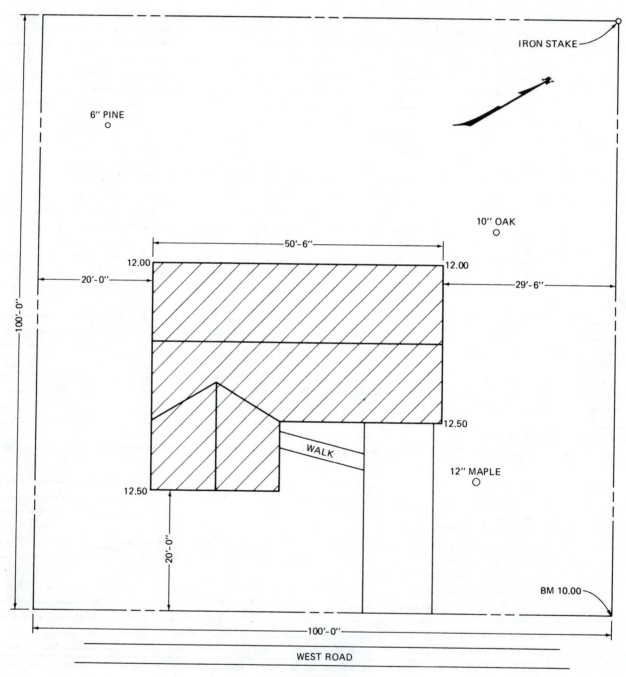

6" PINE

IRON STAKE

10" OAK

50'-6"

12.00   12.00

20'-0"   29'-6"

100'-0"

12.50

WALK

12" MAPLE

12.50

20'-0"

BM 10.00

100'-0"

WEST ROAD

PLOT PLAN
SCALE 1/16" = 1'-0"

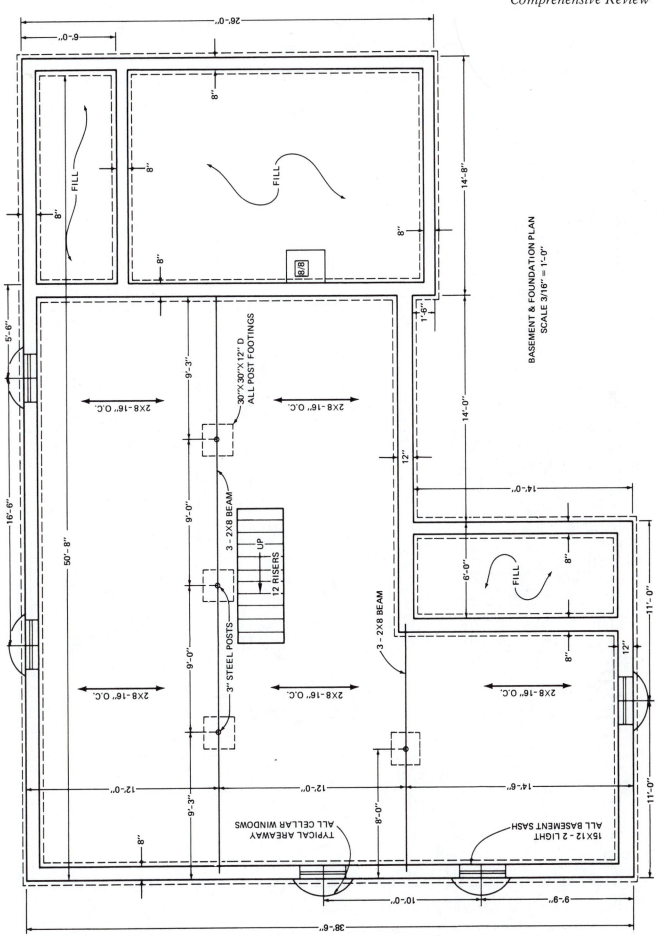

BASEMENT & FOUNDATION PLAN
SCALE 3/16″ = 1′-0″

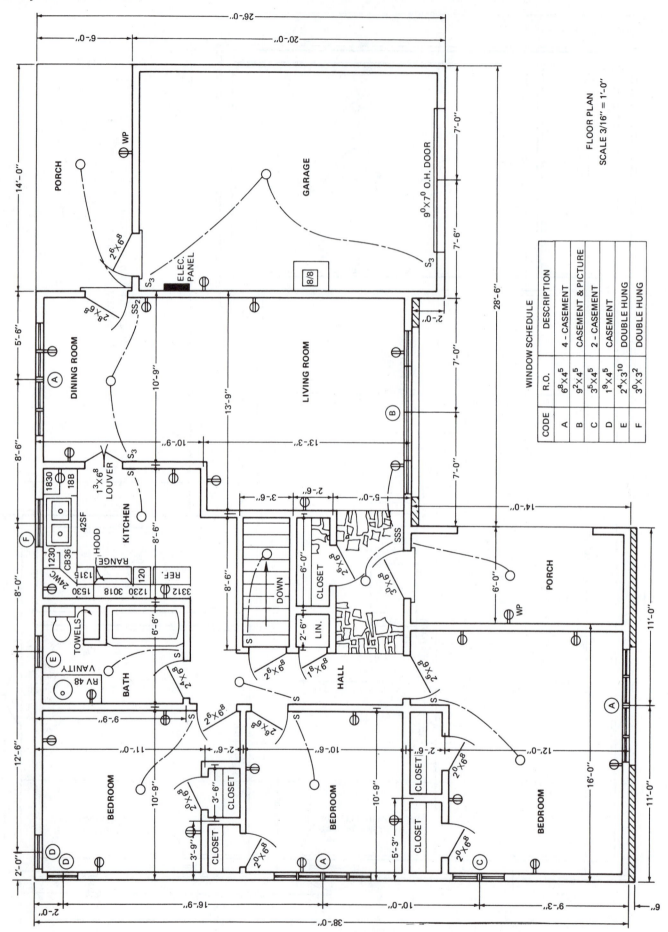

FLOOR PLAN
SCALE 3/16" = 1'-0"

WINDOW SCHEDULE

CODE	R.O.	DESCRIPTION
A	$6^8 \times 4^5$	4 - CASEMENT
B	$9^2 \times 4^5$	CASEMENT & PICTURE
C	$3^5 \times 4^5$	2 - CASEMENT
D	$1^9 \times 4^5$	CASEMENT
E	$2^4 \times 3^{10}$	DOUBLE HUNG
F	$3^0 \times 3^2$	DOUBLE HUNG

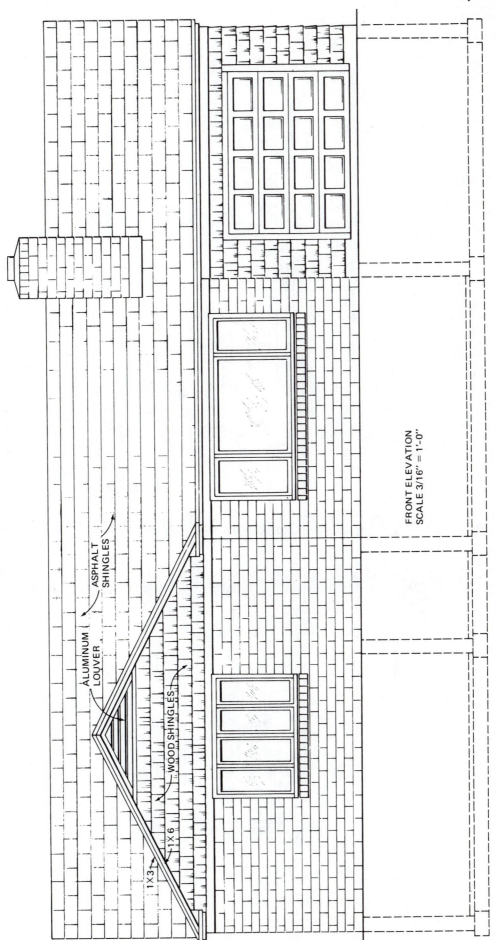

ASPHALT SHINGLES

ALUMINUM LOUVER

WOOD SHINGLES

1 × 6

1 × 3

FRONT ELEVATION
SCALE 3/16" = 1'-0"

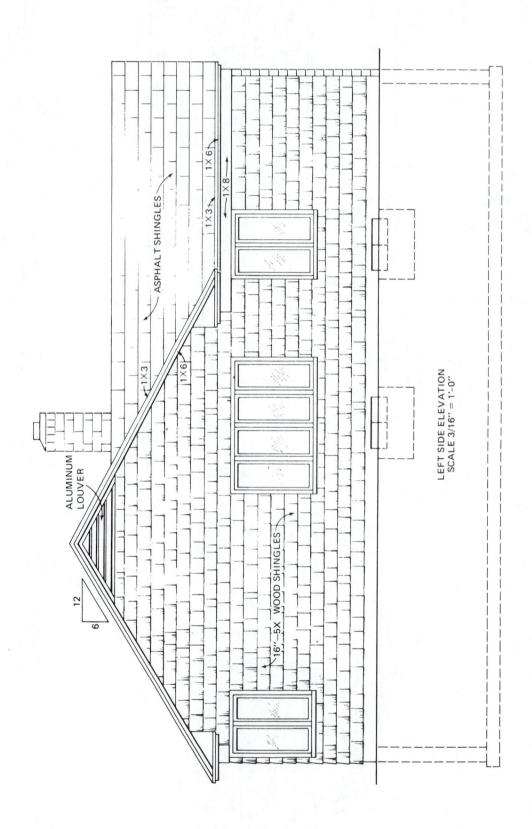

ASPHALT SHINGLES

1×6

1×8

1×3

1×3

1×6

ALUMINUM
LOUVER

12

6

16"—5X WOOD SHINGLES

LEFT SIDE ELEVATION
SCALE 3/16" = 1'-0"

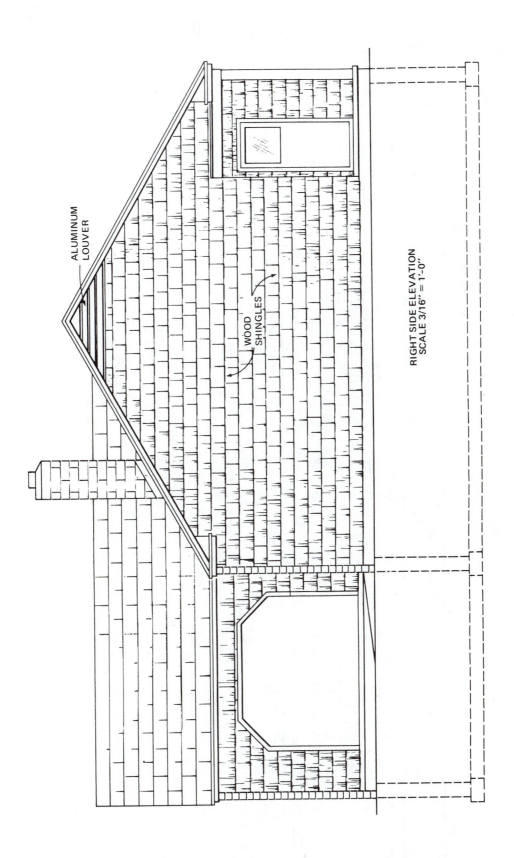

ALUMINUM LOUVER

WOOD SHINGLES

RIGHT SIDE ELEVATION
SCALE 3/16" = 1'-0"

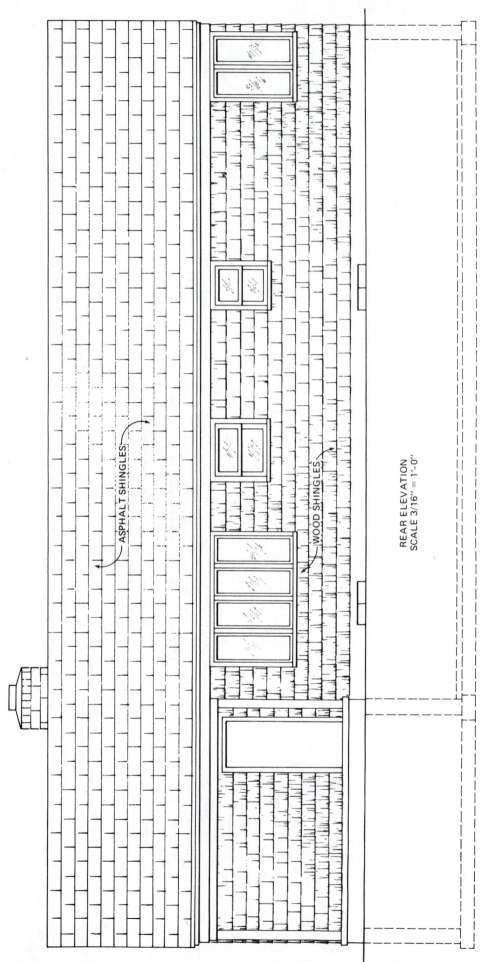

ASPHALT SHINGLES

WOOD SHINGLES

REAR ELEVATION
SCALE 3/16″ = 1′-0″

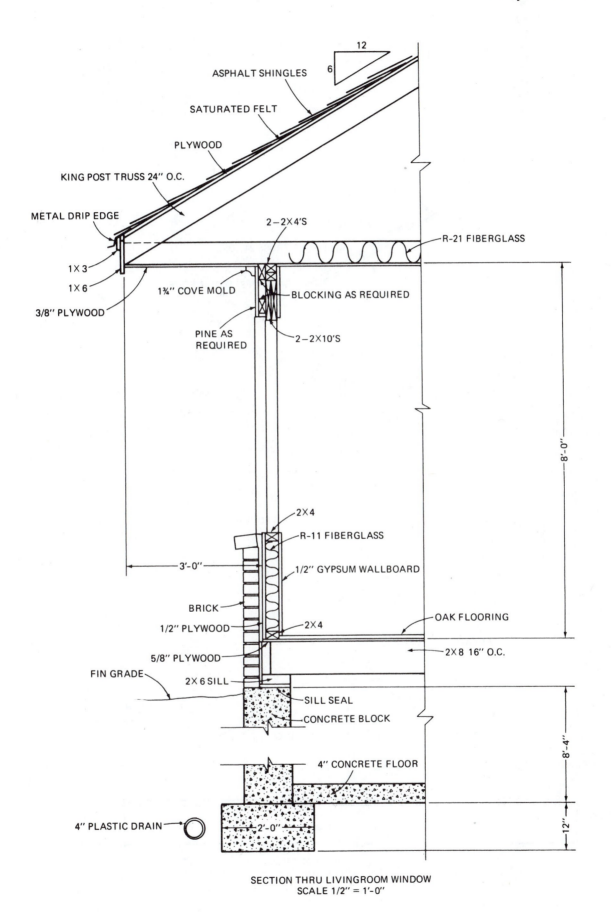

ASPHALT SHINGLES

SATURATED FELT

PLYWOOD

KING POST TRUSS 24″ O.C.

METAL DRIP EDGE

1 × 3

1 × 6

3/8″ PLYWOOD

1¾″ COVE MOLD

PINE AS
REQUIRED

2 − 2 × 4'S

BLOCKING AS REQUIRED

2 − 2 × 10'S

R-21 FIBERGLASS

2 × 4

R-11 FIBERGLASS

1/2″ GYPSUM WALLBOARD

3'-0″

BRICK

1/2″ PLYWOOD

5/8″ PLYWOOD

FIN GRADE

2 × 6 SILL

2 × 4

OAK FLOORING

2 × 8 16″ O.C.

SILL SEAL

CONCRETE BLOCK

4″ CONCRETE FLOOR

4″ PLASTIC DRAIN

2'-0″

8'-0″

8'-4″

12″

SECTION THRU LIVINGROOM WINDOW
SCALE 1/2″ = 1'-0″

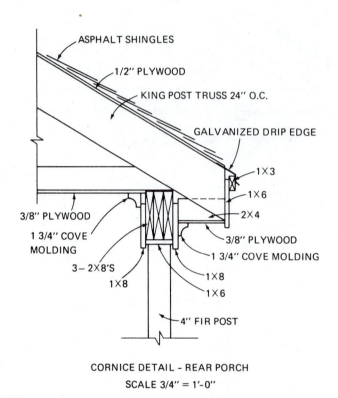

ASPHALT SHINGLES

1/2" PLYWOOD

KING POST TRUSS 24" O.C.

GALVANIZED DRIP EDGE

1×3

1×6

2×4

3/8" PLYWOOD

3/8" PLYWOOD

1 3/4" COVE MOLDING

1 3/4" COVE MOLDING

3—2×8'S

1×8

1×8

1×6

4" FIR POST

CORNICE DETAIL - REAR PORCH
SCALE 3/4" = 1'-0"

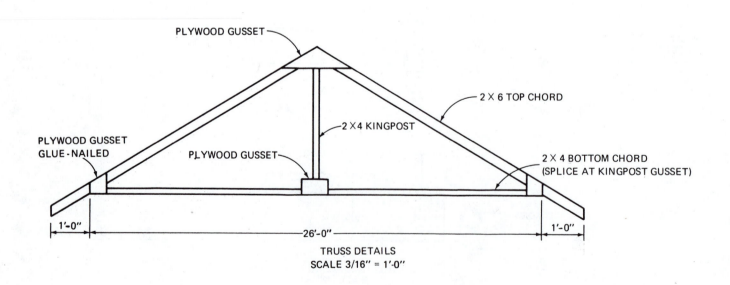

PLYWOOD GUSSET

2 × 6 TOP CHORD

2 × 4 KINGPOST

PLYWOOD GUSSET
GLUE-NAILED

PLYWOOD GUSSET

2 × 4 BOTTOM CHORD
(SPLICE AT KINGPOST GUSSET)

1'-0"

26'-0"

1'-0"

TRUSS DETAILS
SCALE 3/16" = 1'-0"

Building _____

Location _____

Architect _____

Estimator _____

Description		Unit	Quantity	Unit Price	Total Material Cost	Labor	Total
FTG. FORM LUMBER	1 x 8						
WOOD STAKES - 12/BUNDLE							
TRANSIT MIX CONCRETE - FTG.							
TRANSIT MIX CONCRETE - FLOORS							
PORCHES & WALK							
CONCRETE BLOCKS-FOUNDATION	8 x 8 x 16						
CONCRETE BLOCKS-FOUNDATION	8 x 12 x 16						
SOLID BLOCKS-FOUNDATION	4 x 8 x 16						
REINFORCEMENT-MASONRY WALLS							
HALF SASH BLOCKS							
FULL SASH BLOCKS							
STEEL CELLAR SASH	16 x 12						
GALVANIZED AREAWAYS							
MASONRY CEMENT - BLOCKS		BAGS					
PLASTIC DRAIN PIPE	4"	FT.					
#2 CRUSHED STONE							
PORTLAND CEMENT - PARGING		BAGS					
MASON'S LIME- PARGING		BAGS					
ASPHALT FOUNDATION COATING		GAL.					
MASONRY SAND							
SISALKRAFT VAPOR BARRIER							
WIRE MESH -CONC. FLOORS							
LALLY COLUMNS							
FLUE LINERS							
CHIMNEY BLOCKS							
COMMON BRICKS- CHIMNEY							
COMMON BRICKS - VENEER							

Building _____

Location _____

Architect _____

Estimator _____

Description					Unit	Quantity	Unit Price	Total Material Cost	Labor	Total
MASONRY CEMENT -										
BRICK MORTAR					BAGS					
GALVANIZED WALL TIES										
FIR - FRONT BEAM										
FIR - MAIN BEAM										
FIR - STAIR STRINGERS										
FIR - STAIR TREADS										
#1 COM. PINE - STAIR										
RISERS	1	x	8	x 12						
PINE HANDRAIL MOLDING										
SILL SEAL										
FIR - SILL										
FIR - BOX SILL										
FIR - JOISTS										
FIR - JOISTS										
FIR - STAIR HEADERS										
GALVANIZED BRIDGING					PCS.					
CD PLYWOOD W/EXTERIOR										
GLUE - SUBFLOOR	5/8									
FIR - PLATES										
FIR - STUDS HOUSE										
FIR - STUDS GARAGE & PORCH										
FIR - GARAGE DOOR HEADER										
FIR - WINDOW & DOOR HEADERS	2	x	10		LIN. FT					
FIR - BEAM FRONT PORCH	2	x	10	x 10						
FIR - BEAM REAR PORCH										
FIR - BEAM REAR PORCH										
FIR - POST REAR PORCH										
$\frac{6}{12}$ x 26' TRUSS - GARAGE & L.R.										
$\frac{6}{12}$ x 26' TRUSS - (no tail										
ONE END)										

Building _____

Location _____

Architect _____

Estimator _____

Description	Unit		Quantity	Unit Price	Total Material Cost	Labor	Total
6/12 x 22' TRUSS - FRONT WING							
FIR-JACK RAFTERS	2 x 6	LIN. FT.					
FIR RIDGE AT JACK RAFTERS	2 x 8 x 10						
9/12 x 26' GABLE TRUSS							
9/12 x 22' GABLE TRUSS							
ALUMINUM LOUVERS 6/12 PITCH							
CD EXTERIOR GLUE PLYWOOD ROOF SHEATHING	1/2"						
15 LB. SATURATED FELT	500'	RLLS.					
GALVANIZED DRIP EDGE							
240 LB. ASPHALT ROOF SHINGLES		SQ.					
28 GA. ALUMINUM FLASHING	20" x 50'	RLL.					
PINE CORNICE	1 x 3						
PINE FASCIA	1 x 6						
AC EXT. PLYWOOD - SOFFIT	3/8"						
PINE FRIEZE	1 x 6						
COVE MOLDING - FRONT CORNICE	1 3/4"						
CD EXT. GLUE PLYWOOD - WALL SHEATHING	1/2"						
BUILDING PAPER - WALLS		RLL.					
5x CEDAR SHINGLES	16"						
PINE - REAR PORCH BEAM							
PINE - REAR PORCH BEAM							
PINE - REAR PORCH BEAM							
PINE - REAR PORCH BEAM							
PINE - FRONT PORCH TRIM	1 x 6	LIN. FT.					
AC EXT. PLYWOOD - PORCH CEILINGS	3/8						

Building _____

Location _____

Architect _____

Estimator _____

Date _____

Description	Unit	Quantity	Unit Price	Total Material Cost	Labor	Total
COVE MOLDING - PORCHES $1^{3}/_{4}$						
4-CASEMENT WINDOW $6^{8} \times 4^{5}$						
CASEMENT - PICTURE WINDOW $1^{9} \times 4^{5}/5^{8} \times 4^{5}$						
2 CASEMENT WINDOWS $3^{5} \times 4^{5}$						
CASEMENT WINDOWS $1^{9} \times 4^{5}$						
D. H. WINDOW $2^{4} \times 3^{10}$						
D. H. WINDOW $3^{0} \times 3^{2}$						
ALL WINDOWS w/ INSULATED GLAZING & SCREENS						
PRE-HUNG EXT. DOOR $3^{0} \times 6^{8} \times 1^{3}/_{4}$ PINE, LEFT-HAND SWING						
PRE-HUNG EXT. DOOR $2^{8} \times 6^{8} \times 1^{3}/_{4}$ PINE, RIGHT-HAND SWING						
PRE-HUNG EXT. DOOR, $2^{6} \times 6^{8} \times 1^{3}/_{4}$ PINE, RIGHT-HAND SWING						
ALUMINUM COMBINATION- $3^{0} \times 6^{8} \times 1^{1}/_{8}$ STORM & SCREEN DOOR						
ALUMINUM COMBINATION- $2^{8} \times 6^{8} \times 1^{1}/_{8}$ STORM & SCREEN DOOR						
PRE-HUNG INTERIOR FLUSH DOOR $2^{6} \times 6^{8} \times 1^{3}/_{8}$ BIRCH, LEFT-HAND SWING COLONIAL CASING						
PRE-HUNG INTERIOR FLUSH DOOR BIRCH, RIGHT-HAND SWING $2^{6} \times 6^{8} \times 1^{3}/_{8}$ COLONIAL CASING						
PRE-HUNG INTERIOR FLUSH DOOR $2^{0} \times 6^{8} \times 1^{3}/_{8}$ BIRCH, RIGHT HAND SWING COLONIAL CASING						
PRE-HUNG INTERIOR FLUSH DOOR $1^{9} \times 6^{8} \times 1^{3}/_{8}$ BIRCH, LEFT HAND SWING COLONIAL CASING						

Building _____

Location _____

Architect _____

Estimator _____

Description	Unit	Quantity	Unit Price	Total Material Cost	Labor	Total
PRE-HUNG INTERIOR FLUSH DOOR $2^4 \times 6^8 \times 1^{3}/8$						
BIRCH, LEFT HAND SWING,						
COLONIAL CASING						
INTERIOR DOOR JAMBS $2^6 \times 6^8$	SCT.					
LOUVERED DOORS $1^3 \times 6^8 \times 1^{1}/8$						
DOUBLE-SWING HINGES	PR.					
OVERHEAD GARAGE DOOR						
w/ HARDWARE $9^0 \times 7^0$						
OVERHEAD DOOR FRAME $9^0 \times 7^0$						
R-21 FIBERGLASS INSULATION						
R-11 FIBERGLASS INSULATION						
GYPSUM WALLBOARD $1/2"$						
WALLBOARD COMPOUND 5 GAL.	CAN					
PERFORATED TAPE 250'	ROLL					
DEADENING FELT-FLOORS 500'	ROLL					
OAK STRIP FLOORING 1×3						
UNDERLAYMENT PLYWOOD $1/2"$	SHEET					
VINYL TILE - KITCHEN $12" \times 12"$	SQ.FT.					
ADHESIVE - VINYL TILE	GAL.					
CERAMIC FLOOR TILE $1" \times 1"$	SQ.FT.					
SLATE FLOORING	SQ.FT.					
ADHESIVE FOR TILE & SLATE	GAL.					
GROUT FOR TILE & SLATE 5 LB.	PKG.					
AC EXT. PLYWOOD -						
TUB AREA $1/2"$						
CERAMIC WALL TILE $4^1/4 \times 4^1/4$	SQ.FT.					
CERAMIC CAP TILES 2×6	PC.					
MARBLE THRESHOLD - BATH $2'-4"$						
COLONIAL BASE MOLDING $3"$						
STOOL - WINDOW TRIM						
APRON - WINDOW TRIM						

Building _____

Location _____

Architect _____

Estimator _____

Description	Unit	Quantity	Unit Price	Total Material Cost	Labor	Total
COLONIAL CASING - WINDOWS						
CHROME ADJUSTABLE CLOSET RODS						
#2 PINE SHELVING	1×12 UN. FT.					
AD INT. PLYWOOD - LIN. CLOSET & BATH						
NAIL & FASTENER ALLOWANCE						
1830 WALL CABINET						
1230 WALL CABINET						
24 WC CORNER WALL CABINET						
1530 WALL CABINET						
3018 WALL CABINET						
3312 WALL CABINET						
D12 BASE CABINET W/ DRAWERS						
B15 BASE CABINET						
CB 36 BASE CABINET						
SF 42 SINK FRONT						
B18 BASE CABINET						
LAMINATED PLASTIC COUNTER TOP						
VANITY W/ LAMINATED PLASTIC TOP	48"					
FINISH HARDWARE ALLOWANCE						

Two Story Garrison House

As an estimator salesman working in a lumber yard, a contractor or a prospective homeowner may bring you a set of plans for a house and want you to make an estimate of the cost. You wouldn't have a set of specifications such as you had in this textbook but only a list of what the contractor wants you to figure. If it is a prospective homeowner, you would have to ask them the questions to know what to figure. The plans give a lot of information but not all that is necessary to make an accurate estimate. A two story garrison type house was picked, because if you can accurately estimate this, you can easily estimate a ranch, split level, or any size garage.

For our review example, let's say you have been given the following set of plans and a list of what the contractor wants you to figure. With this information, make a material list of the house.

Figure the following:

- 8 and 12 inch block foundation
- Andersen basement windows #2817
- 10′ perforated drain pipe for foundation
- asphalt foundation dampproofing

- 2 x 8 joists, plywood sub flooring and sheathing
- conventional roof framing as shown
- asphalt shingle roofing
- 1/2 x 8 cedar bevel siding
- 6″ fiberglass insulation ceilings
- 3″ fiberglass insulation sidewalls
- 3/8″ gypsum wallboard
- ranch style trim
- 1 x 3 oak flooring
- ceramic tile in tub areas
- pre-hung door units
- Andersen windows with thermopane glass-removable grilles
- oak kitchen cabinets and vanities

Don't figure painting, electrical, plumbing or heating. Now make a material list of the two story garrison house with this information. You will have to refer to this list and study the plans as well as constantly referring to the plans as you make the estimate. You will find that most of the time, this is how you will be called upon to make an estimate.

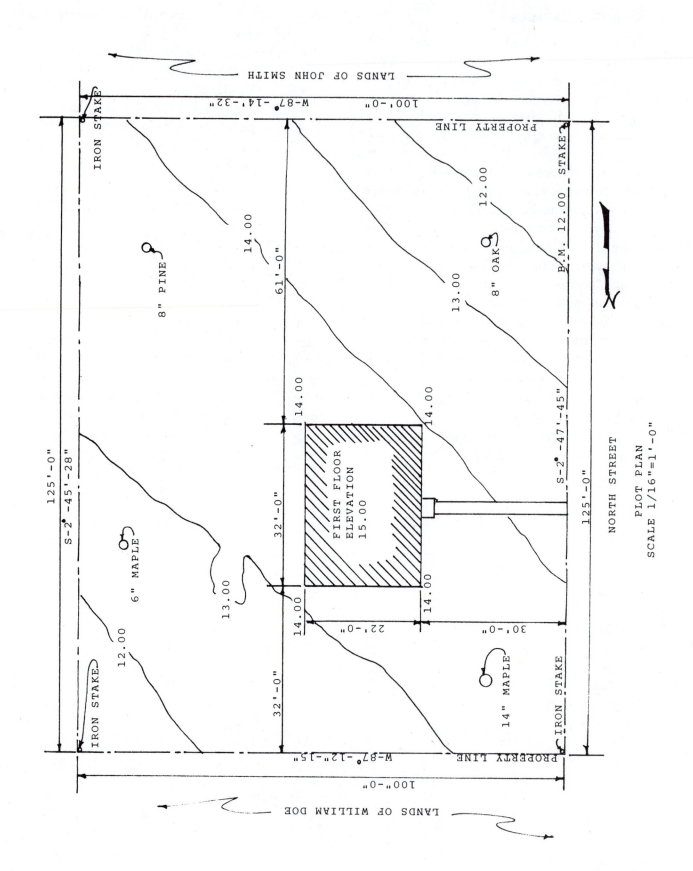

PLOT PLAN
SCALE 1/16"=1'-0"

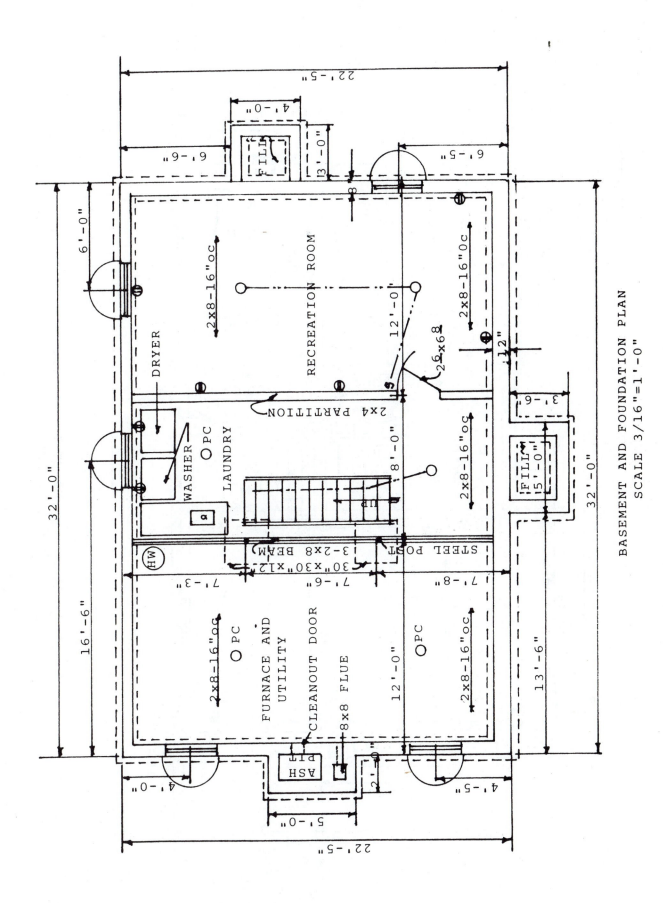

BASEMENT AND FOUNDATION PLAN
SCALE 3/16"=1'-0"

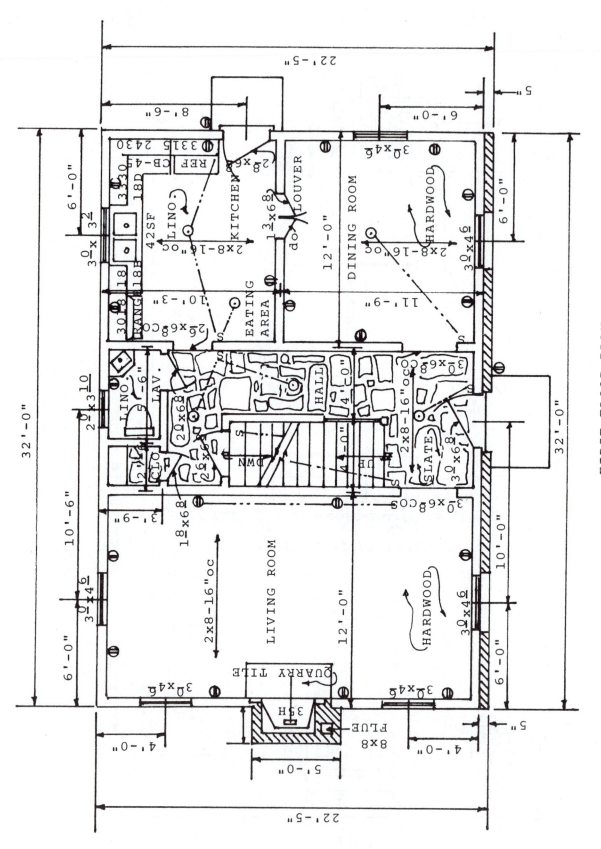

FIRST FLOOR PLAN
SCALE 3/16"=1'-0"

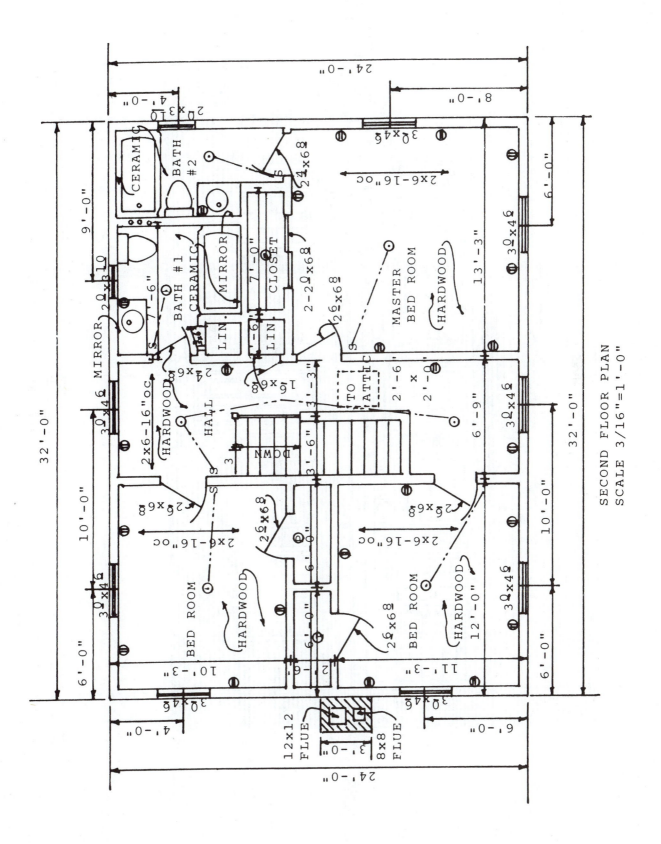

SECOND FLOOR PLAN
SCALE 3/16"=1'-0"

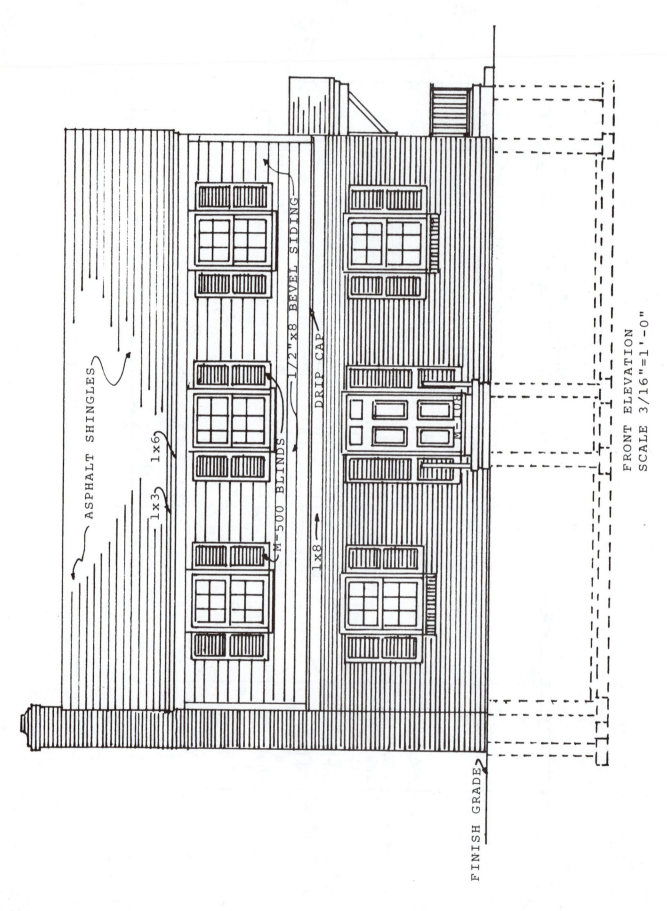

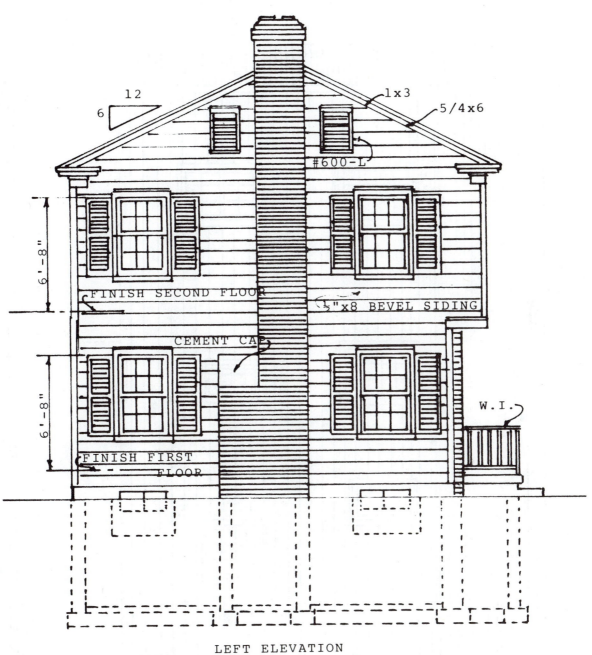

12

6

1 x 3

5/4 x 6

#600-L

6'-8"

FINISH SECOND FLOOR

½"x8 BEVEL SIDING

CEMENT CAP

6'-8"

FINISH FIRST FLOOR

W.I.

LEFT ELEVATION
SCALE 3/16"=1'-0"

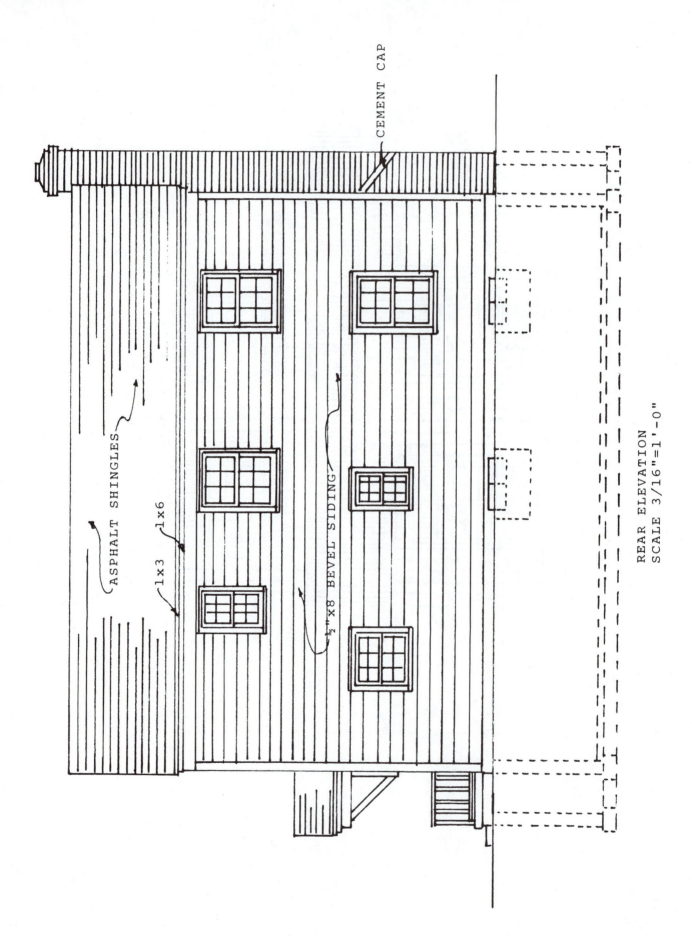

CEMENT CAP

ASPHALT SHINGLES

1x3

1x6

½"x8 BEVEL SIDING

REAR ELEVATION
SCALE 3/16"=1'-0"

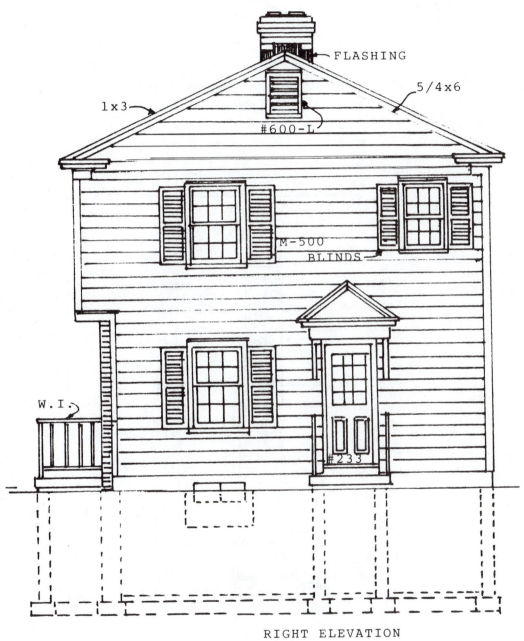

RIGHT ELEVATION
SCALE 3/16"=1'-0"

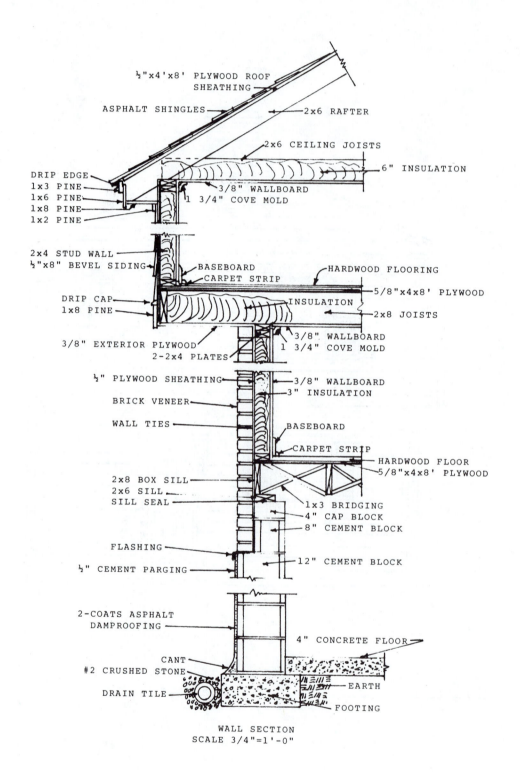

½"x4'x8' PLYWOOD ROOF SHEATHING

ASPHALT SHINGLES

2x6 RAFTER

2x6 CEILING JOISTS

6" INSULATION

DRIP EDGE
1x3 PINE
1x6 PINE
1x8 PINE
1x2 PINE

3/8" WALLBOARD
1 3/4" COVE MOLD

2x4 STUD WALL
½"x8' BEVEL SIDING

BASEBOARD
CARPET STRIP

HARDWOOD FLOORING

5/8"x4x8' PLYWOOD

DRIP CAP
1x8 PINE

INSULATION

2x8 JOISTS

3/8" EXTERIOR PLYWOOD
2-2x4 PLATES

3/8" WALLBOARD
1 3/4" COVE MOLD

½" PLYWOOD SHEATHING

BRICK VENEER

WALL TIES

3/8" WALLBOARD
3" INSULATION

BASEBOARD

CARPET STRIP

HARDWOOD FLOOR
5/8"x4x8' PLYWOOD

2x8 BOX SILL
2x6 SILL
SILL SEAL

1x3 BRIDGING
4" CAP BLOCK
8" CEMENT BLOCK

FLASHING

½" CEMENT PARGING

12" CEMENT BLOCK

2-COATS ASPHALT DAMPROOFING

4" CONCRETE FLOOR

CANT
#2 CRUSHED STONE

DRAIN TILE

EARTH

FOOTING

WALL SECTION
SCALE 3/4"=1'-0"

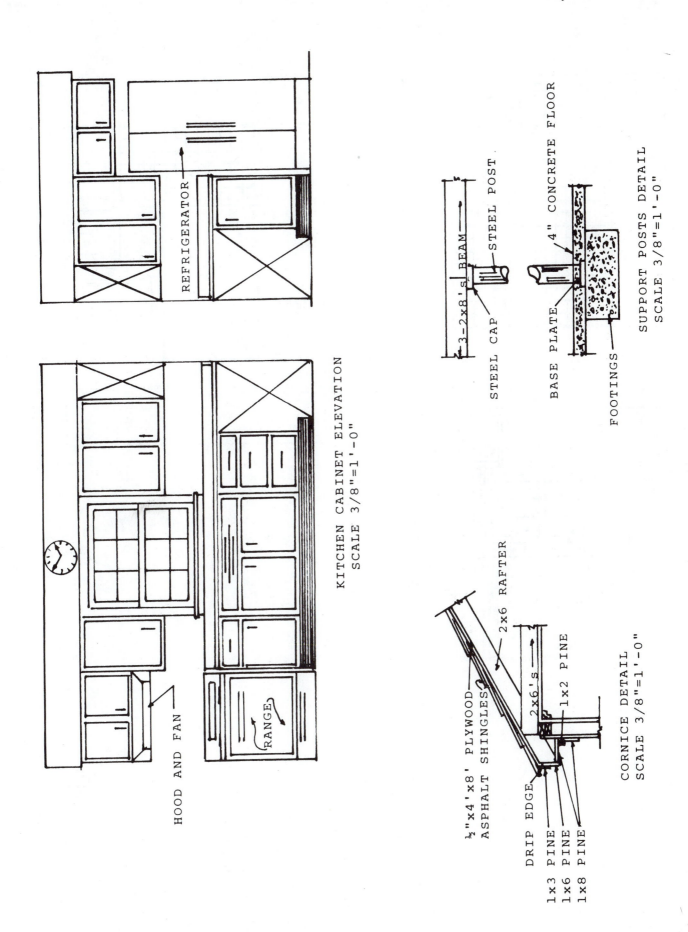

KITCHEN CABINET ELEVATION
SCALE 3/8"=1'-0"

HOOD AND FAN

REFRIGERATOR

RANGE

STEEL POST

3-2x8's BEAM

STEEL CAP

4" CONCRETE FLOOR

BASE PLATE

FOOTINGS

SUPPORT POSTS DETAIL
SCALE 3/8"=1'-0"

2x6 RAFTER

½"x4'x8' PLYWOOD

ASPHALT SHINGLES

2x6's

1x2 PINE

DRIP EDGE

1x3 PINE
1x6 PINE
1x8 PINE

CORNICE DETAIL
SCALE 3/8"=1'-0"

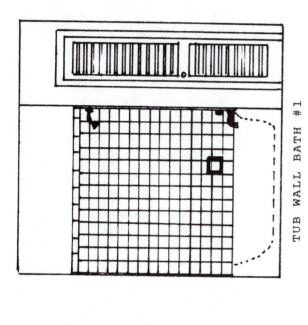

TUB WALL BATH #1

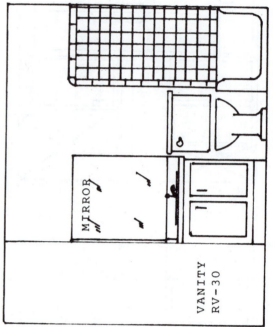

BATH #2   MASTER BEDROOM BATH

VANITY
RV-30

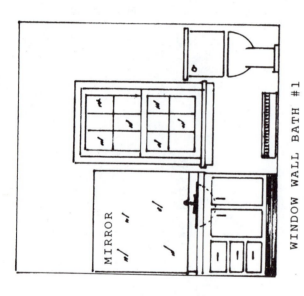

WINDOW WALL BATH #1

VANITY
RV-36

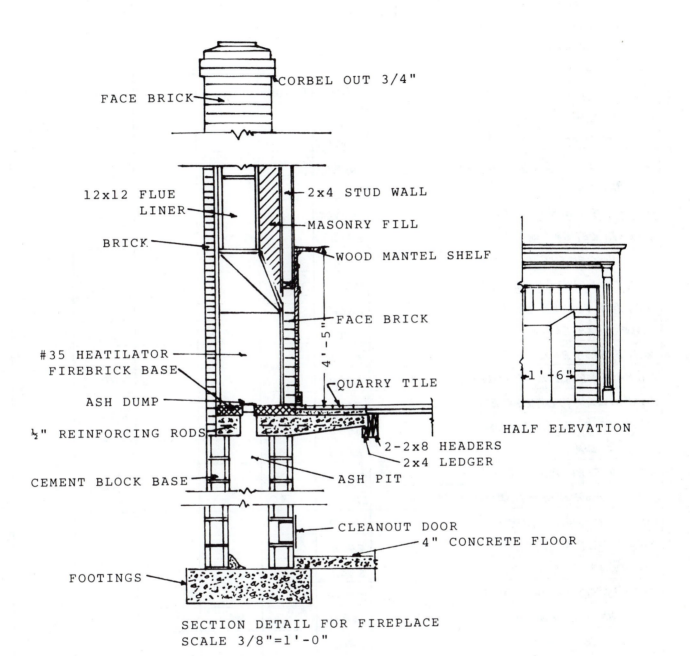

FACE BRICK

CORBEL OUT 3/4"

12x12 FLUE LINER

2x4 STUD WALL

MASONRY FILL

BRICK

WOOD MANTEL SHELF

FACE BRICK

4'-5"

#35 HEATILATOR FIREBRICK BASE

QUARRY TILE

ASH DUMP

½" REINFORCING RODS

2-2x8 HEADERS

2x4 LEDGER

CEMENT BLOCK BASE

ASH PIT

CLEANOUT DOOR

4" CONCRETE FLOOR

FOOTINGS

HALF ELEVATION

1'-6"

SECTION DETAIL FOR FIREPLACE
SCALE 3/8"=1'-0"

Building TWO STORY GARRISON

Location _____

Architect _____

Estimator _____

Description	Unit	Quantity	Price	Cost	Labor	Total
1X8 SHIPLAP FORMS FOOTING	LIN. FT.					
WOOD STAKES (12 EACH)	BUNDLES					
CONCRETE 1-3-5 MIX FOOTINGS	CU. YDS.					
GRAVEL FILL	CU. YDS.					
CONCRETE 1-2-4 MIX FLOOR + WALK	CU. YDS.					
8 X 8 X 16 CEMENT BLOCK	EACH					
12 X 8 X 16 CEMENT BLOCK	EACH					
4 X 8 X 16 SOLID CAP BLOCK	EACH					
8 X 8 X 16 CORNER BLOCK	EACH					
HALF SASH BLOCK	EACH					
WHOLE SASH BLOCK	EACH					
ANDERSEN BASEMENT WINDOWS #2817	EACH					
GALVANIZED 10 GAUGE AREAWAYS	EACH					
REGULAR CEMENT-MORTAR	BAGS					
MASONS LIME	BAGS					
WASHED SAND-MORTAR	CU. YDS.					
8" DUR-O-WALL - RODS REINFORCING	LIN. FT.					
12" DUR-O-WALL - RODS	LIN. FT.					
REGULAR FIREBRICK	EACH					
ASH DUMP	EACH					
12" X 15" CLEANOUT DOOR	EACH					
6" THIMBLE - FURNACE	EACH					
NO. 35 HEATILATOR	EACH					
8 X 8 FLUE LINER	EACH					
8 X 12 FLUE LINER	EACH					
$\frac{1}{2}$ X 3 $\frac{1}{2}$ X 40" ANGLE IRON	EACH					
ANTIQUE COMMON BRICK CHIMNEY	EACH					
REGULAR CEMENT HEARTH + CAP	BAGS					
QUARRY TILE HEARTH 4" X 4" X $\frac{1}{2}$"	SQ. FT.					
REGULAR CEMENT - CHIMNEY	BAGS					
MASONS LIME - CHIMNEY	BAGS					

Building _TWO STORY GARRISON_

Location _____

Architect _____

Estimator _____

Date _____

Description	Unit	Quantity	Price	Cost	Labor	Total
ANTIQUE COMMON BRICK (VENEER)	EACH					
GALV. WALL TIES - BRICK	EACH					
MASONRY CEMENT - VENEER	BAGS					
$3\frac{1}{2} \times 3\frac{1}{2} \times 4' - 6''$ ANGLE IRONS VENEER WORK	EACH					
14 GAUGE ALUMINUM FLASHING 28" WIDE	LIN. FT.					
4 x 10' PERFORATED PIPE	PCS					
90 DEGREE ELBOWS	EACH					
TEE - PVC PIPE	EACH					
#2 CRUSHED STONE - DRAIN	TON					
REGULAR CEMENT - PARGING	BAGS					
MASONS LIME - PARGING	BAGS					
MASONRY CEMENT PLASTERING TOP - FOUNDATION	BAGS					
REGULAR CEMENT - PLASTER	BAGS					
ASPHALT COATING GALLON CANS	CANS					
LONG HANDLE BRUSH	EACH					
WASHED SAND - PARGING	CU. YDS.					
4" STEEL POSTS WITH WELDED BASE + CAP	EACH					
6 x 8 x 22' BUILT UP BEAM	PCS.					
2x4 PLATE + SHOE BASEMENT PARTITION	LIN. FT.					
2x4 x 14' STUDS BASEMENT PARTITION	PCS.					
2x10x14' STRINGERS BASEMENT STAIRS	PCS.					
2x10x12' STAIR TREADS	PCS.					
2x4x12' PINE HANDRAIL	PCS.					
1x8x12' PINE RISERS	PCS.					
6" SILL SEAL	LIN. FT.					
2x6x12' SILL	PCS.					
2x8x12' BOX SILL	PCS.					
2x8x12' FLOOR JOISTS	PCS.					
2x8x8' JOISTS-HALL	PCS.					
1x3 SPRUCE BRIDGING	LIN. FT.					
2x8x12' TRIMMERS FIREPLACE	PCS.					

Building _TWO STORY GARRISON_

Location _____

Architect _____

Estimator _____

Sheet _3_ of _7_ _____

Date _____

Description	Unit	Quantity	Price	Cost	Labor	Total
2 X 8 X 8' TRIMMERS STAIRWELL	PCS.					
2 X 8 X 12' HEADER - FIREPL.	PC.					
2 X 8 X 14' HEADER - STAIRS	PCS.					
$\frac{5}{8}$" X 4' X 8' PLYWOOD SUB-FLOOR	PCS.					
2 X 4 PLATE + SHOE	LIN. FT.					
2 X 4 X 8' STUDS 1ST FLOOR	PCS.					
2 X 4 HEADERS 1ST FLOOR	LIN. FT.					
2 X 8 BOX SILL 2ND FLOOR	LIN. FT.					
2 X 8 X 12' JOIST 2ND FLOOR	PCS.					
2 X 8 X 8' JOISTS HALLWAY	PCS.					
2 X 8 X 12' JOISTS OVER KITCHEN	PCS.					
2 X 8 X 14' JOISTS OVER DINING ROOM	PCS.					
2 X 8 X 8' JOISTS FRONT OVERHANG	PCS.					
2 X 8 X 12' HEADER	PC.					
2 X 8 X 8' HEADER HALLWAY	PCS.					
2 X 8 X 14' TRIMMERS	PCS.					
1 X 3 BRIDGING 2ND FLOOR	LIN. FT.					
$\frac{5}{8}$" X 4 X 8' PLYWOOD 2ND	PCS.					
2 X 4 PLATE + SHOE 2ND	LIN. FT.					
2 X 4 X 8' STUDS 2ND FLOOR	PC.					
2 X 4 HEADERS 2ND FLOOR	LIN. FT.					
2 X 6 X 12' CEILING JOIST	PCS.					
2 X 6 X 14' CEILING JOIST	PCS.					
2 X 6 X 14' RAFTERS	PCS.					
2 X 8 X 16' RIDGE BOARD	PCS.					
2 X 8 X 10' COLLAR TIES	PCS.					
2 X 4 X 12' GABLE STUDS	PCS.					
$\frac{1}{2}$ X 4 X 8' PLYWOOD ROOF	PCS.					
ASPHALT SHINGLES	SQUARE					
15 # ASPHALT FELT	ROLLS					
GALV. DRIP EDGE 10'	PCS.					

Building __TWO STORY GARRISON__

Location _____

Architect _____

Estimator _____

Date _____

Description	Unit	Quantity	Price	Cost	Labor	Total
$\frac{5}{4}$ X 6 X 14' PINE RAKE	PCS.					
$\frac{5}{4}$ X 6 X 10' PINE RETURNS	PCS.					
1 X 6 PINE CORNICE	LIN. FT.					
1 X 8 PINE CORNICE	LIN. FT.					
1 X 3 PINE CORNICE	LIN. FT.					
1 X 2 PINE CORNICE	LIN. FT.					
$\frac{1}{2}$" X 4 X 8' SHEATHING	PCS.					
KRAFT BUILDING PAPER	ROLLS					
$\frac{1}{2}$" X 8" BEVEL SIDING	BD. FT.					
$\frac{5}{4}$ X 5 PINE CORNER BOARDS	LIN. FT.					
$\frac{3}{8}$" X 4 X 8 EXT. PLYWOOD	PCS.					
1 X 8 PINE PROJECTION FRONT	LIN. FT.					
WOOD DRIP CAP	LIN. FT.					
WOOD LOUVERS #600L	EACH					
2 X 6 X 10' BEAM SIDE ENT.	PCS.					
2 X 4 X 12' SIDE ENTRANCE CEILING JOIST + RAFTERS	PCS.					
$\frac{1}{2}$ X 4 X 8 PLYWOOD ROOF	PC.					
ASPHALT SHINGLES	BUNDLE					
GALV. DRIP EDGE 10'	PC.					
1 X 6 X 10' PINE BEAM	PC.					
1 X 4 X 10' PINE BEAM	PC.					
1 X 6 X 12' PINE CORNICE	PC.					
1 X 6 X 8' RAKE	PC.					
1 X 3 PINE CORNICE	LIN. FT.					
1 X 2 X 10' PINE CORNICE	PC.					
2 X 4 X 8' PINE BRACKETS	PC.					
$\frac{3}{8}$" X 4 X 8' EXT. PLYWOOD	PC.					
WROUGHT IRON RAIL 3' SIDE	PCS.					
WROUGHT IRON RAIL 4' FRONT	PCS.					
6" FIBERGLASS INSULATION	BAGS					
3" FIBERGLASS INSULATION	ROLLS					

337

Building **TWO STORY GARRISON**

Location

Architect

Estimator

Sheet **5** of **7**

Date

Description	Unit	Quantity	Price	Cost	Labor	Total
$\frac{3}{8}$" GYPSUM WALLBOARD	SQ. FT.					
50' ROLLS TAPE	ROLLS					
5 GALLON CANS JOINT CEMENT	CANS					
1$\frac{3}{4}$" COVE MOLDING	LIN. FT.					
BASE + CARPET STRIP	LIN. FT.					
$\frac{5}{8}$" x 4 x 8' UNDERLAYMENT	PCS.					
INLAID LINOLEUM	SQ. YD.					
MULTI COLORED SLATE	SQ. FT.					
SLATE ADHESIVE - HALLWAY	QTS.					
WHITE CEMENT GROUT	LBS.					
1 x 3 SELECT OAK FLOORING	SQ. FT.					
DEADENING FELT	ROLLS					
1" x 1" CERAMIC TILE FLOOR	SQ. FT.					
TILE FLOOR ADHESIVE	GAL.					
WHITE CEMENT GROUT	LBS.					
BATHROOM DOOR 2'-4" MARBLE THRESHOLDS	EACH					
$\frac{1}{4}$" x 4 x 8' PLYWOOD (TUB) WATERPROOF GLUE	PCS.					
4$\frac{1}{4}$ x 4$\frac{1}{4}$ CERAMIC TILE	SQ. FT.					
2 x 6 CERAMIC CAP TILE	LIN. FT.					
INSIDE CORNER CAP	EACH					
OUTSIDE CORNER CAP	EACH					
CHINA SOAP + GRAB	EACH					
CERAMIC WALL ADHESIVE	QTS.					
WHITE CERAMIC GROUT	LBS.					
EXT. READI - HUNG DOOR COMPLETE WITH SILL 3-0 x 6-8 x 1$\frac{3}{4}$"	UNIT					
EXT. READI-HUNG DOOR 2-8 x 6-8 x 1$\frac{3}{4}$"	UNIT					
CASING STYLE 37 - DOORS	LIN. FT.					
JAMBS 3-8 x 6-8 HALL	SET					
CASING STYLE 37 C.O.	LIN. FT.					
JAMBS 2-6 x 6-8 KITCHEN	SET					
LOUVER DOORS 1-3 x 6-8 M-500 KITCHEN	EACH					

Building _TWO STORY GARRISON_

Location _____

Architect _____

Estimator _____

Date _____

Description	Unit	Quantity	Price	Cost	Labor	Total
CASING STYLE 37	LIN. FT.					
STOP STYLE 93	LIN. FT.					
READI-HUNG DOOR UNITS 2-6X6-8 1ST FLOOR	UNIT					
READI-HUNG DOOR UNITS 2-0X6-8 1ST FLOOR	UNIT					
READI-HUNG DOOR UNITS 1-8X6-8 1ST FLOOR	UNIT					
READI-HUNG DOOR UNITS 2-6X6-8 2nd FLOOR	UNIT					
READI-HUNG DOOR UNITS 2-0X6-8 2nd FLOOR	UNIT					
READI-HUNG DOOR UNITS 2-8X6-8	UNIT					
READI-HUNG DOOR UNITS 1-3X6-8	UNIT					
READI-HUNG DOOR UNIT 2-0X6-8 (BY PASS DOORS)	UNIT					
ADJUSTABLE CLOSET RODS	EACH					
1X3 PINE ROD SUPPORTS	LIN. FT.					
1X12 PINE SHELVES CLOSET	LIN. FT.					
ANDERSEN NARROWLINE WINDOWS 3046W 1ST FLOOR	EACH					
ANDERSEN NARROWLINE WINDOWS 3032W 1ST FLOOR	EACH					
ANDERSEN NARROWLINE WINDOWS 2032W 1ST FLOOR	EACH					
ANDERSEN NARROWLINE WINDOWS 3046W 2nd FLOOR	EACH					
ANDERSEN NARROWLINE WINDOWS 24310W 2nd FLOOR	EACH					
ANDERSEN NARROWLINE WINDOWS 20310W 2nd FLOOR	EACH					
WINDOW CASING NO. 37	LIN. FT.					
WINDOW STOP NO. 90	LIN. FT.					
WINDOW STOOL NO. 71PL	LIN. FT.					
WINDOW APRON NO. 35	LIN. FT.					
FIREPLACE MANTEL M1448	EACH					
BLINDS 3-0X4-6 M500	PAIR					
BLINDS 2-4X3-10 M500	PAIR					
2-6X6-8X1$\frac{3}{8}$" M510 LOUVER DOORS	EACH					
HOUSED STRINGERS - PINE 14 RISERS 13 TREADS	SET					
NEWEL POSTS M769 3"X3"X4-2" BIRCH	EACH					
BALUSTERS M836 BIRCH 33	EACH					
HANDRAIL M720 BIRCH	LIN. FT.					

Building _TWO STORY GARRISON_

Location _____

Architect _____

Estimator _____

Date _____

Description	Unit	Quantity	Price	Cost	Labor	Total
SUB-RAIL M834	LIN. FT.					
FILLET M835	LIN. FT.					
TREADS, OAK 1-$\frac{1}{2}$" X 3-6	EACH					
PINE RISERS 7$\frac{5}{8}$" X 3-6"	EACH					
LANDING TREAD OAK 8'	EACH					
COVE MOLD 8062 OAK	LIN. FT.					
HANDRAIL BRACKETS	EACH					
WEDGES, PINE	EACH					
WHITE GLUE	PT.					
WALL CABINET 3018 OAK	EACH					
WALL CABINET 1830 OAK	EACH					
WALL CABINET 3330 OAK	EACH					
WALL CORNER WC 2430	EACH					
WALL CABINET 1530 OAK	EACH					
WALL CABINET 3315 OAK	EACH					
BASE CABINET B-18 OAK	EACH					
BASE CABINET SF-42 OAK	EACH					
BASE CABINET D-18 OAK	EACH					
BASE CABINET CB-45	EACH					
LAMINATED PLASTIC COUNTER TOP	LIN. FT.					
VANITY RV-36 COMPLETE WITH TOP BATH #1	EACH					
VANITY RV-30 COMPLETE WITH TOP BATH #2	EACH					
$\frac{1}{4}$" SILVER PLATE MIRROR 36"x 30" BATH #1	EACH					
$\frac{1}{4}$" SILVER PLATE MIRROR 30"x 30" BATH #2	EACH					
MISC. PINE ALLOW						

CONSTRUCTION CONTRACT

STATE _____

COUNTY _____

This CONTRACT, made this _____ day of _____ 19 _____ by

_____ of _____ (hereinafter called

the "Owner"), and _____ of

_____ (hereinafter called the "Contractor"),

WITNESSETH that the parties hereto agree as follows:

(A)   The Contractor will furnish all labor and materials and perform such work as required according to

the plans and specifications for the construction of a new residence for the consideration of _____

dollars $ _____ in accordance with the "General Conditions" shown

in this contract and the specifications and drawings as follows:

(B)   The Contractor will start work by _____ , 19_____ and will complete the work

by _____ , 19_____ .

(C)   The Owner will make payments as follows:

Payments shall be made by the Owner on a basis of ninety percent (90%) of the estimated amount of work completed and materials furnished at the end of each month and the balance shall be paid within thirty (30) days after the work herein contemplated shall be completed and materials in every particular according to the plans and specifications and after final inspection and acceptance by the Owner and the Architect.

Requisition for payment shall be submitted to the Architect for his approval on the tenth of each month covering work performed up to the first.

When application for final payment is made, the Contractor must furnish certified affidavits stating all obligations for labor and material used on the project have been paid for in full. Such affidavits to be in such form as is acceptable to the Architect and/or Owner.

IN WITNESS THEREOF, the parties hereto have executed this CONTRACT as of the date first above written.

_____   _____
              (witness)   (contractor)

                                          _____
                                                       (owner)

GENERAL CONDITIONS

I.   **VISITING THE SITE:**

The Contractor and each subcontractor shall visit the site and ascertain the existing conditions and shall also carefully study the plans and specifications as no allowance will be made for failure to do this.

II.   **SCOPE OF CONTRACT:**

The work will be erected under one "GENERAL CONTRACT" which shall include General Construction work, Heating, Plumbing, Electrical, and Painting.

III.   **CONTRACTOR'S BOND:**

The Contractor must deliver to the Owner an executed bond in the amount of the contract price, guaranteeing the faithful performance of the contract and for the payment of all persons performing labor and furnishing materials in connection with this contract. Bonds shall be furnished and executed by an accepted Surety Company approved, and authorized by and under the Laws of the State of _____ .

IV.   **CONTRACTOR'S INSURANCE:**

The Contractor shall not commence work under this contract until he has obtained all insurance required under this paragraph and such insurance has been approved by the Owner, nor shall the Contractor allow any subcontractor to commence work on his sub-contract until all similar insurance required of the subcontractor has been obtained and approved.

Certificates of insurance for all contractors must be filed with the Architect.

All policies must include completed operations coverage for one year after date of Final Certificate.

COMPENSATION INSURANCE:

The Contractor shall take out and maintain during the life of this contract Workman's Compensation Insurance for all of his employees employed at the site of the project, and in the case any work is sublet, the Contractor shall require the subcontractor similarly to provide Workman's Compensation Insurance for all of the subcontractors' employees unless such employees are covered by the protection afforded by the Contractor.

PUBLIC LIABILITY AND PROPERTY DAMAGE INSURANCE:

The Contractor shall take out and maintain during the life of this contract such Public Liability and Property Damage Insurance as shall protect him and any subcontractor performing work covered by this contract, from claims for damage for personal injury, including accidental death, as well as claims for property damage, which may arise from operations under this contract whether such operations be by himself or by any subcontractor or by anyone directly or indirectly employed by either of them.

CONTINGENT LIABILITY:

The above policies for Public Liability and Property Damage Insurance must be so written as to include Contingent Liability and Contingent Property Damage Insurance to protect the Contractor against claims arising from the operations of subcontractors.

V. CODES, REGULATIONS, AND PERMITS:

Each Contractor shall inform himself fully of all Codes, Regulations, and Conditions relating to the construction and labor under which the work is being or will be performed. No allowance will be made for failure to do this.

Each Contractor shall obtain and pay all costs for temporary and permanent permits (except building permit), give all legal notices and pay all legal fees, secure certificates of inspection that may be required by authorities having jurisdiction over the work, all as required for work of this contract except where otherwise specified. The Owner will secure and pay for the building permit.

VI.   CHANGES IN THE WORK:

When the value of the change in the work is determined by "Cost and Percentage" the charge or credit for the work covered by the approved change shall be determined by the actual cost of:

1. Labor, including foreman.

2. Materials entering permanently into the work.

3. Rental cost of construction plant and equipment during the time of use on the work.

4. Insurance.

To the above cost shall be added ten percent (10%) for overhead including applicable taxes, and six percent (6%) for profit.

VII.   STATE LABOR LAWS:

All Contractors on this project must comply with the Labor Laws of the State of _____ as pertaining to this work.

VIII.   MATERIAL AND LABOR:

All materials and workmanship shall in every respect be in accordance with the best modern practices and wherever the contract drawings, specifications or directions of the Architect admit of a doubt as to what is permissible and/or fail to note the quality of materials and workmanship conformity with modern practice is to be followed. The Contractor shall perform the work indicated in the specifications and/or shown on the drawings and shall furnish all materials, equipment, and incidentals necessary to complete the work in accordance with the specifications and drawings.

IX.   PROTECTION OF WORK AND PROPERTY:

The Contractor shall at all times safely guard the Owner's property or adjacent property from injury or loss in connection with this contract. He shall at all times safely guard and protect his own work; and that of adjacent property (as provided by law and the contract documents) from damage. The Contractor shall replace or make good any such damage, loss or injury unless such be caused directly by errors contained in the contract document, or by the Owner, or his duly authorized representatives.

All passageways, guard fences, lights and other facilities required for protection by local authorities, or local conditions must be provided and maintained.

X.   PLANS AND SPECIFICATIONS - INTERPRETATIONS:

The plans and specifications are complementary. Anything shown on the plans and not mentioned in the specifications or mentioned in the specifications and not shown on the plans, shall have the same effect as if shown or mentioned, respectively, in both. In the case of any conflict or inconsistency between plans and specifications, the specifications shall govern. Any discrepancy between the figured dimensions and the drawings shall be submitted to the Architect, whose decision shall be conclusive.

No change shall be made from the plans and specifications without first receiving written permission from the Architect.

XI.   EQUAL OPPORTUNITY:

The Contractor, in the performance of this contract, will comply with all applicable equal opportunity requirements.

XII.   ALLOWANCES:

All cash allowances shall be for the actual net cost (to include furnishing, packing and shipping to the Contractor of materials only). Allowances do not include installation costs, profits, etc. All allowances or any part thereof which are not expended shall be deducted from the contract price without discount or deduction. All bills for materials furnished under Cash Allowances shall be submitted in triplicate to the Architect for approval.

XIII.   INSPECTIONS AND TESTS:

All material and workmanship (if not otherwise designated by the specifications) shall be subject to inspection, examination and test by the Architect or other representative of the Owner, at any and all times during which the manufacture and/or construction are carried on. Without additional charge the Contractor shall furnish promptly all reasonable facilities, labor and materials necessary to make tests so required safe and convenient.

XIV.   QUALIFYING TERMS AND DEFINITIONS:

Wherever the terms "or equal," "necessary," "suitable," "as directed," "satisfactory," "when directed," "good and sufficient," or other general qualifying terms are used on the drawings or in these specifications they shall be construed as though followed by the words "in the opinion of the Architect," "or by the Architect," as the case may be. Generally, throughout these specifications phrases such as "The Contractor shall," "The Contractor will," "furnish and install," "furnish all labor, materials and equipment necessary to install," etc. and similar phrases have been intentionally omitted for the purpose of brevity and such omitted phrases shall be supplied by inference in the same manner as they are when a "note" occurs on the drawings. Wherever the word "provide" appears in these specifications it shall mean "The Contractor shall furnish all labor, materials and equipment as required to install, etc."

XV.   SUBSURFACE CONDITIONS FOUND DIFFERENT:

Should the Contractor encounter subsurface and/or latent conditions at the site materially differing from those shown on the plans or indicated in the specifications, he shall immediately give notice to the Architect of such conditions, before they are disturbed. The Architect shall thereupon promptly investigate the conditions, and if he finds that they materially differ from those shown on the plans or indicated in the specifications, he shall at once make such changes in the plans and/or specifications as he may find necessary, and any increase or decrease of cost resulting from such changes to be adjusted in the manner provided herein for adjustments as to extra and/or additional work changes.

XVI.   COOPERATION OF CONTRACTORS:

The Contractor shall cooperate with all other contractors and subcontractors employed on the project to the fullest extent to avoid delay and shall promptly render assistance to all of them in every way in which he can reasonably be of service to them in expediting the completion of the work.

XVII. CLEANING UP:

The Contractor shall keep the premises free from the accumulation of waste material and rubbish and at the completion of the work, he shall remove from the premises all rubbish, implements and surplus materials and leave the building broom-clean.

APPENDIX

The following labor estimating tables provide a quick method for determining the amount of time required to complete various types of construction.

Time requirements shown are based on average production per man hour, for quality workmanship, under average conditions. Time requirements do not include supervision.

How to use labor estimating tables

Locate in the tables the type of work to be done. Tabulate skilled and unskilled hours. Add these figures to get the number of hours required per estimating unit, (100 sq. ft., 100 lin. ft., etc.) Multiply this total by the wage rate per hour prevailing in your area. This is the labor cost per estimating unit. Multiply the unit cost by the number of units involved. Repeat this process for all phases of the job to get total labor cost.

If you desire you can also develop a "price per sq. ft." for labor on each type of construction. Or, by combining the various totals of a given job you may develop a "per cu. ft." price for the labor required. Once these figures are prepared they may be used for future projects of the same kind of construction.

EXAMPLE: Extra room, size 10 ft. x 16 ft., 160 square feet area.

Construction, Wood floor with 2 in x 8 in.. joist 12 in. on center with braces. Plywood subfloor with paper, 3/8 in. or 1/2 in. prefinished oak floor.

Labor rates, skilled $13.00, unskilled $9.75 per hr.

From the tables use the following figures:

	Skilled	Unskilled
Joist labor	5 hrs. (one unit)	2 hrs. (one unit)
Subfloor labor	1 hr. "	1/2 hr. "
Finished floor labor	4 hrs. "	1/2 hr. "
(160 sq. ft. =	10 hrs.	3 hrs.
1.6 estimating units) x 1.60 S.F.		x 1.60 S.F
	16 hrs.	4.8 hrs.

16 hrs. skilled   @   13.00   208.00
4.8 hrs. unskilled @   9.75   46.80
                        Total $254.80 labor cost for
                                            160 sq. ft.

Reduced to a square ft. price, the labor cost for this type floor construction is $1.593.

EXCAVATIONS, BACKFILLS, CONCRETE WALLS AND SLABS

	HOURS PER UNIT	
	Machine	Unskilled
Excavation, 100 cubic yards.		
Rates based on average type dry, solid soil.		
Handwork		130
Machine work (3/4 yard dipper continuous operation)		
Power shovel	1	
Backhoe	1 1/2	
Bulldozer (medium size tractor and blade)	3	
Backfilling, 100 cubic yards.		
Loose soil		
Machine work (see above)		
Bulldozer or backhoe	1	
Power tamping	2	

	HOURS PER UNIT	
	Machine	Unskilled
Handwork		65
Hand tamping		80
Ditching, 100 linear feet.		
Based on trench size of 12" x 24" or 2 cubic feet per linear foot.		
Machine work (trencher)	2/3	
Handwork		10
Sewers and Drains, 100 linear feet.		
Laying in 2' ditch and covered		
3' & 4' vitrified, 2' lengths		6
6" vitrified, 2' lengths		7
8" vitrified, 2' lengths		8
10" vitrified, 2' lengths		9
12" vitrified, 2' lengths		10
(Deduct 25% if 4 ft. lengths. Add 10% if asphalt, rubber or cement joints.)		
3-4-6" plastic, fibre 10' joints		4 3/4
3 & 4" drain tile, 1' lengths		5

	HOURS PER UNIT	
	Machine	Unskilled
6″ drain tile, 1′ lengths		5 1/2
8″ drain tile, 1′ lengths		6
Footers (Excavating), 100 linear feet.		
Based on 8″ x 16″ footer.		
Machine work (trencher)	1/3	
Handwork		5
(Add or deduct for other dimensions of ditches or footers.)		
Footers (Placing), 100 linear feet.		
Based on 8″ x 16″ footer.		
Setting forms to level grade. Wood	2	2
Steel	1 1/2	1 1/2
Placing reinforcing rods		1/4
Placing key forms	1	
Placing ready mixed concrete	1/2	4
(Average conditions and wheeling distance to forms)		
If vibrated concrete		1/3
Laying drain tile		1
Placing 12″ porous fill over tile		
Foundations (Concrete), 100 square feet.		
Average type construction to 8′ heights. 8″ to 12″ thick walls. Normal openings included.		
Setting 2′ x 8′ sectional forms. Wood	1 1/4	3/4
Steel	1	3/4
Building forms		
(plywood construction)	2	3/4
(1″ x 8″ sheathing and 2″ x 4″ or 2″ x 6″)	4	1 1/2
Placing reinforcing steel		
(1/2″ rods spaced 12″ x 12″ or wire mesh reinforcing)		1 1/2
Corbeling, chamfering or setbacks (up to 4″ x 6″)	1/2	
Placing concrete		
(Ready mixed concrete under average conditions)		
8″ walls	1	3
If vibrated concrete		1/2
12″ walls	1	4
If vibrated concrete		3/4
Removing forms, ties, etc.		
Sectional forms	1/2	1/2
Built in place forms	1	1/2
Hand rubbing walls, (minor blemishes)		1/2
Cleaning, oiling sectional forms		1
(Adjust rates for extra wheeling distance and handling or foundation heights over 8 Rate may change for unusual type forms or form hardware and vibrators attached to forms.)		
Foundations (Masonry), 100 square feet.		
(Average conditions, struck joints, common bond, openings included)		

	HOURS PER UNIT	
	Machine	Unskilled
Foundations (Masonry) — Cont.		
8 x 8 x 16 concrete masonry units	6	6
10 x 8 x 16 concrete masonry units	6 1/2	6 1/2
12 x 8 x 16 concrete masonry units	7	7
(Rate based on heavy units. Deduct 10% for medium weight and 20% for lightweight units)		
Placing masonry reinforcing		1/2
8″ solid brick walls	11	11
12″ solid brick walls	16	16
Loadbearing structural tile		
5 x 8 x 12 laid flat	5 1/2	5 1/2
8 x 12 x 12	7	7
10 x 12 x 12	7 1/2	7 1/2
12 x 12 x 12	8	8
Waterproofing		
Cement plaster, 1 coat	2	
Membrane (felt, polyethylene cemented to wall)		3
Tar or asphalt, 1 coat Brush coat		1 1/4
Trowel coat		2 1/2
Concrete (Walls), 1 cubic yard.		
Based on sections 12″ to 36″ thick. 20 to 100 cubic yard projects. Average type construction. Ready mixed concrete dumped in forms to 8′ heights.		
Setting 2′ x 8′ sectional forms wood	3/4	1/4
steel	1/2	1/4
Building forms		
Plywood construction	1	1/2
1″ x 8″ sheathing and 2″ x 4″ or 2″ x 6″	2	1
Placing reinforcing steel		
12″ x 12″ spacing 3/4″ rods wired		1/2
Placing concrete	1/4	1 1/2
If vibrated concrete		1/4
(Adjust vibration rate for unusual reinforcing problems and vibrators placed on forms.)		
Concrete Slab Construction, 100 sq. ft.		
Base Preparation		
Handwork after machine grading		1/2
(Not required if hand excavated)		
Placing, grading, tamping base material.		
Stone, Slag, Sand, Gravel, Cinders, etc.		1 1/4
Placing reinforcing wire mesh or rods		1/2
Vapor Barrier—Polyethylene, sisal reinforced paper felt or other non-rigid material		1/2
Perimeter Insulation		
1″ x 12″ x 2″ x 24″ rigid in 8′ strips or sheets		1/2

	HOURS PER UNIT	
	Skilled	Unskilled
Concrete (Slab Construction) — Cont.		
Slab Insulation		
Up to 2″ rigid or semi-rigid insulation laid on base under entire floor area	1/2	1/4
Insulating Concrete, 100 square feet		
Lightweight vermiculite, perlite, pumice (Not finish troweled)		
4″ thick	3/4	3/4
6″ thick	1	1
8″ thick	1 1/2	1 1/2
(Finish troweled floor use figures below for concrete slab installation)		
Concrete Slab Pouring, 100 square feet.		
3″ thick	1 1/2	1 1/2
4″ thick	1 1/2	1 1/2
If vibrated concrete		1/4
5″ thick	1 1/2	1 1/2
6″ thick	2	2
If vibrated concrete		1/3
8″ thick	2 1/2	2 1/2
If vibrated concrete		1/2

(Time based on the use of ready mixed concretes delivered to the site, no forming, finished or troweled surface on prepared base. Includes placing of 1/4″ or 1/2″ asphalt, rubber or fibre expansion joint and normal blocking of 4′ squares. Add for wheeling concrete if required. Adjust for machine finishing.)

CONCRETE FLATWORK, STEPS

Walks, Driveways, Patios, 100 square feet.
Based on hand work under average conditions. Placed on prepared grade.

	Skilled	Unskilled
Walks (4″ concrete) Average 4′ width		
Grading, leveling 4″ to 6″ base materials		1 1/4
Forming: metal or wood and placing joints	3/4	1/2
Pouring and finishing ready mixed concrete	1 1/2	1 1/2
If vibrated concrete		1/4
Removing forms		1/4
For other walk surfaces see brick and patio block under heading of Floors.		
Driveways 6″ concrete, to 16′ widths on prepared grade		
Grading, leveling 4″ to 8″ base materials		1 1/2
Forming: metal or wood & placing asphalt, rubber or fibre expansion joints.	1	1/2
Placing reinforcing mesh	1/2	
Pouring and finishing using ready mixed concrete	2	2

	HOURS PER UNIT	
	Skilled	Unskilled
Walks, Driveways, Patios — Cont.		
If vibrated concrete		1/3
Removing forms		1/2
Adjust for machine work & special curing methods, weather protection, etc.		
Porches and Patios		
Concrete slab construction on prepared base, shored wood formed or metal deck.		
Placing reinforcing mesh		1/2
Placing reinforcing rods		
8″ x 8″ centers 3/8″ rods		1
Pouring ready mixed concrete troweled finish		
4″ thick	1 1/2	1 1/2
5″ thick	1 3/4	1 3/4
6″ thick	2	2
If vibrated concrete		1/4
Placing corrugated metal deck over steel or concrete joists	1/2	1/2
Complete forming and shoring to 8′ heights	8	4
Removing forms	1/2	1
Steps, 10 sq. ft. tread area.		
(On roughed-in concrete base)		
12″ treads, 8″ risers.		
Brick treads and risers, tooled joints		
Treads and risers	5	2 1/2
Treads only	3	1 1/2
Concrete masonry treads and risers (4 x 8 x 12 solids or equivalent)	2	1
Stone or precast concrete treads on concrete base.		
one piece to 4″ thick cut to size	1	1
Rough slate or stone treads to 4″ thick	2	2
Concrete step treads to 12″ wide, 4″ to 6″ thick		
Forming	1	
Pouring ready mixed concrete	1/2	1/2
If vibrated concrete		1/4
Removing forms and finishing	1/2	1/2
Concrete steps, risers & treads (see labor required for cheek walls under Foundations, concrete or masonry)		
Forming steps 8″ rise 12″ treads	2	1

	HOURS PER UNIT	
	Skilled	Unskilled

Steps — Continued

	Skilled	Unskilled
Pouring steps and risers	1	1/2
If vibrated concrete		1/4
Removing forms and finishing	1	1/2

Reinforced concrete steps 8"
rise to 12" treads to 20" length,
4 ft. wide to 8 ft. height

	Skilled	Unskilled
Complete forming and placing steel	6	1 1/2
Pouring ready mixed concrete	2	1
If vibrated concrete		1/3
Removing forms and finishing	1	1 1/2

Wood steps and stairs
(see frame construction)

Masonry, 100 square feet.

(All rates under masonry cover standard
or modular units

Face brick — based on 3/8" flush
joints. Normal openings, sills,
headers. To 16" heights.
4" brick veneer

	Skilled	Unskilled
Stretcher bond	12	8
Stacked bond	15	10
Soldier course	15	10
Header course	18	12
Roman size (12") Stretcher bond	12	18
Stacked bond	15	10
Norman size (12") Stretcher bond	10	7
Stacked bond	13	9
SCR (1/2" joints) Stretcher bond	9	6
Double brick Stretcher bond	8	6

Adjust for special bonds, tooled
raked, struck or special joints,
special color patterns, curved or fancy
walls or brick sizes. Add 10% for
glazed brick.

Glass block — 1/4" joints, 3-7/8"
thick block

	Skilled	Unskilled
5 3/4" x 5 3/4"	20	10
7 3/4" x 7 3/4"	15	8
11 3/4" x 11 3/4"	11	6
4 3/4" x 11 3/4"	20	

Cleaning masonry — brick, tile,
concrete, stone etc. using muriatic
acid and water, (no scaffolding
included).

	Skilled	Unskilled
Rough surfaced walls	1 1/2	3/4
Smooth surfaced walls	1	1/2

Concrete masonry (other than
foundation work) — average weight
regular or patterned units, stretcher
bond hand tooled joints.   Reinforcing mesh. To 16' heights. Normal
openings, sills, headers, control joints
included.

Masonry — Continued

	Skilled	Unskilled
4 x 8 x 12	5	5
2 x 8 x 16	3	3
3 x 8 x 16	3 1/2	3 1/2
4 x 8 x 16	4	4
6 x 8 x 16	4 1/2	4 1/2
8 x 8 x 16	5	5
Add 25% for stacked, ashler or other bonds.		

Masonry fill — Vermiculite insulation

	Skilled	Unskilled
8" walls		1/2
12" Walls		3/4

Figure cavity walls based on
space to be filled at the rate of
5 minutes per 4 cubic feet for
openings over 2" wide.

Stone work—coursed ashler
4" — 5" widths. Random lengths.
End cuts made on job.

	Skilled	Unskilled
2 1/4" thick	16	16
5" thick	12	12
Random widths 2 1/4" to 10 1/2" thick	14	14
Random widths 5" to 10 1/2" thick	10	10

Random ashler cut to size

	Skilled	Unskilled
2 1/4" to 5" thick	10	10
5" to 10 1/2"	7	7

Rustic rubble stone 6"
to 18" size cut and fit on
job

	Skilled	Unskilled
	16	16

Back plastering masonry work

	Skilled	Unskilled
(all types) 1 coat work	1 1/2	1

Backcoating masonry work (all types)

	Skilled	Unskilled
1 coat work asphalt type trowel		2 1/2
brush		1 1/4

Liquid (transparent waterproofing) — brick, concrete
masonry, stone

	Skilled	Unskilled
1 coat application brush coat		1
spray coat		1/2

Painting concrete masonry,
brick, tile, stone, concrete.

	Skilled	Unskilled
Cement base paints 1 coat brush work		1 1/2
spray work		3/4
roller work 9"		1

Back-up walls — normal
openings.

	Skilled	Unskilled
4" Common brick	8	6
8" Common brick	10	8

Concrete masonry

Average weight units Reinforcing mesh included	Skilled	Unskilled
4 x 8 x 12 " " "	5	5

	HOURS PER UNIT	
	Skilled	**Unskilled**

Masonry — Cont.

Reinforcing mesh included

	Skilled	Unskilled
4 x 8 x 16 " " "	4	4
6 x 8 " " "	4 1/2	4 1/2
8 x 8 " " "	5	5
10 x 8 " " "	6	6
12 x 8 " " "	7	7

Structural tile

	Skilled	Unskilled
4 x 5 x 12	5 1/2	5 1/2
5 x 8 x 12	4 1/2	4 1/2
4 x 12 x 12	4 1/2	4 1/2
6 x 12 x 12	5	5
8 x 12 x 12	6	6
10 x 12 x 12	6 1/2	6 1/2
12 x 12 x 12	7	7

Add 10% to all types for cavity walls, adjust for special sizes, shapes, unusual wall designs.

Gypsum partition tile (partition walls) 3/8" joints 12 x 30" units, unusual openings

	Skilled	Unskilled
2" thickness	3	2 1/2
3" thickness	3 1/2	3
4" thickness	4	3 1/2
6" thickness	5	4

Screen or decorative walls, flush joints.

	Skilled	Unskilled
8 x 8 to 8 x 12 Face 4" thick concrete	5	5
8" thick concrete	6	6
8 x 8 to 16 x 16 Face 4" thick concrete	6	6
8" thick concrete	7	7
20 x 20 to 24 x 24 Face 8" thick concrete	6	6
8 x 8 to 8 x 12 Face 4" thick clay	5	5

Adjust for unusual wall designs, joints, patterns or unit types.

Glazed Masonry Work, 100 pieces.

Interior or exterior. Tooled joints. Up to 8' heights.

5-1/16" x 7-3/4" Face

	Skilled	Unskilled
1-3/4" Soap Stretcher	4 1/2	4 1/2
3-3/4" Stretcher	5	5
5-3/4" Stretcher	5 1/2	5 1/2
7-3/4 " Stretcher	6	6

5-1/16" x 7-3/4" Face

	Skilled	Unskilled
1-3/4" Soap Stretcher	6 1/2	6 1/2
3-3/4" Stretcher	7	7
5-3/4" Stretcher	7 1/2	7 1/2
7-3/4" Stretcher	8	8

Glazed Masonry - Cont.

7-3/4" x 15-3/4" Face

	Skilled	Unskilled
1-3/4" Soap Stretcher	10	10
3-3/4" Stretcher	11	11

Glazed concrete masonry

	Skilled	Unskilled
2 x 8 x 16	10 1/2	10 1/2
4 x 8 x 16	11 1/2	11 1/2
6 x 8 x 16	12 1/2	12 1/2
8 x 8 x 16	14	14
12 x 8 x 16	17	17

Both types add 25% if glazed two sides. Double rates for all shapes.

Cutting masonry units (Average power saw) Clay tile, brick, concrete masonry.

	Skilled	Unskilled
To 32" perimeter units	1	
32" to 48" perimeter units	1 1/2	

COPING, LINTELS, BEAMS, COLUMNS

Wall Coping, 100 linear feet.

Vitrified

	Skilled	Unskilled
9" – 13"	6	6
18"	7	7

Includes placing corners and ends.

Precast concrete or cut stone

	Skilled	Unskilled
4" thick to 16" widths; 4–5' lengths	6	5

Sills—Lintels, 100 linear feet.

Door—window. Precast concrete or stone cut to size to 5' lengths.

	Skilled	Unskilled
To 4" x 8"	5	2
6" x 8" – 8" x 8"	5 1/2	2
8" x 10" – 8" x 12"	6	3

Concrete Beams, Lintels, 100 linear feet.

Poured in place to 8' heights

	Skilled	Unskilled
8" x 8" setting forms and placing steel rods	7 1/2	3 1/2
8" x 12" setting forms and placing steel rods	8	4

Pouring ready mixed concrete

	Skilled	Unskilled
8" x 8"	1	3
8" x 12"	1	4
Vibration		1/4
Removing forms	1	1/2
Finishing concrete		2

Concrete Masonry Pilasters, 100 linear feet.

Laying units. Placing reinforcing rods. Filling with ready mixed concrete to 16' heights. Hand work.

	Skilled	Unskilled
8" x 8" Core types	9 1/2	17

	HOURS PER UNIT	
	Skilled	Unskilled

Concrete Masonry Pilasters, Cont.

	Skilled	Unskilled
12″ x 12″ Core types	11 1/2	23 1/2

Adjust for additional heights and
Mechanical filling methods and
vibration if required.

Concrete Columns, 100 linear feet.

Poured in place to 8″ heights,
setting forms and placing steel.

	Skilled	Unskilled
8″ x 8″ — 8″ x 12″ — 12″ x 12″	6	3

Pouring ready mixed concrete

	Skilled	Unskilled
8″ x 8″	1	3
8″ x 12″	1	4
12″ x 12″	1	5
Vibration		1/3
Removing forms	3/4	1/2
Finishing concrete		2

Adjust for additional heights or
vibrators attached to forming.

Foundation Block,
(For slab construction) 100 linear feet.

	Skilled	Unskilled
6″ x 12″ x 16″ U and J types or 8″ x 8″ x 16″ header block	2	3

Concrete Masonry Beams and Lintels, 100 linear feet.

Laying units. Placing reinforcing
rods. Filling with ready mixed
concrete. Hand work to 8′ heights.

	Skilled	Unskilled
8″ x 8″ x 16″ units continuous beams	5	7
8″ x 8″ x 16″ units over openings. Shoring included.	6	8
8″ x 16″ x 12″ units continuous beams	12	16
8″ x 16″ x 12″ units over openings. Shoring included.	13	16 1/2
Vibration		1/3

Adjust for additional heights and
mechanical filling methods or
vibrators attached to forms.

CHIMNEYS, FIREPLACES

Chimneys, per foot of height
with flue liner, to 30′ heights.
Inside chimneys with normal face
brick topping. No scaffolding
provided. Footer or Base (use rates
under Footers).

Brick

	Skilled	Unskilled
1 8″ x 8″ flue	3/4	3/4
2 8″ x 8″ flues	1 1/4	1 1/4
1 8″ x 12″ flue	3/4	3/4
2 8″ x 12″ flues	1 1/2	1 1/2
1 12″ x 12″ flue	1	1
2 12″ x 12″ flues	1 3/4	1 3/4

	HOURS PER UNIT	
	Skilled	Unskilled

Chimneys, Fireplaces Cont.

Adjust for other sizes and
combinations. Add 25% for
average face brick outside
chimneys. Flue liner sizes
may vary. Use nearest size.

Concrete masonry chimney units

	Skilled	Unskilled
8″ x 8″ flue size	1/4	1/4
8″ x 12″ flue size	1/3	1/3

Deduct 10% for units without
flue liner or for unlined round
types. Add 10% for solid
masonry or 2 unit types.

Fireplaces, per fireplace.

Based on average design and type
6′ wide and 5′ high, 36″ x 30″ opening.

	Skilled	Unskilled
Hearth construction		
Concrete base	1/2	1
Brick hearth on concrete base	2	2
Tile hearth on concrete base 2″ x 2″ to 6″ x 6″	2	2
Brick work	10	10
Firebrick lining and damper	4	2
Setting steel circulator	1	1

Double above time for double
faced or 3 way installation. Use
1 1/2 times above for corner
projecting fireplace. Adjust for
unusual brick patterns or
other facing materials.

	Skilled	Unskilled
Tile or glass facing 4″ x 4″ to 8″ x 8″ sizes	4	4
Setting factory mantel (to prepared wall)	1	1/2

FRAMING, SHEATHING, DECKING

Floor Joists, Wood, 100 square feet.

Joist Size	Spacing	Approx. Span*	Skilled	Unskilled
2″ x 6″	12″	8′	4 1/2	1 1/2
	16″	7′	4	1 1/2
	18″	6′–6″	4	1 1/4
	24″	6′	3 1/2	1
2″ x 8″	12″	11′	6	2
	16″	10′	4 1/2	1 3/4
	18″	9′	4	1 1/2
	24″	8′	3 1/2	1
2″ x 10″	12″	14′	6 1/2	2 1/2
	16″	12′	6	2 1/4
	18″	11′	5 1/2	2
	24″	10′	5	1 3/4
2″ x 12″	12″	17′	7 1/2	2 3/4
	16″	15′	7	2 1/2
	18″	14′	6 1/2	2 1/4
	24″	12′	6	2

	HOURS PER UNIT	
	Skilled	**Unskilled**

	HOURS PER UNIT	
	Skilled	**Unskilled**

Framing Cont.

Labor rates cover both platform or balloon type construction, also sill, plate, edge joist and bridging. Add 25% for spans over 12', wood girder if required also if above 1st floor level.

*Spanning based on 100 pounds per square foot total load using Douglas Fir or equal.

Steel Bar Joist, 100 square feet.

Based on 36" spacing to 20' spans. Types 2 & 3 lightweight. To 2nd floor levels.

	Skilled	Unskilled
8"	1	1
10"	1 1/4	1
12"	1 1/2	1 1/4

Steel Beams, 100 square feet.

Light weight type to 20' spans; to 2nd floor levels.

	Skilled	Unskilled
6" – 7" widths	1	1
8" – 10" widths	1 1/4	1
12"	1 1/2	1 1/4

Concrete Joists, 100 square feet.

Based on 24" spacing to 12' lengths, at ground level.

	Skilled	Unskilled
3" x 8"	1	2
3" x 10"	1 1/4	2 1/4
4" x 12"	1 1/2	3

Add 25% for 14 to 20' lengths.
Add 50% for 2nd floor construction.

Precast Concrete Floor Slabs, 100 square feet.

On concrete joist or steel beams.

	Skilled	Unskilled
1" to 2" thick 24" x 30"	1 1/4	1 1/4
1" to 2" thick 24" x 60"	3/4	1

Adjust for various sizes and types.

Precast and Prestressed Concrete Floor Beams, 100 square feet.

Over steel or concrete beams. To 20' lengths. Erected to second floor level. Includes top grouting. Hollow beams.

	Skilled	Unskilled
6" x 12" Beams	1/2	1
6"–8" x 16" Beams	1/3	2/3
6" –8" x 48" Beams	1/4	1/2

Adjust for other sizes. Add 25% for spans 20' to 30'. Add machine or crane time.

Commercial Store Fronts and Window Construction.

Based on total front or window area. Average type construction. Excluding glass.

Commercial Store Fronts Cont.

	Skilled	Unskilled
Rough framing wood	8	2
Metal work over rough framing using rolled light metal stock	20	5
Metal work over rough framing or free standing using medium or heavy extruded stock	30	10

Adjust for unusual types, patterns and designs or extra metal cutting and fitting.

Studding, 100 square feet.

Includes normal openings, average type outside walls, frame or veneer construction. Plates, headers, fillers, bracing, firestops, girts included.

	Skilled	Unskilled
2" x 4" 12" centers 8' to 12' heights	3	1/2
12" centers 12' to 20' heights	4	3/4
16" centers 8' to 12' heights	2 1/2	1/2
16" centers 12' to 20' heights	3 1/2	3/4
24" centers 8' to 12' heights	2	1/2
24" centers 12' to 20' heights	3	3/4
2" x 6" 12" centers 8' to 12' heights	4 1/2	3/4
12" centers 12' to 20' heights	5 1/2	1
16" centers 8' to 12' heights	3 1/2	3/4
16" centers 12' to 20' heights	4 1/2	1
24" centers 8' to 12' heights	3 1/2	1

Add 25% for irregular or cut up walls and shed or gable dormers.

Deduct 25% for interior stud partition walls.

SCR construction (Furring)

	Skilled	Unskilled
2" x 2" 12" centers 8' to 12' heights	2 1/2	1/2
2" x 2" 16" centers 8' to 12' heights	2	1/2

Ceiling Joists, 100 square feet.

Normal type frame construction. 1st and 2nd floor levels. Includes bridging, trimmers to 16' spans. Normal openings.

	Skilled	Unskilled
2" x 6" 12" centers	5 1/2	2

		HOURS PER UNIT	
		Skilled	Unskilled
Ceiling Joists Cont.			
2" x 6"	16" centers	5	2
	18" centers	4 1/2	1 1/2
	24" centers	3	1
2" x 8"	12" centers	7	2
	16" centers	6	2
	18" centers	5 1/2	1 1/2
	24" centers	4	1
2" x 10"	12" centers	8	2 1/2
	16" centers	7	2 1/2
	18" centers	6 1/2	2
	24" centers	5	1 1/2
2" x 12"	12" centers	9	3
	16" centers	8	3
	18" centers	7 1/2	2 1/2
	24" centers	6	2
For open beam ceilings			
3" x 6"-8"	16" centers	7	2 1/2
	18" centers	6 1/2	2
	24" centers	5	1 1/2
4" x 6"-8"	16" centers	7 1/2	2 1/2
	18" centers	7	2
	24" centers	5	1 1/2

All of above add 25% for spans over 16' and heights above normal 2nd floor ceiling level.

Rafters, 100 square feet.

Average type construction, normal pitch and flat, gable ends, to 22' lengths.

		Skilled	Unskilled
2" x 4"	12" centers	3	3/4
	16" centers	2 1/2	3/4
2" x 6"	12" centers	3 1/2	1
	16" centers	3	1
	18" centers	2 1/2	1
	24" centers	2	3/4
2" x 8"	12" centers	4	1
	16" centers	3 1/2	1
	18" centers	3	1
	24" centers	2 1/2	3/4
Open beam rafters			
3" x 6" - 8"	16" centers Flat construction	7	3 1/2
	24" centers Flat construction	6	3
4" x 6" - 8"	16" centers Flat construction	8	4
	24" centers Flat construction	7	3 1/2

(Add 25% for pitched roofs over 2" pitch)

All of above add 50% for cut up and hip roof types.

Rafters Cont.

Add 25% for rafters on shed, gable or hip type roofs on dormers and for rafters over 22' long. Deduct 10% for rafters with short or plain overhang.

Sheathing, 100 square feet.

Average type construction, frame or veneer, normal openings.

		Skilled	Unskilled
1" x 6" – 8" Wood			
sheathing	flat roofs	1	1/2
	pitched roofs	1 1/2	3/4
	steep, cut-up or hip roofs	3	1 1/2
	sidewalls	1 1/2	3/4
(Add 25% for diagonal sheathing on sidewalls)			
2" x 6" – 8" / 3" x 6"			
	flat roofs	3	1 1/2
	pitched roofs	4	2 1/2
(Add 25% if exposed underside on beamed ceiling construction)			
Plywood, rigid insulating sheathing to 25/32" thick			
4' x 8' sheets	flat roofs	1/2	1/4
	pitched roofs	3/4	1/2
	steep, cut-up or hip roofs	1 1/4	1
	sidewalls	1	1/2
Gypsum sheathing			
2' x 8' panels	sidewalls	1 1/4	1/2
Strip sheathing			
1" x 2" – 3" – 4" pitched roofs		1	1/2
(Deduct 33-1/3% for use of automatic or power nailers or staplers.)			
Insulating sheathing			
1" to 3" thick	flat roofs	3/4	1/2
	pitched roofs	1 1/4	1
(Add 25% if finished open beam ceiling construction)			
Asbestos covered insulating sheathing 19/16" to 2" thick, nailed or screwed			
4 x 8 panels. Frame construction	flat roofs	1 1/2	1
	pitched roofs	2	1 1/2
(Adjust for various types of fasteners)			

Bar Rib Lath, 100 square feet

For poured concrete or gypsum over concrete, bar joist or steel beams.

	HOURS PER UNIT	
	Skilled	Unskilled
Bar Rib Lath Cont.		
3/4" rib lath.*	1/2	1/2
Steel Decking, 100 square feet.		
Sheets 2' x 8' for floor or roof construction.		
18 or 20 gauge.*	1/2	3/4
Paper Backed Wire Mesh, 100 square feet.		
For floor construction.		
3" x 4" mesh, 12 gauge, rolls 4' x 125'	1/4	1/4
*(Add 25% for 2nd or 3rd floor construction)		
Stairways (Frame), per stairway		
Average type straight flight to 4' wide and 12' long.		
Rough cutting, framing, placing (Add 50% for 2 flight type)	5	2
Roof Trusses, (wood) per truss.		
Average type 1 floor construction, using precut lumber.		
24" spacing 14' to 20' length	1/2	1/2
22' to 32' length	3/4	3/4
SUBFLOORS, FINISHED FLOORS		
Subfloors on Wood Joists, 100 sq. ft.		
1" x 8" Lumber	1 1/2	1/2
Plywood 3/8" to 3/4"	1	1/2
(Deduct 50% for use of automatic or power nailers or staplers)		
Insulation, 100 sq. ft.		
Blanket type placed between joists	3/4	
Underlayment, 100 sq. ft.		
Board-type (Hardboard, Fiber Board, Plywood, etc.) 1/4" to 1/2" thick	1	
Wood Finish Flooring, 100 sq. ft.		
Softwood strip		
2 1/4" face	3	1/2
3 1/2" face	2	1/2
Hardwood strip 7/8"		
1 1/4" face	4 1/2	3/4
2 1/2" face	4	1
Hardwood strip 1/2" - 3/8"		
1 1/2" face	4	1/2
2 1/4" face	3 1/2	1/2
(Deduct 25% for use of automatic nailers)		
(Includes placing paper or felt under floor.)		
Sanding, machine	1	
Finishing		
Three liquid applications to all unfinished wood floors	1 1/2	
Prefinished hardwood		

	HOURS PER UNIT	
	Skilled	Unskilled
Wood Finish Flooring, Cont.		
(Strip, plank, block)		
1/2" – 3/8" x 1 1/2" face	4	1/2
Plank to 6" wide	2 1/2	3/4
Block 9" x 9"	2	3/4
(Either nailed or cemented on prepared subfloor)		
Resilient Type Flooring, 100 Sq. ft.		
Roll linoleum (plain)	3	1/2
Patterns, borders	4	1
Tile, (asphalt, vinyl, rubber, linoleum, cork, etc.)		
4" x 4" size	5	1
6" x 6"	4	1
9" x 9"	3	1
6" x 12"	3	1
12" x 12"	2 1/2	1
9" x 18"	2 1/2	1
Strip (to 8")	3	1
(Includes laying felt or underlayment)		
Ceramic Tile on Concrete, 100 sq. ft.		
1/2" to 2", paper backing	4	4
2" x 2"	5	5
4" x 4"	4	4
6" x 6"	3	3
(Includes laying the necessary underlayment on any base.)		
Slate on Concrete, 100 sq. ft.		
Up to 1" thick.		
Random cut sizes to 12" x 12"	2	2
Rough uncut slabs	1 1/2	1 1/2
Marble on Concrete, 100 sq. ft.		
4 x 6 to 12 x 12 tile		
Cement bed preparation	2	4
Setting tile	8	8
Machine finishing	8	
Brick Floor on Concrete, 100 sq. ft.		
(Standard size on edge)		
Laid in mortar, mortared joints, basket weave or common bond.	12	6
Herringbone or fancy patterns	16	8
(Standard size laid flat)		
Basket weave or common bond	8	4
Herringbone or fancy patterns	12	6
(Use 1/3 of above rates if laid in sand without mortar.)		
Adjust rates for other brick sizes.		
Patio Blocks on Concrete, 100 sq. ft.		
Laid in mortar, mortared joints up to 4" thick. Square, rectangular, octagonal or odd shapes.		

	HOURS PER UNIT Skilled	Unskilled
Patio Blocks Cont.		
6 x 12 to 8 x 16	4	2 1/2
12 x 12 to 12 x 18	3 1/2	2
12 x 24	2 1/2	2
24 x 24	2	2
Use 1/3 of above rates if laid in sand base without mortar.		
ROOFING		
Asphalt Roofing, 100 sq. ft. (Square)		
Roll, Plain Surface Pitched roofs	3/4	1/4
Mineral Surface Pitched roofs	1	1/4
Felt Underlayment		1/4
Shingles, Individual type	3	1
Individual type Dutch lap	1 3/4	1
Interlocking	2	1
Strip 10″ x 36″	2	1
Strip 12″ x 36″	2 1/4	
(Deduct 33-1/3% for use of automatic or power nailers or staplers)		
Asbestos Shingles, 100 sq. ft. (Square)		
American method 8″ x 16″	4 1/2	2 1/2
Dutch lap hexagonal 16″ x 16″	2 1/2	1 1/2
Colonial method 10″ x 24″ to 12″ x 30″	2 1/4	1
Wood Shingles, 100 sq. ft. (Square)		
16″ Approx. 5″ exposure	4	1 1/2
18″ Approx. 6″ exposure	3 1/2	1 1/4
24″ Approx. 8″ exposure	3	1
(Add 35% for staggered or thatched butts or double coursing.)		
Slate, 100 sq. ft. (Square)		
16″ x 8″	5	2 1/2
18″ x 9″	4	2
20″ x 10″	3 1/2	2
22″ x 12″	3	1 1/2
Random widths to 3/8″ thick	7	3 1/2
Graduated slate to 3/4″ thick	8	4
Clay Tile, 100 sq. ft. (Square)		
Spanish, Mission or Shingle type	5	2 1/2
Interlocking tile type	6	3
(Add for stripping and underlayment)		
Metal Sheets, 100 sq. ft. (Square)		
Pitched roofs.		
Corrugated aluminum or galvanized steel in sheets. Over wood framing.	1 1/4	3/4
Crimped types, aluminum or steel in sheets over wood framing	1 1/2	3/4
Rolls over flat areas. Copper or tin 14″ widths. Flat seams	2 1/2	2 1/2
Corrugated Asbestos Sheets, 100 sq. ft. (Square) (Screws or Fasteners)		
1/4 x 42–8′ length over open rafters or steel	2	2

	HOURS PER UNIT Skilled	Unskilled
Built-Up, 100 sq. ft. (Square)		
(Includes topping of slag, stone, gravel or chips)		
3 Ply over wood or insulating deck	1 1/2	3/4
3 Ply over gypsum or concrete deck	1 3/4	1
4 Ply over wood or insulating deck	1 3/4	1
4 Ply over gypsum or concrete deck	2	1 1/4
Concrete Slabs, Lightweight, 100 sq. ft. (Square)		
On steel beams or bar joist. 2nd floor levels. Add for mechanical hoisting equipment.		
Flat type 1″ to 2″ thick 24″ x 5′	1	2
Channel type.		
2 3/4″ thick 24″ x 8′	1 1/4	2 1/2
3 3/4″ thick 18″ x 8′	1 1/2	3
3 3/4″ thick 24″ x 8′	1 1/4	2 1/2
NOTE: Add 25% to all roofing estimates for steep pitched or cut-up roofs and unusual conditions. Above rates include time for average type ridges, hips and valleys. Underlayment or scaffolding not included. Add for preparation of surface, eaves, etc. when estimating reroofing jobs.		
Metal Work 100 linear feet		
Eaves and gutters, metal, standard sizes.	5	5
(Deduct 1/3 for interlocking prefit types)		
Downspouts 2″ to 6″ round or square	4	2
Valleys 20″ width metal	4	
Flashing Parapet walls, chimneys and dormer sides, etc.	4	
Raggles for flashing (cutting only in set up masonry walls.)	1	4
Wood Gutters, 100 linear feet		
3″ x 5″ and 4″ x 6″	5 1/2	4
5″ x 7″	6 1/2	5
Cant Strips, 100 linear feet		
3″ x 6″	1 1/2	1/4
SIDING		
Wood Siding (horizontal), 100 sq. ft.		
Shiplap, patterns, rustic types		
1″ x 3 1/2 – 4″	2 1/2	3/4
1″ x 4 1/2 – 5″	2 1/4	3/4
1″ x 6 – 8″	2	1
Lap, bevel or bungalow types		
1/2″ x 8″	2 1/2	3/4
3/4″ x 10″	2 1/4	3/4
3/4″ x 12″	2	3/4
Add 25% for cuts to fit ends and mitered corners. Includes metal corners if required.		

Siding Cont.	Skilled	Unskilled
Vertical patterned types		
1" x 6-8"	3	1 1/2
1" x 10-12"	2 3/4	1 1/4
Board and batten		
1" x 6-8"	3 1/2	1 1/2
1" x 10-12"	3	1 1/4
(Includes horizontal stripping over studs)		
Wood shingles		
16" approximately 5" exposure	5	1 1/2
18" approximately 6" exposure	4 1/2	1 1/2
24" approximately 8" exposure	3 1/2	1 1/2
Add 25% for special patterns or double coursing. Adjust for fasteners or special nailers.		
Asphalt Siding*, 100 sq. ft.		
Brick, Stone, Shingle patterns.		
Roll types (15" widths)	2	1/2
Panel types (insulating)		
10-7/8" x 43"	3	1
14" x 43"	2 1/2	1
(Deduct 25% for use of automatic or power nailers or staplers)		
Composition Siding*, 100 sq. ft.		
Hardboard, wood fibre etc., 1/4"- 5/16", plain or factory painted. 10" - 12" widths. Horizontal.		
Self venting and spacing metal furring.	1 3/4	1/2
10"- 12" widths. Horizontal.	1 1/4	1/2
16" widths. Horizontal	1	1/2
10" - 12" widths. Horizontal Shadow edge furring	2	1/2
16" widths. Horizontal. Shadow edge furring	1 3/4	1/2
10" - 12" widths. Vertical Board & Batten	2	1
16" widths. Vertical Board & Batten	1 3/4	1
(Deduct 25% for use of automatic or power nailers or staplers)		
Asbestos Siding*, 100 sq. ft.		
8" x 32" to 12" x 24"	2	1
Plain Sheet Types*		
3/16" and 1/4" thick, 48" wide 8' long	1 1/2	1 1/2
Corrugated Sheets*		
1/4" x 42" - 8' lengths	1 1/4	1 1/4
* Includes corners, trim, etc.		

Shingle Backer, 100 sq. ft.	Skilled	Unskilled
8 3/4" to 11 3/4" 48" lengths	1 1/4	1/2
13 1/2" to 15 1/2" 48" lengths	1	1/2
(Deduct 25% for use of automatic or power nailers or staplers)		
Lattice strip furring	1/4	1
3/4" x 2" strip furring	1/2	1/4
Paper, Felt, Etc., 100 sq. ft.	**1/4**	**1/4**
Aluminum Siding, 100 sq. ft.		
Clapboard types 8"		
Horizontal plain	1 1/2	1 1/2
insulated	2	2
Vertical plain	2	2
insulated	2 1/2	2 1/2
(Includes felt, paper, foil, moisture barrier, corners, fittings and trim)		
Corrugated sheets, 8' lengths		
32" coverage sheets, over wood frame	1	1
over steel frame	2	1 1/2
(Deduct 10% for 45" coverage sheets)		
Galvanized Steel Siding		
Corrugated Sheets, 100 sq. ft.		
8' sheets, 26" coverage, over wood frame	1 1/4	1
over steel frame	2	2
Adjust for unusual cutting and fitting and heights over 20'.		
To all siding jobs add 25% for unusual conditions, cut up walls, bays and gables.		
Adjust for wall preparation on all types of remodeling projects.		
Stucco, 100 sq. yards.		
Cement stucco 3 coats. Float finish over tile, concrete, brick, metal lath or stucco base. Average openings included to 16' heights. Scaffolding provided.		
Grey, white portland cement or colored prepared stucco	25	18
Add 10% skilled time for textured finish		
Add 15% skilled time for troweled finish, unusual finished surfaces or patterns.		
Based on average conditions, normal openings. New frame construction.		

	HOURS PER UNIT	
	Skilled	Unskilled
Rigid Board Types, 100 sq. ft.		
4' x 8' to 1" thick		
Flat roofs	1/2	1/4
Pitched roofs	3/4	1/2
Steep pitched or cut-up roofs	1	3/4
(Add 10% for 2" & 3" thickness)		
Sidewalls	1	1/2
(Deduct 33 1/3% for automatic or power nailers or fasteners)		
Non-Rigid Types, 100 sq. ft.		
Batt type 2 to 4" thick between studs		
15-19-23" x 24" batts	3/4	1/2
15-19-23" x 48" batts	1/2	1/2
Strip types 2 to 4" thick, between studs		
15" widths	1/2	1/2
Blanket types 1 to 4" thick, between studs		
16-20" widths, 4 to 8' lengths	3/4	1/2
24-33" widths, 4 to 8' lengths	1/2	1/2
Wide stock over studs or sheathing	1/3	1/3
(Add 25% for ceiling installations)		
Reflective Types, 100 sq. ft.		
1 sheet stripped in place between studs	1/2	1/4
Multiple, accordian sheets	3/4	1/4
Pouring Types, 100 sq. ft.		
Between 4" studs when accessible	1	1
Over ceilings 4 to 6" thick	1	1/2
Poured in cavities concrete masonry or cavity type walls to 4" space	1	1/2
Semi-Rigid Types, 100 sq. ft.		
Applied over frame or masonry walls using adhesives or asphaltic cements	2	2
(Add 25% for each additional 2" in thickness)		
(Add 25% for ceiling or unusual wall installations all thicknesses)		
Vapor Seals, Moisture Barriers, 100 sq. ft.		
Roll or sheet types, aluminum, polyethylene, papers		
Tacked in place	1/2	1/4
Cemented on frame or masonry surfaces	3/4	1/2

Adjust all rates for extra time and conditions on unusual projects and remodeling jobs.

WALLS AND CEILINGS

Average type construction, normal openings. Walls and ceilings combined. Necessary working level planking or scaffolding included.

	HOURS PER UNIT	
	Skilled	Unskilled
Plaster Bases, 100 square yards		
Gypsum lath 3/8" - 1/2" 16" x 48"	6	1 1/2
3/8" - 1/2" 24" to 12'	4	1 1/2
Insulating lath 1/2" 18" x 48"	7	2
Metal lath, average size sheets, over studs	4	1 1/2
(Adjust for staples or mechanical fasteners)		
Metal lath, average size sheets, over masonry	5	1 1/2
Paper backed wire or expanded mesh	3 1/2	1 1/2
Foamed plastic, 12" x 9' planks, 1-2" thick cemented	8	1 1/2
Plaster, average 1/2" - 5/8" grounds, 100 square yards.		
Scratch coat over gypsum or insulating lath	2	2
Scratch coat over masonry or gypsum tile	2 1/2	2
Scratch coat over metal lath or mesh	3	2
Brown coat over average scratch coats	3	2
Sand finish coat*	4	2
White coat lime finish*	6	2
* Add 10% if color mixed on job.		
Prepared color finish	5	2
Keenes cement finish, smooth	4 1/2	2
Keenes cement finish, 4" x 4" tile pattern	5 1/2	3
Acoustical Plaster (1 coat over suitable base— handwork)	4	2
2 coat work over gypsum or insulating lath	10	5
3 coat work over gypsum or insulating lath	13	6
2 coat work over metal lath, mesh	11	6
3 coat work over metal lath, mesh	14	6
2 coat work over masonry, gypsum tile concrete	10 1/2	5
3 coat work over masonry, gypsum tile concrete	13 1/2	6
2 coat work over foamed plastic base	10	5
3 coat work over foamed plastic base	13	6

Deduct 25% unskilled time for factory mixed plasters. Add 10% for heavily wood fibred plasters or heavy plasters. Adjust for special finishes of all types, machine application and for work over radiant heating cables or tubing, etc.

	HOURS PER UNIT	
	Skilled	Unskilled
Metal Lath and Plaster Partitions,		
100 square yards.		
Hollow 4″ thick installed in masonry		
or concrete work		
2 coat work plastered 1 side only	41	10
2 coat work plastered both sides	52	16
3 coat work plastered 1 side only	44	10
2 coat work plastered both sides	55	18
(Add for various partition		
widths)		
2 coat work both sides	46	12
3 coat work both sides	52	18
Gypsum Lath and Plaster Partitions,		
100 square yards.		
Solid 2″ thick, includes floor and		
ceiling runners, concrete or masonry		
construction		
2 coat work, 2 sides	50	16
3 coat work 2 sides	53	18
Suspended Ceilings, 100 square yards		
Hung from concrete or steel, 1 1/2″		
main runners with 3/4″ channels on		
12″ centers, metal lath wired or clipped on.		
2 coat work complete ceiling	45	15
3 coat work complete ceiling	48	17
Less 10% if hung from wood joist		
or rafters.		
Plaster Bonding, 100 square yards.		
Over concrete and masonry surfaces		
Asphaltic type brushed on		5
Cement base types brushed on		6
Cement base types troweled on		
1/8″ to 1/2″	4	4
Adjust and add for chipping,		
roughing up or other wall prep-		
arations if required.		
Gypsum Board, Walls, 100 square feet.		
3/8″ - 1/2″ 5/8″ thick		
4′ x 8′ to 10′ panels	1	1
Finishing joints	1	
Finishing joints 2 ply application		
2nd layer cemented in place finished	3 1/2	1 1/2
Gypsum Board, Ceilings, 100 square feet.		
Plain 4′ x 6′ to 12′ lengths	1 1/2	1 1/2
Wood veneer or other finishes	2	1 1/2
Plank type 16″ widths	2 1/2	1 1/2
(Adjust for screws or unusual		
fastening methods)		
Plywood Panels to 4′ x 12′ panels,		
100 square feet.		
Plain joints or tongue and grooved	1	1
Covered joints with moulding	3	1
Fitted joints cemented	4	1

	HOURS PER UNIT	
	Skilled	Unskilled
Plywood Panels, Cont.		
Plank type paneling tongue and		
grooved 6″ to 16″ widths	3	1
Patterned Paneling, 100 square feet		
3/4″ x 6″ to 12″ widths, horiz.		
or vertical	3 1/2	1
Beaded Wood, 100 square feet.		
1/2″ x 3 1/2″	3	1
Insulating Board, 100 square feet.		
Rigid panels 4′ x 8′ to 12′	1	1
Plank type	1 1/2	1
Insulating or Acoustical Tile		
100 sq. ft.		
Plain or decorated (Nailed,		
stapled, clipped to metal strips)		
12″ x 12″	3	1
12″ x 24″	2	1
24″ x 24″	1 1/2	1
16″ x 32″	1	1
(Furring not included)		
Add 25% for patterns and adhesive		
application.		
Wall Tile, 100 square feet.		
Applied with adhesives to prepared		
walls, plain patterns. Includes base		
and cap. Metal trim.		
Plastic 4″ x 4″	5	1
9″ x 9″	3	1
Metal 4″ x 4″	6	1
Ceramic 4″ x 4″	8	2
(Add 50% if applied in		
mortar bed)		
Adjust all work for unusual		
conditions, materials, designs, etc.		
Hardboard, 3/16″ - 1/4″ - 5/16″, 100 sq. ft.		
Plain 4′ x 8′, 8′ to 12′ lengths.		
Nailed	1 1/4	1
4′ x 8′, 8′ to 12′ lengths.		
Nailed with metal fittings	2 1/2	1
Metal Sidewalls and Ceilings,		
100 square feet.		
Patterned types	5	2
(Add for furring and intricate		
designs)		
Curtain Wall Panels, 100 square feet.		
1-9/16″ and 2″ thickness		
4′ x 8′ - 9′ - 10′ - 12′ panels	4 1/2	4
(Includes 2″ x 4″ studs and		
plates floor and ceiling) (Screws		
or nails)		
Adjust for special fastening methods.		
Furring, Plaster Grounds, Up to		
3/4″ x 4″ wood strips, 100 square feet.		

	HOURS PER UNIT			HOURS PER UNIT	
	Skilled	Unskilled		Skilled	Unskilled

Furring, Plaster, Cont.

	Skilled	Unskilled
Frame construction 12" centers	1	
16" centers	3/4	
24" centers	1/2	
Nailable masonry construction		
12" centers	1 1/4	
16" centers	1	
24" centers	3/4	

Adjust for installation on concrete walls and for use of power activated powder or explosive type automatic fasteners. Add 25% all types of work if figuring ceilings only.

Moldings, Trim, and Accessories, 100 linear feet.

	Skilled	Unskilled
Plastering and corner beads, corner lath, metal furring, screeds, picture molding, bullnose beads, etc.	2/3	1/3
Door and window casing beads (metal) metal trim, corners, mitered and fitted before plastering or used with dry wall construction	2 1/2	1

WINDOWS AND DOORS

Window Sizes:

A To 3' x 3'. Single unit

B Over 3' to 3' x 5'-6" Single unit

C Over 3' to 6' x 5'-6" Double unit

D Over 3' to 6' to 8' x 6' Triple unit and picture windows

Wood Windows, per each unit

Window frames, assembled from stock sections

Size	Skilled	Unskilled
A	1/2	1/4
B	3/4	1/2
C	1 1/4	1/2
D	1 1/2	1/2

Setting window frames, includes handling and bracing when required.

Size	Skilled	Unskilled
A	1/2	1/4
B	3/4	1/2
C	1	1/2
D	1 1/2	1/2

Fitting and hanging double hung sash,

Size	Skilled	Unskilled
A	1/2	1/4
B	3/4	1/2
C	1 1/2	1/2
D	2 1/2	3/4

Fitting and hanging casement sash, Per opening.

Size	Skilled	Unskilled
A 1 sash	1/2	1/4
B 2 sash	1	1/4
C 3 sash	1 1/2	1/2
D 4 sash	2	1/2

Setting complete window units. All types glazed. To prepared openings.

Size	Skilled	Unskilled
A	3/4	1/4
B	1	1/2
C	1 1/2	1/2
D	1 3/4	3/4

Metal Windows, per each unit.

All residential types, fins or wood surround attached. Setting sash units. To prepared openings.

Size	Skilled	Unskilled
A	3/4	1/2
B	1 1/4	1/2
C	1 3/4	3/4
D	2	1

(Adjust for unusual installations and glazed units.)

Basement Sash, wood or metal, per each unit

	Skilled	Unskilled
	1/4	
Setting poured in place basement frames & sash	1/2	
Setting utility metal sash		
4 light open	1/2	
6 light open	3/4	

Commercial Projected.

Pivoted or security type metal sash. Bracing and handling to second floor heights. Fixed or vented.

	Skilled	Unskilled
Small units to 2' x 4'	3/4	1/4
2' x 4' to 4' x 8'	1 1/2	1/2
4' x 8' to 6' x 8'	2	3/4
6' x 8' to 8' x 8'	2 1/2	1

(Add 25% for architectural projected or heavy awning types)

Add 25% to all work if second floor installation only.

Adjust for plate glass or twin glass glazing where work is figured glazed.

Trim, Interior, per each unit.

Single member type, casings, stool, apron, stops. Soft wood.

Size	Skilled	Unskilled
A	1	
B	1 1/4	
C	1 1/2	
D	2	

(Add 25% for wood jambs over plaster returns)

Plain wood Stool and Apron.

	Skilled	Unskilled
All sizes	1/2	

	HOURS PER UNIT	
	Skilled	**Unskilled**
Trim, Interior, Cont.		
Plain Wood Stool only. All sizes	1/4	
Adjust for various types trim.		
Metal Stools to 4'	3/4	
Metal Stools 4' to 8'	1	
Glass, Marble, 1 piece precut stools to 4'	1	
Ceramic tile stools to 4' plain		1 1/2
Trim, Exterior, per each window		
Size A	1/2	
B	3/4	1/4
C	1	1/4
D	1 1/2	1/4
(Includes flashing at head if required)		
Shutters, per pair		
Wood, Metal		
Small	1/2	
Medium	3/4	
Large	1	
(Add 50% if over masonry construction)		
Storm Sash & Screens, (Prefit wood comb.) per unit		
Size A	1/4	1/4
B	1/2	1/4
C	3/4	1/4
D	1	1/4
(Add 25% for metal or plastic)		
Doors (Residence types)		
Setting exterior door frames. Wood residential types. Standard sizes	1	1/4
Setting front entrance door frames with patterned side panels to 5' widths	2	1/4
Setting front entrance door frames with side light panels to 5' widths	2 1/2	1/4
Cutting and setting 2" x 6" - 8" frames from stock material 3' x 7'	1	1/4
Same as above 8' x 8' to 10' x 8'	1 1/2	1/2
10' x 10' to 16' x 10'	2	3/4
Interior door jambs and heads. Assembling from stock sections. Standard sizes	3/4	1/4
Setting interior door frames. Standard sizes	1	1/4
Same as above small closet	1/2	
large closet	3/4	
Setting sliding door pockets in stud walls. Standard sized doors	3/4	1/4
Building sliding door pockets & setting track		
1 pocket, single door	2 1/2	
2 pockets, double door	4 1/2	

	HOURS PER UNIT	
	Skilled	**Unskilled**
Glass Sliding Doors, setting complete units.		
To 4' widths	2	
4' to 6' widths	2 1/2	
6' to 8' widths	3	
Metal Door Frames, setting each.		
Standard size	3/4	1/4
Double size	1 1/2	1/2
Outside Door Trim, 1 piece setting each.		
Standard size	3/4	
Inside Door Trim, 1 piece, setting each.		
Standard size doors	1 1/4	
Small closet doors	1/2	
Large closet doors	3/4	
Thresholds, setting each.		
Wood	1/4	
Metal	1/4	
Ceramic, Marble	3/4	
Hanging Doors, per each.		
Exterior, wood, standard sizes.		
All types. 3 butts.		
1 3/4"	1	1/2
metal, prefit	3/4	1/4
Interior metal, prefit 1-3/8"	1/2	1/4
wood 1-3/8"	3/4	1/2
wood sliding 1-3/8"	1/2	1/2
metal sliding 1-3/8"	1/2	1/4
metal bifolding, pair	3/4	1/4
wood bifolding, pair	3/4	1/4
wood french, pair	1	1/4
closet small, 1-1/8"-1-3/8" thick	1/2	
closet large, 1-1/8"-1-3/8" thick	3/4	
folding fabric or slatted types	3/4	
Exterior combination storm/ screen wood complete	1	1/4
Complete prefit metal or plastic	3/4	1/4
(Adjust all door rates for special designs and types)		
Gargage and heavy doors.		
1 3/4" hinged 4' x 8'	2	1/2
sliding 4' x 8'	2 1/2	1/2
sliding 8' x 8'	3	1/2
Setting overhead doors, wood 1 3/8", metal or glass fiber. (complete)		
8' x 6'-6" to 8' x 8'	3	1
to 10' x 7' & 8'	3 1/2	1
12' x 7' & 8'	4	1
12' x 10' & 12'	4 1/2	1 1/2
16' x 6'-6" to 8'	5 1/2	2
(Adjust for metal doors, add 25% for 1 3/4" doors)		

	HOURS PER UNIT	
	Skilled	Unskilled

Weatherstripping, per opening.

	Skilled	Unskilled
Standard type metal for average size door.	3/4	

Louvers, Vents, per each.
Metal, screened or regular.

	Skilled	Unskilled
Small medium sizes	1/4	
Large sizes	1/2	
Half circle types	1	

Access Doors, prefit metal.

	Skilled	Unskilled
To 24" x 30"	1/2	
Over to 36" x 48"	3/4	

Metal-Clad Doors, complete masonry or steel construction. (Underwriters)

		Skilled	Unskilled
Swinging	Single	4	2
	Double	6	2
Sliding	Single	6	2
	Double	8	2

Adjust all window and door rates for transoms if required. Add 25% hardwood trim. Adjust for masonry construction unless indicated and for unusual conditions. Rates do not include hardware unless shown as complete.

MILLWORK AND TRIM

Normal openings. Frame construction. Soft wood. First class work. Rooms 10' x 10' or over. Add 25% for hardwood. Adjust if drilling or screws are required.

Wood Interior Trim, 100 linear feet.

	Skilled	Unskilled
Base — 1 piece 2 1/2	1	
Shoe mold	1 1/2	
Chair rail	2 1/2	1
Picture mold	3	1
Cornices, crown, bed, cornice mold		
single member to 3 1/2" widths	6	2
single member over 3 1/2" widths	7	2
2 to 4 member	12	2
Ceiling beams, build in place		
3" x 6" to 6" x 12"	18	2
Closet shelving 3/4" x 12" stock		
includes cleats	3	1

Metal Interior Trim, 100 linear feet.
Applied either before or after plastering.
Applied with screws or clipped-on.

	Skilled	Unskilled
1 piece base	3 1/2	1
2 piece base	4 1/2	1
Cove mold	2	1
Chair rail	2 1/2	1
Picture mold	3	1
Cornice to 8" width with preformed mitered corners	6	2
Closet shelving. Adjustable lengths 12" and 16" widths	2	1

	HOURS PER UNIT	
	Skilled	Unskilled

Metal Interior Trim Cont.

Adjust time for all trim if applied over masonry walls or over plaster over masonry.

Wood Exterior Trim, 100 linear feet.

	Skilled	Unskilled
Corner board, verge boards, fascia, frieze to 3/4" x 8"	3	2
Shingle mold, bed mold	1 1/2	1
Soffits, 3/4" x 6"-8" to 24" widths	7	1
Plywood, hardboard, asbestos pegboard 1/4" x 12" x 24"	4	1
(Add for screens if used)		
Metal with screens or vents	3	1
Cornices, 2 members to 12" widths	6	1

Porch Rail, 100 linear feet.

	Skilled	Unskilled
Top, bottom, balusters	3	1 1/2

Stairs, Exterior Wood, per flight.
Plain open stairs. 4' widths to 12' lengths,

	Skilled	Unskilled
single flight	4	2
double flights	6	2

Porch Columns to 10' heights with caps and bases where required.

	Skilled	Unskilled
4' x 4"-6" x 6" solid	1/2	1/2
built up square to 8" x 8"	1 1/2	1
round hollow to 12"	1	1
round turned solid	3/4	1/2

Mantels, per unit.

	Skilled	Unskilled
Setting average type factory built mantel units. To prepared walls	2 1/2	1 1/2

Cabinets, Cupboards, 100 sq. ft. face area
Average type work. Includes base and mold as required.

	Skilled	Unskilled
Setting factory built base cabinets, cases, range, oven sections. Also broom and utility cabinets, vanities	4	2
Hanging top cabinets	3	2

Counter Tops, 10 sq. ft. surface area
Placing factory built tops over cabinet bases without sinks

	Skilled	Unskilled
3/4" Plywood, lumber	1/6	1/6
1" to 2" Maple	1/2	1/4
Stainless Steel	1/5	1/5

Covering sink, base, vanity tops, plain type using metal trim.
Cutting, fitting, cementing

	Skilled	Unskilled
Laminated plastic	3/4	
Linoleum	1/2	
Ceramic tile 4" x 4"	1 1/2	
(Add 50% if sinks or lavatories included)		

Misc. Factory-Built Cabinets per each unit.

	HOURS PER UNIT	
	Skilled	**Unskilled**

Factory Built Cabinets, Cont.

Placing or setting to rough openings

	Skilled	Unskilled
Package receivers, through-wall types. Average size.	1	
Medicine and wall cabinets —		
Small	1/2	
Large	3/4	
Ironing board, broom, shoe cabinets, etc.	1/2	
Clothes chutes, complete, (except basement bin)	1 1/2	
Bath and Kitchen Accessories. (Soap dish, towel racks, paper holder, hand bars, etc.)	1/4	
Setting prefit lightweight shower doors — plastic glazing	1	
heavy weight plate glass	2	

GLAZING

Average type work, to second floor heights. Single strength, double strength.

Glazing Wood Sash and Doors.

Using putty, per 10 lights

Approximate glass size

	Skilled	
8″ x 10″ to 12″ x 14″	1	
16″ x 20″ to 20″ x 28″	2	
30″ x 36″ to 36″ x 40″	2 1/2	
40″ x 48″ to 48″ x 60″	4 1/2	

Deduct 1/3 if glazed with wood stops.

Glazing Metal Sash.

Using putty or plastic glazing compound, per 10 lights

Approximate glass size. Square inches per light

	Skilled	
Up to 300 square inches	1 1/4	
300 to 600 square inches	2 1/2	
600 to 900 square inches	3 1/4	
900 to 1200 square inches	4 1/2	
1200 to 1800 square inches	5 1/2	
1800 to 2400 square inches	6 1/2	

Deduct 1/3 if glazed with metal or plastic stops or strips.

Adjust all rates for twin-pane lights, plate, wired, ribbed or special types of glass and other unusual conditions.

Store or Commercial Building Fronts, 100 square feet.

	Skilled	
Setting plate glass over 4′ x 5′ to 8′ x 10′	10	
Over 8′ x 10′ to 10′ x 15′	15	

PAINTING **Skilled**

Based on 1st coat over new work unless indicated otherwise. Surface areas.

Outside Walls, 100 square feet.

Oil, stain or rubberized types.

	HOURS PER UNIT	
	Skilled	**Unskilled**
Wood sidings	1	
Wood shingles (stain)	1 1/4	
Burned smooth brick	1 1/4	
Burned rough brick	1 3/4	
Concrete masonry, concrete, stucco	1	

(Add 10% if masonry paints)

(Reduce 50% if spray work-large projects)

(Reduce 25% if roller work)

Inside Walls and ceilings, 100 square feet.

Flat, casein, or rubberized types.

	Skilled	
Sizings	1/4	
Plaster, white coat, dry wall	1/2	
Insulating plank, panels, tile, rough raw finish	1	
Plywood, lumber, plasterboard,	3/4	
composition board, smooth finishes	3/4	

(Deduct 10% if calcimine or similar types. Add 25% all work if ceilings only)

(Reduce 50% if spray work, large projects)

(Reduce 25% if roller work)

Cabinets, Cupboards.

	Skilled	Unskilled
Vanities, cabinet or closet doors, bookcases. 1 side	1	1/2

Wallpaper, 100 square feet.

Ordinary type walls.

	Skilled	
Butt joint work, Average type paper	2	
Special papers and coated fabrics, etc.	3	

Floors, 100 square feet.

	Skilled	
Filling, wiping	2/3	
Shellac, varnish, stain	1/2	
Painting — Wood types	1/2	
Concrete	1/3	

(Reduce 25% if roller work)

Roofs, 100 square feet.

Average conditions, smooth surfaces.

	Skilled	
Asphalt, aluminum and other free flowing types	1/2	
Fibred semi-plastic types	1	
Wood shingles — stained	3/4	
painted	1	

(Adjust for unusual conditions, spray work etc.)

Trim, 1 coat per opening.

Inside and Outside

	Skilled	
Windows and doors, average size	1/4	
Picture mold, chair rail, base, cornice, etc. Figure each item for each average size room as 1 opening	1/4	

	HOURS PER UNIT	
	Skilled	Unskilled

	HOURS PER UNIT	
	Skilled	Unskilled

Doors, 1 coat per opening.
All types, inside, outside 2 sides average size — 1/2
Combination storm doors — 3/4
Inside, medium size closet & cupboard — 1/3
Inside, small size closet & cupboard — 1/4

Windows — Based on wood double hung, casements, sliding, projected, awning, window-wall types with large lights. 1 side.
Small sizes — 1/4
Medium sizes — 1/3
Large sizes — 1/2

Metal, Residential types average sized lights. 1 side. All types.
Small size units — 1/5
Medium size units — 1/4
Large size units — 1/3
Adjust both wood and metal residential rates for small lights of glass where found.

Metal, Commercial types, average sized lights. 1 side.
Small openings — 1/4
Medium openings — 1/3
Large openings — 1/2
All window units figure each unit and combine time when two or more units are combined by mullions.

Shutters. All types, per 2 sides — 1/4
Stairways. Complete open types, per stairway — 4
Wood Mantels, per unit — 1/2
Caulking, 100 linear feet.
Average type work using gun, around windows, door trim, etc. — 1 1/4
Adjust all painting time for enamels, special type paints, finishes and unusual conditions, old work, and spray painting. Deduct 10% for successive coats on large areas of inside walls, ceilings and outside walls.

HARDWARE
Lock Sets, per each unit.
Outside doors, plain type sets — 1/2
Outside doors, front, fancy type sets — 3/4
Outside storm doors — 1/2
Outside door closers — 1/2
Inside doors softwood — 1/5

Lock Sets, Cont.
Inside doors hardwood — 1/4
Inside closet, cupboard, cabinet, etc. — 1/10
Inside sliding doors, swinging, single — 1/5
Inside sliding doors, swinging, double — 1/4
Door Accessories, per 10 units.
Door bumpers, stops, surface bolts, night latches, closet hooks, handles, pulls, catches, and misc. small items. Simple installation — 1/2
Sash Hardware, per each unit.
Double hung, locks and lifts per window — 1/10
Casement sash, locks and operators per sash — 1/4
Closet Rods, per rod — 1/6
Garage or heavy swinging doors, latches and lock sets, holders and top or bottom bolts, per door — 1/2

WORK DESCRIPTION	LABOR	MATERIAL	TOTAL
Site/Subgrade Prep.			
Construction utilities			
Heat			
Water			
Electricity			
Engineering Fees			
Temporary construction			
Clearing site			
Grading			
Fill material			
Foundation			
Layout			
Stakes			
Batter boards			
Line			
Nails			
Excavation			
Pier footings			
Continuous foundation			
Dimension			
Boards			
Stakes			
Nails			
Ties			
Wire			
Reinforcing steel			
Column reinforcement			
Concrete			
Curing			
Foundation drainage			
Drain tile			

WORK DESCRIPTION	LABOR MATERIAL TOTAL	WORK DESCRIPTION	LABOR MATERIAL TOTAL
Porous fill	_____	Nails	_____
Waterproofing	_____	Reinforcing steel	_____
Concrete Floor Const.		Control joint	_____
Backfill	_____	Expansion joint	_____
Top soil	_____	Screeds	_____
Aggregate	_____	Concrete	_____
Fill and tamp	_____	Curing	_____
Aggregate	_____	**Exterior/Interior Walls**	
Dirt	_____	Wall panels	_____
Vapor Barrier	_____	Dimension	_____
Concrete forming	_____	Nails	_____
Dimension	_____	Reinforcing steel	_____
Board	_____	Chairs	_____

GLOSSARY

Aggregate — Materials such as sand, rock and vermiculite used to make concrete.

Ampacity — The current-carrying capacity of an electrical device or conductor.

Apron — Trim placed against the wall directly below the window stool.

Architect's scale — A flat or triangular scale used to make and read architectural drawings.

Areaway — Metal, concrete, or masonry construction to hold back earth and provide an opening around basement windows and doors.

Ash dump — A hatch placed in the bottom of a fireplace which disposes ashes into the ash pit.

Awning window — A window that is hinged near the top so that the bottom opens outward.

Backfill — Earth placed against a building wall after the foundations are in place.

Backsplash — The raised lip on a countertop to help prevent water from running down the back of the cabinet.

Balloon framing — Type of construction in which the studs are continuous from the sill to the top plate, under the roof rafters. The second-floor joists are supported by a strip or ribbon, which is cut into the studs.

Balusters — Vertical pieces which support a handrail.

Balustrade — An assembly of balusters (vertical supports) and a handrail; commonly found on open stairways.

Barefoot — Method of construction in which the rafters rest on top of the ceiling joists.

Base cabinets — Any kitchen cabinet that rests on the floor.

Bid bond — Guarantees that the contractor will not withdraw the bid for a specified period of time.

Board foot — (board measure) One hundred forty-four cubic inches of wood, or the quantity contained in a piece which measures 12″ x 12″ x 1″ thick.

Box sill — The header joist placed at right angles to the ends of the floor joists.

Boxed rake — A rake cornice that extends beyond the end of the building and is enclosed with a fascia and soffit.

Branch — (electricity) Circuits that carry the current to various parts of the building.

Branch — (plumbing) The piping from a fixture to the point in the plumbing system where it joins piping from other fixtures.

Bridging — The lateral bracing of the floor joists.

Btu — British thermal unit is the amount of heat required to raise the temperature of one pound of water one degree Fahrenheit.

Btuh — British thermal unit per hour.

Butt hinges — The hinges on which the doors swing (sometimes referred to as *butts*).

Casement window — A window that is hinged at one side so that the opposite side opens outward.

Centerline — An actual or imaginary line that equally divides the surface or sides of something in half.

Chimney cap — Special masonry construction at the top of a chimney to protect the masonry from the elements.

Collar beams — Horizontal members which tie opposing rafters together, usually installed approximately half way up the rafter.

Common brick — Clay brick which is formed and then baked to a hard material. Common bricks generally used as structural material where appearance is not important.

Concrete — Building material consisting of fine and coarse aggregates bonded together by cement.

Conductor — Electrical wire; a cable may contain several conductors.

Consolidate — Refers to concrete and means to work to remove voids or air pockets.

Contour lines — Lines on a map or plot plan to identify the elevation. These lines describe the slope of the ground.

Contract — An agreement in which one party agrees to perform certain work and the other party agrees to pay for the work and services.

Cornice — The assembly of boards and moldings used in combination with each other to provide a finish to the ends of the rafters which extend beyond the face of the outside walls.

Cornice return — Where the level cornice at the eaves turns the corner at the end of the house.

Course — A single row of building materials such as concrete blocks or roofing shingles.

Damper — A door installed in the flue of a fireplace or furnace to regulate the draft.

Dampproofing — Construction technique which helps prevent moisture from seeping through to the inside of a foundation wall.

Datum — The permanent points in a city or county establishing elevations above sea level.

Double-hung window — A window consisting of two sash which slide up and down past one another.

Drywall — A type of interior wall covering using gypsum wallboard and special compound to conceal the joints.

Face brick — Shale brick which is formed and then baked to a hard material. They are dense and will not absorb water. Face brick is usually used for exterior facing.

Fascia — The part of a cornice that covers the ends of the rafters. Usually a 1 x 8 board.

Finish hardware — All of the hardware which is exposed to view in the completed house.

Float — To smooth concrete with a tool called a float. Floating is done before the concrete cures and does not produce as smooth a surface as trowelling.

Floor plan — An architectural drawing showing the layout of rooms, location of windows and doors, and the location of special equipment.

Flue — A duct for carrying off smoke and gases of combustion. Chimneys normally have a fire clay flue lining.

Flush doors — A door of any size having both surfaces smooth and flat.

Footing — Concrete construction which may be found under foundation walls, columns, or chimneys and which distribute the weight of these structures to prevent settling.

Frieze — A horizontal board directly below the cornice and above the siding. A frieze is not used on all construction.

Gable — (gable end) The triangular area between the rafters and top wallplate at the ends of a building with a gable roof.

Gable studs — Vertical framing members placed between the end rafters on a gable roof and the two plates of the end wall.

Grout — Special cement for tile joints.

Gypsum wallboard — Drywall material composed of reconstituted gypsum encased in paper.

Headers — Used to support the ends of framing members which have been cut to make an opening for stairways, chimneys, fireplaces, doors, and windows.

Hip rafter — The rafter running from the corner of a building to the ridge of a hip roof.

Insulated glazing — Two pieces of glass with air space between them.

Jack rafter — Spans the space between the wall plates and the hip rafters, or the ridge and valley rafter.

Jambs — Side members of a window or door frame.

Joists — Horizontal members that support the floor.

K factor — The rate at which a material conducts heat.

Lally column — Steel pipe used as a support post.

Lintel — Structural member that spans a clear opening to support the structure above the opening. Can be found above doors, windows, and fireplace openings.

Lock set — An assembly consisting of a latching mechanism.

Masonry cement — Cement which is specially prepared to make mortar.

Miter joint — A joint with pieces cut at 45° angle to form a right angle corner.

Mortar — Material made from portland cement and additives to bond masonry units together.

Mullion — The vertical piece between two windows when they are mounted side-by-side. Windows with mullions are called mullion windows.

Nominal size — The size by which materials are specified. Actual size is usually slightly smaller.

Oakum — A fibrous rope-like material used for caulking joints in cast iron soil pipe.

Orthographic-projection — A method of drawing used by draftsmen to show three-dimensional objects on two-dimensional paper. (also referred to as third-angle projection.)

Panel doors — Any door having decorative panels on either surface.

Parging — Thin coat of cement plaster used to smooth rough masonry walls.

Payment bond — Guarantees that all the contractor's debts relating to the job will be paid so there will be no liens against the property.

Penny size — (abbreviated d) Penny refers to the length of a nail.

Performance bond — Guarantee issued by a bonding company stating that the work will be done according to the plans and specifications.

Perimeter drain — An underground drain pipe around the footings.

Plate — A horizontal piece, the same width as a stud placed at the bottom and top of a stud wall.

Platform framing — A method of framing in which each floor is framed separately, with the subfloor in place before the wall and partition studs of that floor are erected.

Plot plan — A drawing that provides all of the information necessary to clear the lot, the size of the lot, and the location of the building.

Plywood — A wood material made up of an odd number of layers laminated into a sandwich-like panel.

Polyurethane varnish — A clear plastic coating material which is extremely durable.

Portland cement — Finely powdered limestone material which is mixed with water to bond aggregates in concrete. Also can be used to mix mortar.

Primer — Type of paint used for the first coat on surfaces that do not hold paint well.

Quantity takeoff — List of materials required to complete the construction of a building.

R value — Thermal resistance: ability to resist the flow of heat.

Rafter — The framing members of a roof. The rafters extend from the wall to the ridge of the roof.

Rake — The inclined portion of a cornice at the ends of a roof.

Resilient flooring — vinyl, vinyl-asbestos, linoleum, or other synthetic floor covering which provide a smooth surface.

Ridge board — The top member of the roof framing which runs the length of the roof.

Rough plumbing — The concealed portion of the plumbing system. This includes all piping and fitting up to the finished wall or floor surface.

Rowlock — Pattern of bricklaying in which bricks are laid on edge.

Run of rafter — The horizontal distance covered by one rafter.

Run of stairs — The horizontal distance covered by a set of stairs.

Sash — Frame holding the glass of a window.

Saturated felt — Paper-like felt which has been treated with asphalt to make it waterproof.

Screed — A straight board used to level concrete.

Sectional view — A drawing which shows what one would see after making a vertical cut through the building.

Sheathing — Boards applied horizontally or diagonally, plywood, or fiber board applied to the outside face of the studs.

Sill — A single piece of wood laid flat on the top of the foundation wall.

Sill cock — (sometimes called a hose bibb). A faucet mounted outside of the building to which a hose may be attached.

Sill seal — Man-made material which is laid on top of the building sill to prevent small gaps between the foundation and the sill.

Site-constructed — Built on the job.

Sliding window — A window consisting of two sash which slide from side to side past one another.

Soffit — In cornice construction, the surface of the bottom or underside is called the soffit.

Soil pipe — Cast iron pipe used for waste plumbing.

Soldier — Pattern of bricklaying in which bricks are stood on end.

Span of Roof — The horizontal distance covered by a roof; the width of the building.

Specifications — A document which conveys important information to the builder that cannot be shown on the working drawings. Specifications explain quality, color, and finishes of materials to be used, as well as the workmanship to be expected.

Square — 100 square feet.

Stack — The large vertical pipe in the main house drain into which the branches run.

Stair carriage — The supporting frame member under the treads and risers.

Stair risers — The vertical pieces between the stair treads.

Stair stringers — The inclined pieces at the ends of the treads and risers which support the stairs.

Stair tread — The top surface of a step in stair construction.

Stair winders — Tapered treads used to make a turn without a landing.

Standard glazing — Single piece of glass in a window.

Stool — Trim which forms the finished window sill.

Stop molding — Small molding that stops the door from swinging through the opening as it is closed. Also used to hold window sash in place.

Stud — Vertical framing member that supports the weight of the ceiling upper floors, and the roof.

Subfloor — The layer of the floor which is applied to the joists. The subfloor is covered with finished flooring or underlayment.

Sweat — Method of soldering used in plumbing. The fittings are called sweat fittings.

Termite shield — Metal shield installed at the top of the foundation to prevent termites from entering the wood framing.

Thermostat — A switch which is activated by changes in temperature. All heating systems use thermostats.

Trap — U-shaped fitting which prevents sewer gas from entering the house.

Trimmers — The joists at the ends of the headers.

Truss — A factory-build assembly used to support a load over a wide span. Trusses are often used in place of rafters.

U factor — The combination of all of the K factors in any given type of construction.

Underlayment — Any material installed on the subfloor that will give a smooth surface to receive the finish floor covering.

Valley rafter — Carries the ends of the jack rafters where two roof surfaces meet to form a valley.

Vanity — A bathroom vanity is a cabinet in which a lavatory is installed.

Vapor barrier — Material used to help prevent the movement of water vapor through building surfaces.

Varnish — Combination of resin, oil, thinner, and dryer which makes a transparent coating. Most varnish does not contain pigment.

Vent pipe — Allows atmospheric pressure into the plumbing system and prevents vacuum from building up as the waste water is discharged.

Vertical contour interval — The vertical distance between the contour lines which show the change in elevation.

Wall cabinets — Any cabinet that is mounted on a wall above a countertop or base cabinet.

Waste plumbing — Includes the pipes and fittings that carry the waste water from the fixtures to the house septic system or municipal sewer and the pipes and fittings that make up the vent system.

Wood stain — Similar to paint in the fact that it contains a vehicle and pigment. It is not opaque, however.

Working drawings — Drawings containing all dimensions and information necessary to carry a job through to a successful completion.

REFERENCES

There are numerous sources available to the estimating instructor. A few are listed with their addresses.

ASSOCIATIONS

American Association of Cost Engineers
308 Monongahela Building
Morgantown, WV 26505

American Institute of Architects
1735 New York Avenue, N.W.
Washington, DC 20006

International Conference of Building Officials
5360 S. Workman Mill Road
Whittier, CA 90601

National Association of Home Builders
1625 L Street, N.W.
Washington, DC 20036

PUBLICATIONS

Books

ARCHITECTURAL GRAPHIC STANDARDS,
   Ramsey and Sleeper
John Wiley & Sons, Inc.
605 Third Avenue
New York, NY 10016

BLUEPRINT READING AND SKETCHING FOR
   CARPENTERS – RESIDENTIAL, McDonnell and
   Ball
Delmar Publishers Inc.
2 Computer Drive West, Box 15-015
Albany, New York 12212

BUILDING COST FILE
Construction Publishing Company
2 Park Avenue
New York, NY 10016

BUILDING ESTIMATOR'S REFERENCE BOOK
Frank R. Walker Company
Chicago, IL 20656

BUILDING TRADES BLUEPRINT READING
Delmar Publishers Inc.
2 Computer Drive West, Box 15-015
Albany, New York 12212

CONSTRUCTION DICTIONARY
National Association of Women in Construction
Greater Phoenix Chapter #98
Phoenix, AZ

CONSTRUCTION ESTIMATING, Jones
Delmar Publishers Inc.
2 Computer Drive West, Box 15-015
Albany, New York 12212

GUIDE TO INFORMATION SOURCES IN THE
   CONSTRUCTION INDUSTRY, Godel
Construction Publishing Company
2 Park Avenue
New York, NY 10016

SWEET'S CATALOG FILE
McGraw-Hill Information System, Inc.
1221 Avenue of the Americas
New York, NY 10020

Periodicals

Architectural Design, Cost, and Data
Allan Thompson (Publishers)
P.O. Box 796
Glendora, CA 91740

Home Improvements
Peacock Business Press
200 S. Prospect Avenue
Park Ridge, IL 60068

Homebuilding Business
Gralla Publications
1501 Broadway
New York, NY 10036

House & Home
McGraw-Hill Publications
1221 Avenue of the Americas
New York, NY 10020

index